AF856143

Adventure Guide

Costa Rica

5th Edition

Adventure Guide

Costa Rica

5th Edition

Bruce & June Conord

HUNTER

HUNTER PUBLISHING, INC,
130 Campus Drive, Edison, NJ 08818
☎ 732/225-1900; 800/255-0343; fax 732/417-1744
www.hunterpublishing.com

Windsor Books
The Boundary, Wheatley Road, Garsington
Oxford, OX44 9EJ England
☎ 01865-361122; fax 01865-361133

Ulysses Travel Publications
4176 Saint-Denis, Montréal, Québec
Canada H2W 2M5
☎ 514/843-9447; fax 514/843-9448

Printed in the United States

ISBN 1-58843-502-4

© 2005 Hunter Publishing, Inc.

This and other Hunter travel guides are also available as e-books in a variety of digital formats through our on-line partners, including NetLibrary and Amazon.com.

All rights reserved. No part of this publication may be reproduced, stored in a retrieval system, or transmitted in any form, or by any means, electronic, mechanical, photocopying, recording, or otherwise, without the written permission of the publisher.

This guide focuses on recreational activities. As all such activities contain elements of risk, the publisher, author, affiliated individuals and companies disclaim responsibility for any injury, harm, or illness that may occur to anyone through, or by use of, the information in this book. Every effort was made to insure the accuracy of information in this book, but the publisher and author do not assume, and hereby disclaim, liability for any loss or damage caused by errors, omissions, misleading information or potential travel problems caused by this guide, even if such errors or omissions are the result of negligence, accident or any other cause.

Cover: Golden Toad © Robert Fried/robertfriedphotography.com
Index by Nancy Wolff
All other images © Bruce and June Conord

Maps by Kim MacKinnon & Kim André, © 2005 Hunter Publishing, Inc.

1 2 3 4

www.hunterpublishing.com

Hunter's full range of guides to all corners of the globe is featured on our exciting website. You'll find guidebooks to suit every type of traveler, no matter what their budget, lifestyle, or idea of fun.

Adventure Guides – There are now over 40 titles in this series, covering destinations from the Virgin Islands and the Yucatán to Tampa Bay & Florida's West Coast, Panamá and the Alaska Highway. They offer complete information on what to do, as well as where to stay and eat. *Adventure Guides* are tailor-made for today's active traveler, with a focus on both cultural and physical adventures such as cooking and language classes, hiking and biking, skiing and more.

Alive Guides – This ever-popular line of books takes a unique look at the best each destination offers: fine dining, jazz clubs, first-class hotels and resorts. In-margin icons direct the reader at a glance. Top-sellers include *The Cayman Islands, St. Martin & St. Barts,* and *Aruba, Bonaire & Curaçao.*

One-of-a-kind travel books available from Hunter include *A Traveler's Guide to the Galapagos Islands, The Virginia Handbook; Cruising Alaska* and many more.

Full descriptions are given for each book on the website, along with reviewers' comments and a cover image. You can also view book pages and the table of contents. Books may be purchased on-line via our secure transaction facility.

Dedication

Never doubt that a small group of thoughtful, committed citizens can change the world. Indeed, it's the only thing that ever has.

~ Margaret Mead, American anthropologist

To the men, women, and children around the world who do their best to conserve and protect our precious environment for future generations – theirs is the silent green army; and to them, this book is dedicated.

Acknowledgments

Although Costa Rica is a small country, its diverse geography and plentiful eco-adventure tourism opportunities made writing a well-researched guidebook quite a challenge. Readers who write to us are deeply appreciated, especially ones such as Kathleen Stewart, Cindy Kippley, Carol Fahey, who alerted us to changes, and Rachelle Burk, who provided new info on the golden toad. Individual parks and attractions made themselves available to us on short notice and we thank them. We are also grateful to all the Ticos and expats who gave freely of their time and made us feel so very welcome wherever we went. We must single out Allan Templeton in Manuel Antonio and Thomas Douglas in San José for particular generosity, and our personal thanks for years of friendship and hospitality go to Hilda Castro and family in Cedros.

Then, of course, our gratitude goes to our editor, Kim André, whose support and encouragement – and fine editing – keeps us on track. Lastly, we would like to thank you, our readers, for choosing this guidebook – it's the highest kind of praise! *Gracias*.

Author Profiles

A wandering minstrel I –
A thing of shreds and patches,
Of ballads, songs, snatches,
And dreamy lullaby!
~ *The Mikado*, Gilbert and Sullivan, 1885

The Conords have been traveling and writing about their journeys for years. They enjoy adventure and ecological travel, as well as historical and cultural forays in foreign lands. Their other Hunter titles, *Adventure Guide to the Yucatán* and *Cancún, Cozumel & The Riviera Maya Alive!* garnered critical acclaim and won several prestigious travel writing and photography awards from the North American Travel Journalists' Assoc.

Bruce went to Rutgers University and has written biographies of John Lennon, Bill Cosby and Cesar Chavez. He has worked as an import manager, teacher, advertising executive, copywriter and publisher's representative. Writing, photography, soccer and travel are his passions.

June grew up in southwest England and went to Plymouth Art College. Whenever she gets near the ocean, she feels happy. Her professional credits include numerous newspaper and magazine articles and stock photography. She has edited, photographed and worked alongside Bruce in many of their joint efforts to eke out a living and travel more. For relaxation, she could happily spend all day on a pebble beach, poking a stick in rock pools and collecting shells.

Their individual work is represented by a stock photography agency and together their photographs of Mexico have been featured at an exhibit in the prestigious New York Arts Club.

Visit their website at www.adventureguides.com.

READ WHAT REVIEWERS SAY:

"I read your guide book cover to cover when planning our last trip to Costa Rica (Dec. 2004 - Jan. 2005). It was an excellent source for planning a trip with 19 relatives. Thank you for a most informative book." Emilie Zouhar, La Mesa, CA

"... packed with solid information. [The book] sports a fresh modern design, easy-to-read typeface... [is] fun to read and easy to use." *Tico Times*

"[Adventure Guides] aim to deliver content... Moderately-priced and truly user-friendly, they are packed with information that other series rarely cover." *Library Journal*

"Thank you for a wonderfully written travel book.... This is one of the best I have seen on any country..." Dinesh Khera, traveler

"... a comprehensive travel guide... it's a must-read. This land of rustic beauty is explored to the hilt, complete with colorful photos and maps and full of easy-to-read practical tips." Anton Community Newspapers

"These useful guides are highly recommended... " *Library Journal*

"[Adventure Guides] direct you away from the theme parks and into the great outdoors... the information on trekking routes, canoeing, wildlife refuges - even golf courses - is well researched." *The Sunday Telegraph*

"...intended for the adventure-minded travelers with special affection for the outdoors and nature. Each *Adventure Guide* packs in outdoor-oriented activities set in different regions. There's something for nearly everyone." *Midwest Book Review*

Contents

■ Maps

Original luggage tag from Hotel Europa,
San José's oldest hotel.

Introduction

Never a ship sails out of the bay
But carries my heart as a stowaway
~ Roselle Mercier Montgomery, writer, 1874-1933

Costa Rica is a magical land known for its beautiful scenery and appealing people. Geographically close to the US yet far different, its historic political stability and intriguing Spanish culture make it an important draw for all kinds of vacationers.

IN THIS CHAPTER

Don't expect either the obvious poverty or indigenous customs of Mexico or Guatemala here. Costa Rica is a cosmopolitan country blessed with an exotic landscape and an uncommon landscape perfect for eco-tourism. It is not as inexpensive as the rest of Central America, or even as cheap as it was before eco-tourists discovered its beauty. We overheard a casual remark that Costa Rica is a "third-world country with first-world prices." While in some ways that is true, your vacation dollars will still go a long way here, and even the higher-priced itineraries offer tremendous value. In addition, thousands of expatriates and retirees have settled here for the climate, culture and lower cost of living, making Costa Rica a very North American-friendly destination.

But it is the host of natural wonders and ecological diversions that draws most visitors – and vacation adventure opportunities abound.

Whitewater raft through pristine jungles. Bathe in natural hot water springs. Slide through a canopy of trees on a high wire. Visit coffee plantations, where some of the finest coffee in the world is grown. Sunbathe on a beach of black volcanic or powdery white sand. Watch a nighttime volcano put on a fireworks display of ruby-red hot lava. Hide on a darkened beach to watch giant sea turtles as they dig nests and lay their eggs. Climb to the top of a high volcano and look down into its crater filled with a pea green lake. Visit one of the many butterfly farms. Swim, snorkel, hike, bike, camp, eat, sleep, read a book – do whatever you want. Costa Rica offers a smorgasbord of pastimes for your pleasure. We've been many times and find more to do and see each time we visit.

How to Use this Book

The writer that does the most is the one who gives his reader the most knowledge – and takes from him the least time.
~ Charles Caleb Colton, English writer, 1780-1832

You know the old saying. If you find fault with our book, please tell us – but if you like it, tell a friend!

We have arranged this guide in a way that makes it easy for readers ether to plan a logical itinerary or get right into the details of specific adventures. Unless you're just going to one area, travel in Costa Rica always involves passing through the Central Valley. The best way to think of it is as a spoked wheel, with San José as the hub. You can see a lot of the country if you are based in the capital. Costa Rica is small enough to get from coast to coast in one long day of driving, or half an hour in a plane. (Then again, some areas have such bad roads you're lucky to get a few kilometers in a half-hour – we'll steer you away from those!) We divided our guide into regions: San José; the Central Valley; North by Northwest; the Nicoya Peninsula; Pacifica (the Central Pacific Coast); the Caribbean Coast; and Zona Sur, the South Pacific. Each offers its own diverse enjoyments.

We wrote this *Adventure Guide* with an ecologically responsible slant. Costa Rica is an environmental paradise. In fact, 2002 was declared the Year of Eco-tourism by the United Nations. Don't be a tourist, be an eco-tourist.

Flexible travelers can always get a lot out of their vacations, and that's especially true in Costa Rica. Starting in San José and wandering off to find your pleasures is a great way to go. The main north-south road is the Inter-American Highway, which connects the towns of La Cruz (near the Nicaraguan border) with Canoas (on the border with Panama).

Nearly every major establishment we mention in this guide has e-mail and a website; we've included the web address after the phone number. In the interest of space, we usually give just the website when you can use that to send them an e-mail. Those establishments without their own websites are still likely to have e-mail; we've listed only the e-mail in these cases. Because Internet cafés come and go faster than we can type, we haven't bothered to list them. Just ask. There will be one nearby.

Remember, new places pop up overnight, places that are good go bad, formerly lousy restaurants clean up their act, and hotels remodel. If you find things are different than we have suggested in the pages of this book, please let our publisher know or send us an e-mail directly (brucewrite@aol.com or junioc@aol.com). Same goes if you've found something good that we didn't include – let us know. Our goal is to make your vacation the best ever!

■ Prices

Although we made every effort to be as thorough, complete and accurate as possible, things change in Costa Rica – sometimes *muy rapido*. We tried to use a system of dollar-sign symbols.

Hotel Prices	
Prices are per night for two people and do not include 16% tax	
[No $]	less than US $20
$	US $20-$40
$$	US $41-$80
$$$	US $81-$125
$$$$	US $126-$200
$$$$$	over US $200

Restaurant Prices	
Prices are for an average entrée	
$	less than US $5
$$	US $5-$10
$$$	US $11-$20
$$$$$	over US $20

■ Very Useful Web Information

There's tons of information available. Make sure you read our *Handy Hints* section on page 113. And, if you plan to rent a car, please read our driving tips on page 97.

There are several sources of Costa Rican information on the web. The government's official site is **www.tourism-costarica.com**. If you'd like to have a Costa Rica tourism brochure mailed to you, call them at ☎ 800/343-6332. Some good maps can be found for purchase or download at **www.mapcr.com**.

Other good resources include **www.costarica.com, www.central-america.com** and **www.bruncas.com**. Gay and lesbian travel info can be found at **www.hometown.aol.com/gaycrica/guide.html**. For a round-up of hotels, log on to **www.hotels.co.cr** or the useful **www.yellowweb.co.cr**. Check bus routes and schedules online at **www.costaricabybus.com**. A new website, **www.cupotico.com**, has car rental, restaurant and hotel discount coupons you can download and print.

If you want to keep up with news in Costa Rica, look every Friday at **www.ticotimes.com** and check daily on **www.insidecostarica.com** and **www.amcostarica.com**, where you can sign up for daily headlines by e-mail. *La Nacion* newspaper has a good site: **www.incostarica.net/centers/visitor**.

For eco-tourism discussions look to **www.planeta.com**. Devoted eco-tourists should check out the eco-cultural offerings of Cooperena cooperatives at **www.turismoruralcr.com** or **www.ecotourism.co.cr**. The first site is most important for community-based ecological and cultural tourism. Scattered across Costa Rica are rustic rural lodges that provide local guides, typical family food, eco-tourism activities and cultural interchange with local communities. The simple life. Sports folk should check out **www.costaricaoutdoors.com**.

Smiling Tica.

Locally run websites with accommodations and general information include **www.monteverdeinfo.com** (Monteverde), **www.maqbeach.com** (Manuel Antonio & Quepos), plus Pacific coast beach destinations: **www.nosara.com**, **www.samarabeach.com**, and **www.tamarindo.com**. For the Osa Peninsula and southern part of the country from Dominical to Panama, look to the Pacific region's **www.southerncostarica.com**, or on the southern Caribbean coast, **www.greencoast.com, www.purtoviejo.net** and **www.puertoviejoweb.com** (Puerto Viejo). Surfers can check **www.crsurf.com** in Dominical. Sports and cultural info are available at **www.internet.co.cr.**

■ We Love To Get Mail!

A traveler has a right to relate and embellish his adventures as he pleases, and it is very impolite to refuse that deference and applause they deserve.

~ Rudolf Erich Raspe, 1737-1794

Hunter Publishing makes every effort to ensure that its travel guides are the most current sources of information available to travelers. If you have any information that you feel should be included in the next edition of this guide, please write to us at 130 Campus Drive, Edison, NJ 08818, or send an e-mail to kim@hunterpublishing.com.

Feel free to include your opinion or experiences of the places you've visited, as well as price updates and suggestions for ways in which we could improve this book for future readers. If you'd like to contact the authors directly, e-mail them t book@adventureguides.com.

Costa Rica At A Glance

On a Clear Day You Can See Forever.
~ 1965 musical

WHEN TO VISIT: The high season in Costa Rica, December through April, is the dry season. The rainy season, which lasts from May to November, usually sees sunny mornings, with rain showers in late afternoon and evening. Secondary roads can become rutted during those months, and four-wheel-drive vehicles are strongly recommended. Overall, the climate is tropical, with an average temperature of 72°F (22°C). It can be much hotter along the coastal areas of the country, and cooler in the mountains.

MONEY: Costa Rican currency is the **colon** (co-LOAN). It floats daily against the dollar and can be exchanged at banks and change booths. American dollars and major **credit cards** are acceptable almost everywhere, except in small business establishments or hotels and restaurants in remote locations. **Travelers' checks** are not exchanged as favorably as cash. If you pay by credit card, a small surcharge is sometimes added. **ATMs** are available in most cities and towns with bank offices.

PEOPLE: The population of Costa Rica was 3,622,171 as of January 2000, which includes 40,000 native people who belong to eight different cultural groups. The official language is Spanish, but many of the people speak some English, a required course in all schools. Costa Ricans are affectionately known as *Ticos* (TEA-coes) – and you would be hard pressed to find a more friendly and welcoming culture.

EDUCATION: Costa Rica's constitution requires 6% of its Gross Domestic Product be dedicated to education – and as a result it has a higher literacy rate (95%) than the United States. All post offices have computers for general use, and Internet connections are also available there. Costa Rica also imports students from overseas who come to the Spanish-language schools that abound throughout the country.

RELIGION: Catholicism is the dominant religion, as it is in most of Latin America. Consequently, nearly all major holidays are religious in nature. The government and popular culture is secular, though still conservative.

MAJOR CITIES: San José, population one million, is the capital and cultural heart of Costa Rica. Other major cities (by population) are: **Alajuela**, **Cartago**, **Heredia**, **Liberia**, **Limón** and **Puntarenas**.

WEATHER: Costa Rica is a tropical country with two seasons – dry and wet. Temperature in the Central Valley is spring-like all year long. It's colder at higher altitudes in the mountains and hotter in the lowlands and along the shore.

ENTRY REQUIREMENTS: Canadians and North Americans are not required to have a visa and may visit Costa Rica for a maximum of 90 days without one. To enter the country you must have a valid passport, or a photo ID (like a driver's license) and a copy of your birth certificate. We strongly recommend having a passport. All Europeans need a valid passport.

ELECTRICITY: The voltage throughout the country is 110, the same as in North America. However, three-prong outlets are scarce, so bring along an adapter if you need one. Travelers with appliances set for 220 will need an adapter that changes the voltage and allows for use of a different plug.

TIME ZONE: Costa Rica is on **Central Standard Time**, six hours behind Greenwich Mean Time and one hour behind EST in the States. It does not currently use daylight saving time, although the idea has been considered.

HEALTH: No shots are required, but we always suggest having a **Hepatitis A** shot as a precaution. The water in the major cities of Costa Rica is safe and most hotels and restaurants offer purified tap water. You might prefer to drink bottled water (*agua purificada*) or seltzer (*agua mineral*) to be sure. Costa Rica has excellent, low-cost medical care and well-qualified practitioners. Many North Americans come to Costa Rica for cosmetic surgery or dental work.

SAFETY & CRIME: Costa Rica is a safe destination for 99% of its tourists, but it's always a good idea to exercise caution whenever one travels. In general, the country has a low crime rate, but in recent years there have been increasing instances of tourists and expatriates being robbed, as well as several murders. Additionally, most eco-adventures involve some sort of danger, so be sure to use less testosterone and more common sense when deciding on your level of participation in these activities.

THINGS TO BUY: Choose from coffee and coffee-related products, reproduction pre-Columbian jewelry, craftily carved wooden boxes, attractive Chorotegan pottery, leather goods, hand-painted art (on bird feathers), guitars and other musical instruments or painted oxcarts. There's also an abundant selection of clothes and crafts imported from Panama, Ecuador and Guatemala available.

DRIVING/CAR RENTAL: Rental cars are expensive, but a good way to see Costa Rica outside of San José. You should buy all the insurance offered – and then some. Drivers in Costa Rica are maniacs – worse than Bostonians – and, for a non-confrontational people, very aggressive behind the wheel. Combine that with unpredictable road conditions and there can be "awkward" moments. Drive very cautiously. In rainy season, make sure that you rent a four-wheel-drive auto. Think mass transit – buses are a good alternative and very reasonably priced.

■ Costa Rica's Top 20 Spots

Adventure today means finding one's way back to the silence and stillness of a thousand years ago.
~ Pico Iyer, *NY Times* magazine

Our list of "Top 20 Things to Do or See" is a guide to the best that Costa Rica has to offer. It's not in any order, nor does it cover anywhere near all of the country's attractions. But it should give you some food for thought in planning your vacation. Happy trails to you!

1. ARENAL VOLCANO: Famous for its nighttime lava fireworks, Arenal Volcano towers above a lovely lake of the same name. The area has plenty of natural activities, eco-adventures, and the lake is particularly popular with fishermen and windsurfers. The thermal springs at the base of the mountain offer a refreshing dip any time of the day. The volcano itself rumbles frequently and, if not socked in by clouds, is very impressive – and just a tad exciting. See page 197.

2. MONTEVERDE: The 10,526-hectare/26,000-acre Monteverde Cloud Forest Reserve, nestled in moisture-filled hanging clouds, provides a home to thousands of species of plants, animals and insects. It offers a unique opportunity to experience, up close, the beauty of nature unspoiled – the reason we all come to Costa Rica. See page 212. If you can't make it here, try the Los Angeles Cloud Forest or Tapantí-Macizo de la Muerte Cloud Forest.

3. MANUEL ANTONIO: This is prime real estate. Manuel Antonio National Park has three white sandy connecting beaches and a forest filled with a variety of monkeys. It sits at the base of a mountain on a peninsula that eagerly stretches out to greet the Pacific Ocean beyond and boasts magnificent flora and fauna, as well as fantastic views both in and out of the park. See page 293.

4. TORTUGUERO: Bordered by the Caribbean Sea, Tortuguero National Park contains an incredible network of navigable canals, and boats are the only way to get around. Mangrove forests that edge the canals are the place to look for wildlife. The park's 37 km/23 miles of beaches are protected nesting grounds for the green sea turtle. In season (July to October), you can accompany a guide to the beaches at night and watch turtles lay their eggs. It is a memorable experience. Other seasons offer plenty of non-turtle related nature sightseeing. See page 332.

5. CANOPY TOURS: Most of the rainforest's life dwells in the canopy, high above ground. Consequently, there are a large number of "canopy tours" offered. Most are geared to having a wild ride rather than studying wildlife – but what fun it is. In the popular and ubiquitous versions you slide from tree platform to tree platform, safely attached in a harness, on a thick cable. It is thrilling to zip across the forest like a high-tech Tarzan, and kids love it. Check out a complete list and description of tree-top tours at www.mapcr.com/costaricacanopytour.html. If you would like more cerebral and sedate fun, take the "Aerial Tram." It offers a monkey's view of the surrounding jungle at various levels in the canopy, while seated in a slow-moving cable car and accompanied by a naturalist guide. There is also a "Sky Walk" that features incredibly high wire bridges between trees and above deep gorges. Our favorite places to go in search of such thrills are "Kazm Cañon," which offers rappelling and slides over the Río Colorado next to Rincón de la Vieja National Park; and the waterfalls near Manuel Antonio, where you can rappel down the falls. These types of activities are referred to as "canyoning."

> **WARNING:** Many of these adventures contain an element of danger, so be selective.

6. WHITEWATER RAFTING: Costa Rica is famous for its whitewater river raft trips through the rain forest. Rafting is one of the best ways to spend a day. You can even opt for an overnight trip with a stay in a riverside lodge deep in the rain forest. Our favorite ride is on the Pacuare River (see page 134), where you are rushed through primary and secondary forests and an impressively deep gorge. The Class III and IV rapids will thrill experienced rafters. Safety-oriented professional guides provide a trip well suited for everyone from Grandma to Junior. Exciting, exhilarating, and intoxicating. Other regional rivers also offer challenging rides.

7. LANKESTER GARDENS: Wander along the 17 km/10.6 miles of trails that lead throughout the Lankester Gardens outside of Cartago. Among colorful open garden beds and a cool shady forest, enjoy the countless varieties of orchids, bromeliads, cacti and palms that are grown here for botanical study as well as your viewing. Guides are available. See page 170.

8. IN SEARCH OF THE QUETZAL: Everyone heads to Monteverde to look for the quetzal, the colorful native bird that is an enduring symbol of freedom. If you're going north, by all means keep an eye out. But we think the best chance to find the elusive quetzal is off the Inter-American Highway that connects San Isidro and Cartago. This highway crosses Costa Rica's highest mountains along the Cerro de la Muerte and there are several lodges here that cater to birdwatchers. Practically all mountain nature lodges have guides for hire and offer the opportunity to stay and watch for quetzals. Nesting season for these magnificent birds is March through May, and that's the best time for a sighting. But it is possible, if you're lucky, to catch a glimpse of one any time of the year.

9. IRAZU VOLCANO: The 11,260-foot-tall Irazú Volcano is the highest in Costa Rica and still considered active, even though its last major eruption was on March 19, 1963, the day that President John F. Kennedy arrived in Costa Rica. Today, a few puffs of steam and smoke are the most activity you're likely to see. To get here you'll cross a wide expanse of dark gray barren land resembling a moonscape. At the rim, the sight of the crater filled with pea-green sulfur-laden water – surrounded by a rugged rocky cliff face – leaves one breathless. See page 169.

10. POAS VOLCANO: A long, panoramic, twisting road wends its way through fertile farmland and forest stands, leading up the mountainside to the Poás Volcano. Its crater, filled with turquoise water, is over a mile across and is said to be the second-largest active volcano crater in the world. Occasionally, the crater still boils and shoots steam geysers. Poás Volcano is situated in a protected national park of the same name that offers visitors an information center, a colorful slide show about the volcano, and is the starting point for several well-marked trails around the moonscape crater's rim. See page 184.

11. CORCOVADO: This national park is the largest tract of virgin rain forest in Costa Rica and covers more than half of the Osa Peninsula. Famous for its scarlet macaws and a multitude of other wildlife, it is a colorful must-see for adventure naturalists. See page 362.

12. CAHUITA & PUERTO VIEJO: The southern beach town of Cahuita is laid back and friendly, a place to rest and rejuvenate during the day – and the hot spot for dancing at night. It contains two inviting beaches, one white and one black. Cahuita National Park offers it all – camping, biking, hiking, snorkeling, and more – and is guarded from the treetops by vociferous howler monkeys. Travel farther south to Puerto Viejo to find yet another inviting small beach town with an even more relaxed beat. With a large selection of accommodations and eateries, it makes a good base to cover the whole coast down to the Panamanian border. Worth all the time that you can afford. See pages 343-353.

13. NATIONAL THEATER & GOLD MUSEUM: Right in the heart of San José, alongside the Plaza de la Cultura, is the imposing National Theater (see page 121). Designed by Belgian architects and decorated by Italian artists, the 1,000-seat stone and metal structure offers performances by the world's most famous artists. Don't miss seeing its fabulous gold gilt interior and be sure to have afternoon tea in the theater's café. The Gold Museum (page 122) is below the Plaza de la Cultura, to the rear of the theater. A permanent display of more than 2,000 pre-Columbian gold artifacts and temporary art exhibits make for a rich experience.

14. JADE MUSEUM: Eye-popping pre-Columbian jade, gold and stone art are featured at the Jade Museum, located at the top of the INS building in San José. In addition, frequently changing exhibitions of artists' works will enhance your pleasure. See page 124.

15. BUTTERFLY FARM & INSECT MUSEUM: The Butterfly Farm near Alajuela is Latin America's first and largest exporter of farm-raised butterflies. All visitors get a guided two-hour tour, which fits in well with the Café Britt Coffee Tour during a day of sightseeing. There are now many smaller butterfly farm imitators, but one that's impossible to imitate is Dr. Richard Whitten's **Jewels of the Rainforest Exhibit**. Headquartered at Selvatura park in Monteverde, his extensive collection of weird and wonderful insects, butterflies, bugs and beetles are artistically displayed and accompanied by educational videos, brochures and, if you're lucky, the personal attention of Dr. Whitten or his charming Scottish wife (both accomplished organ players). Another biological attraction is **InBio Parque**, a private project to categorize Costa Rica's diverse insect and plant life, as well as educate and entertain visitors.

16. RINCON DE LA VIEJA: This park surrounds the flanks of the Rincón de la Vieja Volcano and its active crater. Many excellent hiking trails traverse this diverse landscape with its hot springs, geysers, mud pots, waterfalls, volcanic craters and a lake. This is a favorite destination for birdwatchers, offering fabulous views of the lowland pastures and rich wildlife. Hard to get to, but a fascinating place. See page 229.

17. SARCHI & COFFEE TOWNS: If you like to shop, Sarchí is your town. Filled with handicraft stores and small factories, it is best known for its colorfully hand-painted Costa Rican oxcarts (*carretas*), which can be purchased in all sizes (and shipped home). But the town also offers fine furniture and other wooden articles, as well as leather, metal and fabric creations. Enjoy an oxcart painting demonstration at the Plaza de la Artesania shopping mall, then dine at one of the many restaurants. When you're shopped out, take a short side trip to nearby Zarcero. In the center of town you'll enjoy the whimsical and photogenic topiary featuring animal figures from sculpted cypress. A long tunnel of connecting bushes that look like melted Hershey's kisses, lead to an inviting red and white church beyond. Living art. See pages 186-195.

18. WATER, WATER, EVERYWHERE: A plethora of beaches await you on the Caribbean and Pacific Ocean coasts of Costa Rica. They come in a variety of colors with sands that run from white to black and textures ranging from powder soft to coarse and gritty. Public beaches, isolated beaches, great surfing beaches, laze-around beaches – whatever you prefer, it's there. If sand between your toes is not your style, then swim, sail, whitewater raft, kayak or windsurf on the many rivers and lakes between coasts. Or choose the ultimate way to relax – soak in hot springs then cool down beneath a waterfall.

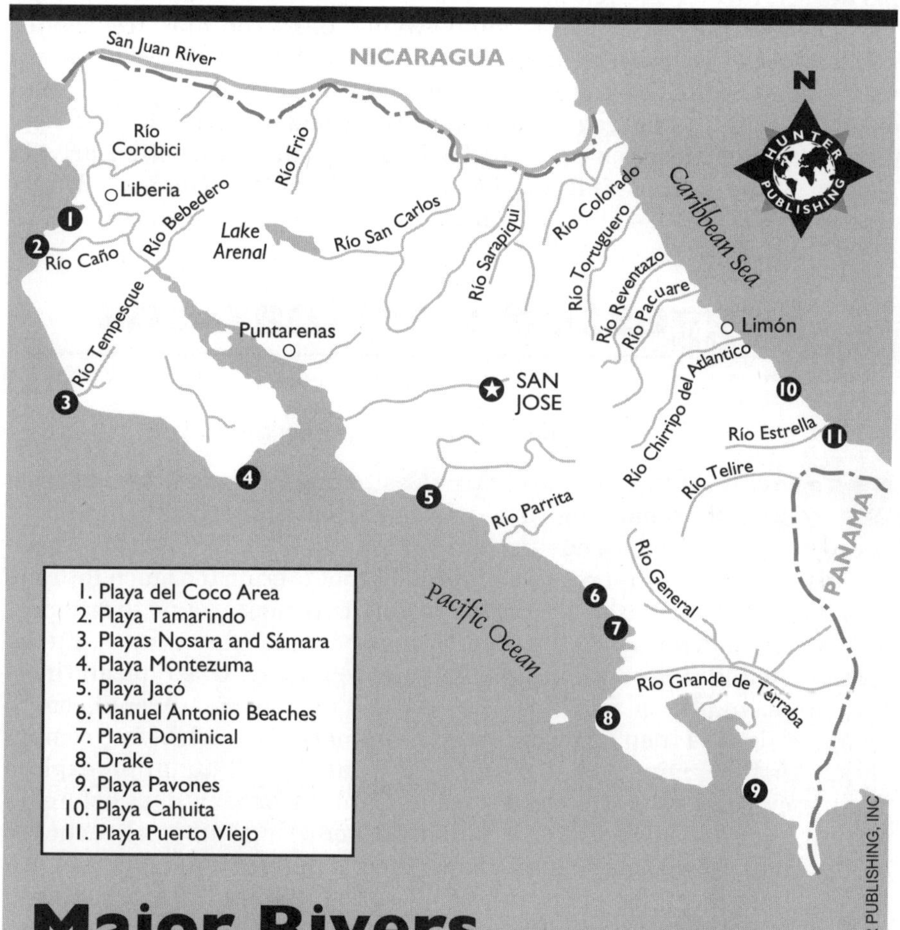

19. NATURE LODGES: If you're looking for a unique experience that your cruise ship friends couldn't imagine, spend at least a night in one of the many ecological nature lodges that offer rustic accommodations. You owe it to yourself to experience the diverse unspoiled countryside, away from the traffic and noise pollution of the cities and towns.

20. FISHING: A growing number of qualified operators offer competitive, world-class sport-fishing charters on both coasts. In the Pacific, marlin, sailfish, tuna and dorado lure the enthusiastic angler. In the turquoise waters of the Caribbean, at the mouth of the Barra del Colorado in particular, tarpon and snook are yours for the catching (and releasing). The beautiful 35-km-long (22-mile) Arenal Lake contains freshwater rainbow bass (*guapote*). In their eagerness to get hooked, they will practically pull you into the water. In the mountains, a short drive from San José, you will find fishing locations for trout and other freshwater species. A fishing license is required for inland angling (except on private property). For detailed information, contact *Costa Rica Outdoors Magazine* at ☎ 506/282-6743 or by e-mail, jruhlow@sol.racsa.co.cr.

Suggested Destinations

And away we go.
~ Jackie Gleason, *The Honeymooners*

We have had the luxury of visiting Costa Rica for much longer periods of time than most tourists dream of. Consequently, at the risk of suggesting too much, we have to bite our tongue when it comes to preparing itineraries. Our first recommendation is not to cram too much into one trip. It's better to come back again. If you are traveling around to different regions, think in terms of a trip from San José that touches what you're interested in and loops back in time for your departure. One typical itinerary, for example, is San José up to La Fortuna and Arenal, on to Monteverde, and then a day or two on a Guanacaste beach before coming back. Another might change directions at La Fortuna to explore Tortuguero, and follow the coast south to Cahuita for a few days before returning to San José. We found a delightful loop along Nicoya's beaches to Montezuma. How about heading down the Pacific coast from Manuel Antonio to Dominical, then back inland along the "Ridge of Death highway" to Mt. Chirripó? Or catch a flight to Osa. The possibilities and combinations are endless. There is even a small cruise ship line, **Cruise West** *(☎ 800/888-9378, www.cruisewest.com)*, that offers a fun, seven- or 12-day cruise along the Pacific coast of Costa Rica, including Coiba Island and the Panama Canal.

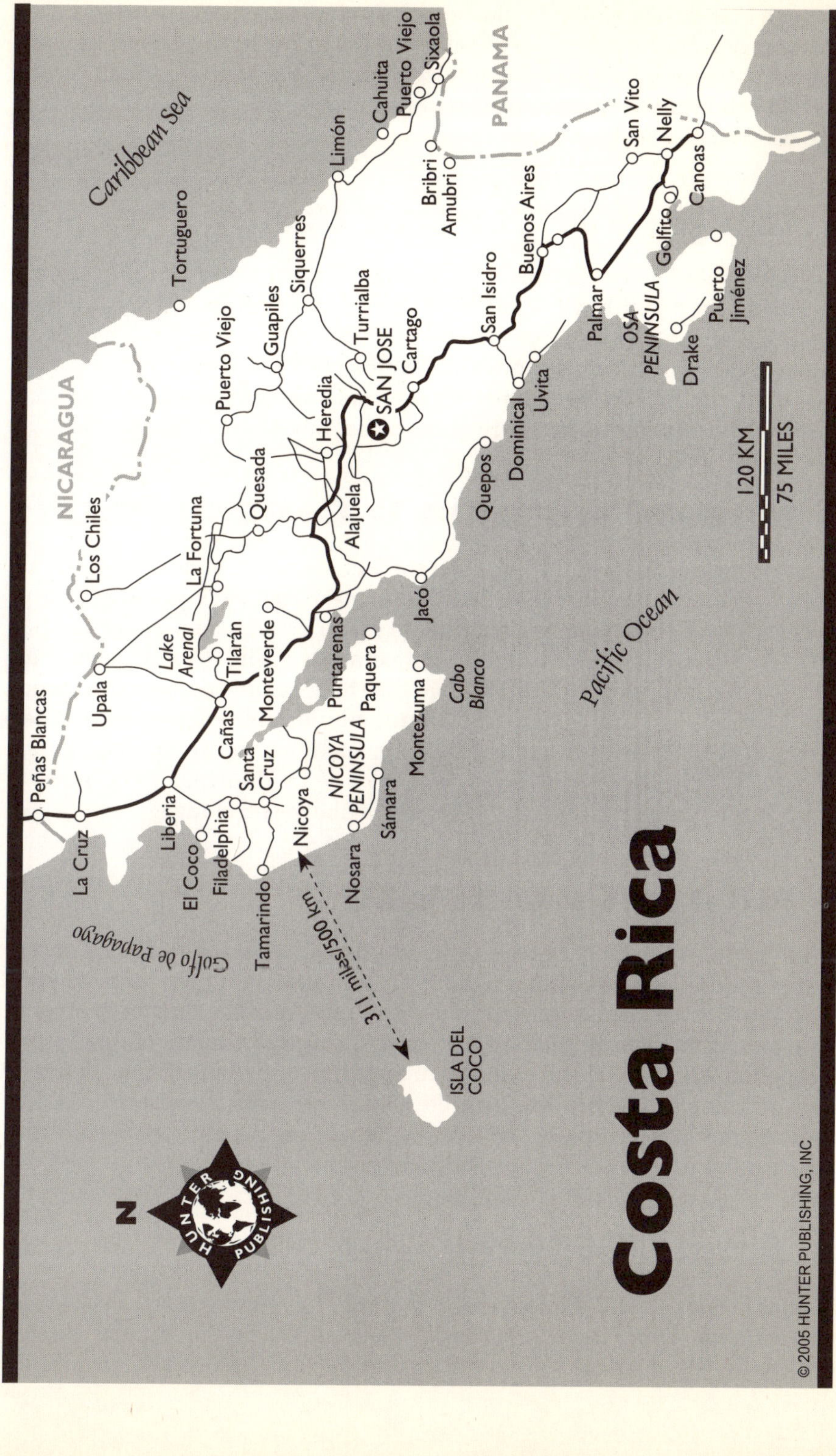

Introduction

Because Costa Rica is small, you can easily combine several areas and attractions in a one-week trip. If this is your first time here, you might want to pick just one or two areas and base yourself out of them. Each region covered offers adventure travel, recreational vacation and eco-tourism opportunities.

■ San José & Central Valley

The metropolitan area of San José, Escazú, Cartago and Heredia can fill more time than you have. It offers museums, shopping, nature and culture, as well as day-trips to nearby attractions such as the volcanoes, whitewater rafting, canopy tours and coffee plantations. San José is a big city and is fun for those who like what cities offer, but less metropolitan-oriented visitors might prefer to stay in the suburbs or nearby towns.

■ Manuel Antonio & Central Pacific

Manuel Antonio National Park is on the Pacific coast, about three or four hours overland from San José. It's the most-visited of Costa Rica's natural reserves. Besides the natural beauty and eco-tourism opportunities you'll find there, the drive itself is spectacular. The stretch from Jacó (where most residents of San José go for the weekend), through Quepos and Manuel Antonio down to Dominical, features long, palm-lined beaches great for swimming and surfing. Included in this trip could be a drive back from Dominical along the Cerro de la Muerte, a twisting part of the Inter-American Highway that rises above the clouds.

■ Nicoya/Guanacaste

Guanacaste is the drier, cattle-ranching province in Costa Rica's northwest shoulder. A new airport in Liberia has opened up the area to even more tourism as people come to visit the fabulous beaches that stretch down from the Nicaraguan border to Cabo Blanco at the tip of the Nicoya Peninsula. The Pacific water is warm and inviting. Some of the country's most attractive beaches are located along this coast, and surfers, sun-worshippers and swimmers gravitate here. Plus, it gets more sun and less rain, which makes for perfect beach weather.

■ Arenal/Monteverde

Although they stand almost back-to-back, the active Arenal Volcano, with its large, windy lake, and the Monteverde Cloud Forest, maintained in part by the original Quaker community that founded it, are hours apart by car. Many visitors combine them for their diverse pleasures, and some smart travelers take the shortcut across the lake. Arenal offers hot springs, lava flows, a hip town (La Fortuna), plus the thrill of real danger. Monteverde, on the other hand, is surreal. Clouds shroud the forest's nature walks. Scientists and students relax in nearby quiet Santa Elena.

■ Tortuguero

The remote corner of northeastern Costa Rica is famous for its deep-sea fishing (charters depart from the mouth of the Barra Colorado River), and the Tortuguero canals, a complex of inland rivers that criss-cross mangrove forests. Here, on Tortuguero's Caribbean beaches, endangered turtles come to nest. A great eco-activity is to watch the turtles dig their nests and lay eggs, or watch the hatchlings struggle down to the sea – all with the accompaniment of a guide, of course. Although some companies now offer day-trips from San José, we recommend at least one overnight in the area.

■ Atlantic Caribbean

The shore south from Limón to Cahuita and Puerto Viejo is what is generally referred to as Costa Rica's Caribbean or Atlantic coast. The atmosphere is closer to that of a Caribbean island than to the hustle and bustle of the Central Valley. That's partly because it looks more Caribbean – beaches are lined with palms – but mostly it's because of the Caribbean culture that survives with the Afro-Caribbeans who settled the area 150 years ago. We find it very loose and lovely. Indigenous reserves and nature lodges can be found inland.

■ Osa Peninsula & The South

The south and southwest corner of Costa Rica contains the largest tracts of unspoiled virgin rainforest. Although a fair amount of this area is accessible by hard four-wheel driving, it's remote enough that most visitors arrive by boat and plane. Corcovado's rich expanse of primary forest covering most of the Osa Peninsula features some wonderful nature lodges. The gulf gets visits by dolphins and whales and is popular with fishermen. Most tourists don't get this far, but it is absolutely worth it.

Surfing Safaris

Let's go surfing now, everybody's learning how,
come on and safari with me.
~ *Surfing Safari*, Beach Boys, 1962

Long known as a surfer's paradise – for more reasons than just waves – Costa Rica's beaches continue to attract foreign surfers who come for tournaments or just for fun. Since both coasts have exceptional curls, more locals are also enjoying this imported sport. Check out **www. crsurf.com** for the latest updates.

Surfing the waves in Costa Rica.

■ North Pacific Coast

PLAYA NARANJO: Also known as **Witch's Rock**, Playa Noranjo is one of the most famous breaks in Costa Rica with near-perfect tubular waves. Hard to reach – four-wheel-drives routinely get stuck – it is located in a wilderness area of Santa Rosa National Park, way up north. Playa Naranjo was made famous in the film *Endless Summer.* There are no facilities here – you have to camp. **Siesta Campers** *(☎ 506/289-3898, www.edenia.com/campers)* rents fully equipped VW Westfalia camper vans. Just to the south of Witch's Rock is a spot known as **Ollie's Point**, named after Lt. Col. Oliver North. A secret airstrip nearby was once used to supply Contra rebels across the Nicaraguan border. It's nice to know that something good has come from Ollie's misdeeds – one of the two cargo

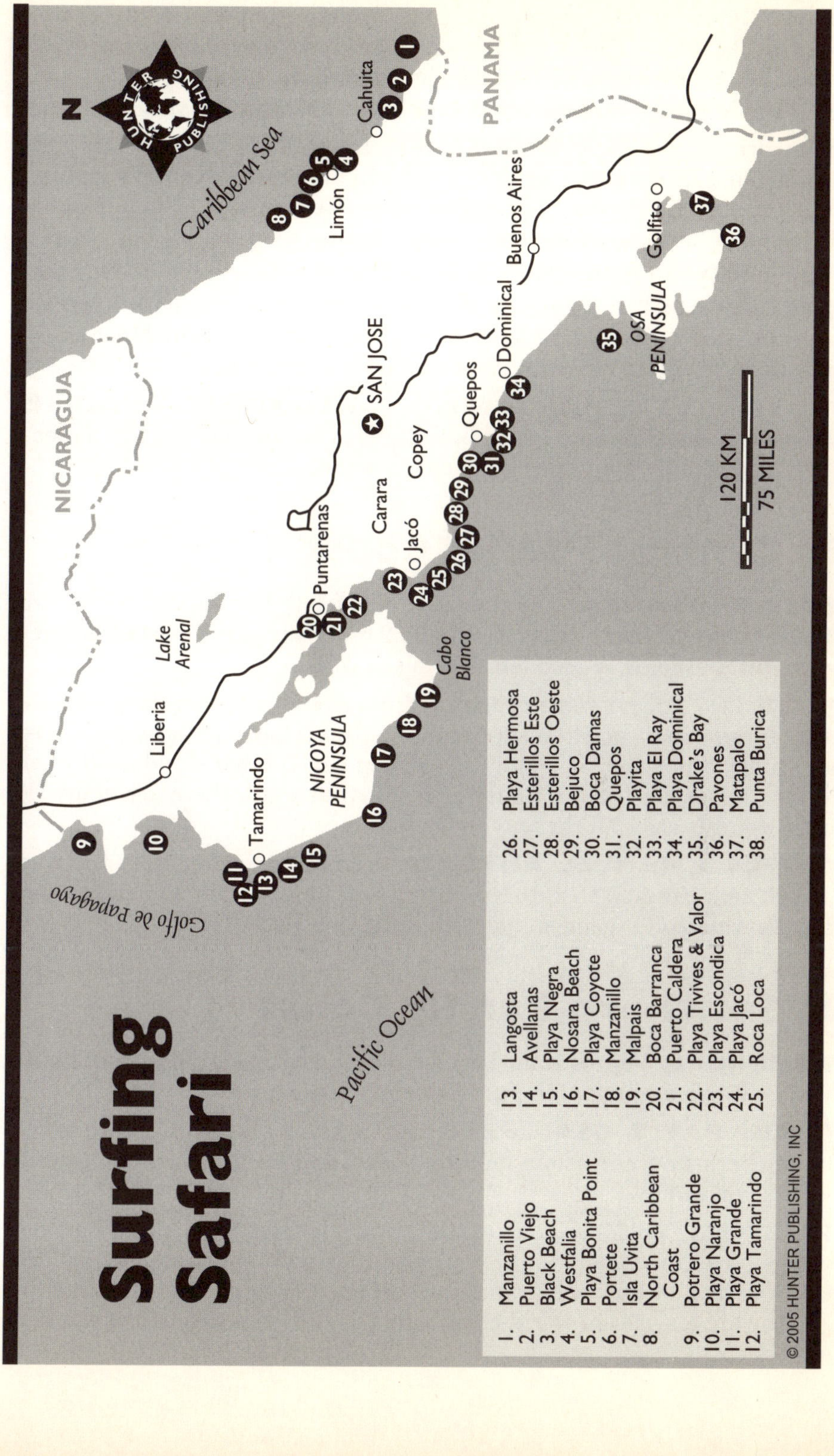
Surfing Safari
NICARAGUA
PANAMA
Caribbean Sea
Pacific Ocean
Golfo de Papagayo
Lake Arenal
NICOYA PENINSULA
OSA PENINSULA
Cabo Blanco
Liberia
Tamarindo
Puntarenas
Carara
Jacó
Copey
SAN JOSE
Quepos
Dominical
Buenos Aires
Golfito
Limón
Cahuita
HUNTER PUBLISHING
N
120 KM
75 MILES
1. Manzanillo
2. Puerto Viejo
3. Black Beach
4. Westfalia
5. Playa Bonita Point
6. Portete
7. Isla Uvita
8. North Caribbean Coast
9. Potrero Grande
10. Playa Naranjo
11. Playa Grande
12. Playa Tamarindo
13. Langosta
14. Avellanas
15. Playa Negra
16. Nosara Beach
17. Playa Coyote
18. Manzanillo
19. Malpais
20. Boca Barranca
21. Puerto Caldera
22. Playa Tivives & Valor
23. Playa Escondica
24. Playa Jacó
25. Roca Loca
26. Playa Hermosa
27. Esterillos Este
28. Esterillos Oeste
29. Bejuco
30. Boca Damas
31. Quepos
32. Playita
33. Playa El Ray
34. Playa Dominical
35. Drake's Bay
36. Pavones
37. Matapalo
38. Punta Burica
© 2005 HUNTER PUBLISHING, INC

planes flown out of here has become a restaurant at Costa Verde in Manuel Antonio. (The other one was shot down over Nicaragua, which made US involvement in the Iran-Contra scandal public.)

PLAYA TAMARINDO: Easy to get to and very popular. The three main breaks are **Pico Pequeño**, a rocky point in front of the Hotel Tamarindo; **El Estero**, a good river-mouth break; and **Henry's Point**, a rocky break in front of the Zullymar Restaurant. Each August, Tamarindo hosts a season-end pro-am competition with a big purse. If you're interested in taking part, check www.crsurf.com for dates.

AVELLANAS: The **Guanacasteco** break features very hollow left and right breaks, 10 km/6.2 miles south of Tamarindo. Nearby **Nosara** is a nice little seaside town with right and left beach breaks.

PLAYA COYOTE, MANZANILLO, SANTA TERESA & MALPAIS: These somewhat remote beach breaks, with several points, are increasingly popular for their speed and consistency.

■ Central Pacific Coast

BOCA BARRANCA: The closest surf beach to San José, this is a river mouth with excellent access. Two km/1.2 miles south is **Puerto Caldera**, with **Jetty Break**, a good left near a sea wall.

JACÓ: Jacó can get some rough surf and closes out when it is over five feet. Inconsistent good beach break. South of Jacó is **Playa Hermosa** and a few other nearby spots with strong beach breaks and good waveforms. Waves are best on rising tides. Farther south there are many isolated point breaks all the way to Quepos.

QUEPOS & MANUEL ANTONIO: A small left point is at the river mouth near the city. Next to the park itself there are left and right beach breaks which offer good shape with larger swells.

■ Southern Pacific Coast

DOMINICAL: Very strong beach breaks with good lefts and rights. Windy, warm and popular, but frequently has riptides.

DRAKE BAY & OSA: Located next to the beautiful Corcovado National Park and accessible only by boat, this break features powerful waves and swells.

PAVONES: Considered one of the world's longest left points, Pavones' waves have an international reputation for good shape and speed. Get there by boat or bus, bring camping equipment or stay in rustic *cabinas*. Nearby are a series of world-class rights. Situated just north of the Panama border.

■ Caribbean Coast

TORTUGUERO: Besides the area's fishing, turtle watching and natural canals, the Tortuguero beaches feature good break surf. The best is near **Puerto Moín**, 15 km/9.3 miles north of Playa Bonita.

PLAYA BONITA: Its point/reef, left break has been described as "thick, powerful and dangerous." Five km/3.1 miles north of Limón.

ISLA UVITA: A little island off Limón, Uvita is the spot where Columbus first anchored in Costa Rica. It has a good but dangerous left break.

CAHUITA: The park has waves year-round. Ask about **Black Beach Cahuita**, a little-known spot with an excellent beach break.

PUERTO VIEJO: The shallow reef here makes the voluminous deep-water wave that passes over it very big and tubular. This juicy and powerful big wave is known as the **Salsa Brava**. Good surfers only.

MANZANILLO: The last surf beach before the Panama border, Manzanillo features a fast beach break.

Wild Windsurfing

For what is it to die but to stand naked in the wind and to melt into the sun?
~ Kahlil Gibran, prophet

Windsurfers are a unique breed, a hybrid breed of surfers and sailors. Although there are a few ocean areas for windsurfing (best in the Pacific northwest and Golfo Dulce), **Arenal Lake** offers ideal inland windsurfing conditions with 55-70 knot winds! Outside of Cartago there's **Cachí Reservoir**. The wind is not always consistent here, but its location close to San José and its quiet beauty make it popular for weekend warrior windsurfers.

Fantastic Fishing

El pez muere por la boca.
(The fish dies because he opens his mouth.)
~ Spanish proverb

Costa Rica's rich waters boast giant marlin, super sailfish, vigorous tarpon, record-class snook and more than a dozen hard-hitting freshwater species. On the Pacific, it is not uncommon to catch and release 30 sailfish and even a marlin or two – in just one day. Boat operators release all billfish that are not record contenders; the waters hold many world records. Even deep-sea fly-fishing nets a record number of fish.

THE FISH CALENDAR	
Here are some of the fish and their seasons, listed geographically. Keep in mind though, fish don't use calendars.	
NORTH PACIFIC WATERS	**Marlin:** Caught 12 months a year. Peak periods are from November to early March and August and September. **Sailfish:** Caught year-round, with May through August the top months. **Tuna:** Available year-round, peaking between August and October. Several fish between 200 and 400 lbs caught annually. In addition to these, there are plenty of **dorado**, **wahoo** and **roosterfish**.
CENTRAL PACIFIC WATERS	**Marlin:** Caught 12 months a year, but October and its shoulder months are best. **Sailfish:** The middle of December to the end of April is best rated, but sometimes October has big schools. **Tuna:** Peak months are June through September, but tuna is available year-round. A dozen or more 200-lb-plus fish are taken every year. **Snook:** The rainy season seems to be best; a world record Pacific black snook was caught near Río Naranjo.
SOUTH PACIFIC WATERS	The Golfito area is famous for its big **roosterfish** year-round. **Marlin:** August through December is peak. **Sailfish:** December to the end of March is best, then again in August and September. **Tuna:** August through March is the best time for the 100 pounders, but fish of up to 30 lbs are caught year-round. **Snook:** The rainy season seems to be best; a world record Pacific black snook was caught near Río Naranjo.
CARIBBEAN WATERS	**Tarpon:** Traditionally these fish are caught most often during the dry season, December through May. **Snook & Fat Snook:** Snook catches generally peak March through May, September and November. Fat snook (Calba) become plentiful November through January. **Billfish:** Out in the deep blue water are Atlantic blue marlin and Atlantic sailfish. Most are caught between February and September. Also caught in great numbers here are **wahoo**, **dorado**, **tripletail**, **kingfish**, **Spanish** and **cero mackerel**, **jack crevalle** and **barracuda**.

The Caribbean coast offers fine lodges for those in pursuit of tarpon, snook and other sport species. There is no greater thrill than hooking an acrobatic tarpon, having it sail 12 feet above the water, only to return, twist, and leap again. Plus, there is always action somewhere along Costa Rica's shore. The best time to come is whenever you can!

Living or Retiring in Costa Rica

Live long and prosper.
~ Mr. Spock, *Star Trek*

At least 25,000 North Americans and thousands of Europeans have chosen the "paradise" of Costa Rica as their home. Many North Americans and Europeans pull up stakes and begin new lives in countries such as Costa Rica and Mexico. There are many reasons to consider either living or retiring there. Moving to a beautiful country with a low cost of living can be perfect, for some, but for others it can disastrous. Do a great deal of in-depth research before you decide to move to Costa Rica.

■ Considerations

? Costa Rica is cheaper than the US for everyday living. The cost of things – fruits, vegetables, meats and most unprocessed food, mass transportation, medical care and most housing – is lower. For fixed-income retirees, dollars can stretch farther in Costa Rica than in the US. At the same time, automobiles and insurance can be more expensive, restaurant costs are comparable, and housing in especially desirable enclaves can be just as costly as some communities in the States. Low-cost housing is found in the *barrios* and neighborhoods where locals live. If your choice is based solely on monetary concerns, some places in Florida can be just as affordable. However, if you are independently wealthy, a life in Costa Rica can be very attractive.

Those expats most satisfied with their decision to live in Costa Rica have a universal trait – flexibility. They must be willing to learn the language and share the customs. In exchange for a lower cost of living in a wonderful place, expatriates must give up many things that they now take for granted; good libraries, art, cinema, and musical and cultural diversity are just a few of a long list of little perks that satellite TV won't replace. Of course, you can choose to live in neighborhoods where English is commonly spoken, but no large complexes exist in Costa Rica that are exclusively American. Besides the need to learn a fair amount of "survival

Spanish," you need to adapt to – and accept – the cultural differences. For example, Ticos operate on what we call "Tico time," which can be frustrating to those who are used to punctuality. If you've ever called a repairman and had to wait for a return call, or for them to show up for a job, you've probably experienced a similar frustration. Multiply that several times and you'll get an idea of what it is like dealing with the bureaucracies of the government, police, phone company, banks, and so on, in Costa Rica. The reluctance of Costa Ricans to be confrontational can also perplex newcomers. Ticos may seem to agree to one thing, but then do another. It's not a unique Tico trait; we know many people who do the same here in the States. Overall, the people of Costa Rica are open, friendly, honest, caring, peace-loving and fun-loving.

■ Real Estate Concerns

Foreigners can own property and invest in Costa Rica, but real estate laws and practices are different than those you are used to. If you are thinking of buying a place, find a good lawyer. Consult the list at the US Embassy or the **American-Costa Rican Chamber of Commerce** *(☎ 506/220-2200, www.amcham.co.cr)* for recommendations. Also see the Association of Costa Rican Residents, below.

Real estate agents are not required to have a license. You should choose one that is registered with the **Costa Rican Chamber of Realtors** *(☎ 506/283-0191, caccbr@sol.racsa.co.cr)*. Unfortunately, Costa Rica has more than its fair share of scam artists that have ripped off trusting souls willing to invest in anything from teak farms and beachfront property to bars and hotel projects.

> **AUTHOR TIP:** If you leave your property vacant or undeveloped, you may lose your land to squatters, who have archaic rights to take unused land. Hire a local caretaker.

Rent First

Before you buy your dream house in Costa Rica, we recommend that you rent for awhile. This will allow you to experience the different seasons and also get to know the lifestyle and the immediate community. However, it's often the case that people decide they want to live in Costa Rica *because* they find a property that is absolutely perfect. Resist the urge. Plan on living here a year before you commit to a large financial outlay. That once-in-a-lifetime opportunity will come again, and we speak from personal experience.

Some people make many trips here before deciding to make a permanent move. Though not technically legal, people commonly overstay their 90-day visa in order to experience more of Costa Rica, paying the US $50 fine (plus a small amount for every month past the 90 days) upon departure. This works best for those who stay less than one year. There are also many "perpetual tourists" that leave the country every three months for 72 hours (often to Panama or Nicaragua) and, upon re-entry, obtain a new visa. However, officials have been cracking down on perpetual tourists, so don't count on it as an option.

■ Master Plan for Your Visa

There are several categories under which you can become a lawful Costa Rican resident. If you are planning to move here, consult a lawyer who specializes in immigration.

Pensionado or Rentista

As a retiree (*pensionado*) or temporary resident (*rentista*) you can own a business, but you are not permitted to work for someone else. It's common for foreign residents living in Costa Rica to create a business as an easy way to obtain legal resident status. As a *pensionado* you must provide proof of a fixed income from abroad of at least US $600 per month; as a *rentista*, you have to show a guaranteed income of US $1,000 per month for the next five years.

> **AUTHOR TIP:** Hire an immigration lawyer to help you navigate the bureaucracy.

Your permanent resident ID Card is valid for one year and renewed at the cost of about US $100. Dependents of *pensionados* and *rentistas* may be included under your application. You are required to live in Costa Rica for four months (not necessarily consecutively) per annum.

Inversionista

An *inversionista* is an investor who lives in Costa Rica at least six months of the year and invests under one of the following three options: US $50,000 in tourism or export businesses; US $100,000 in reforestation projects; or US $200,000 in any other business. Again, use an immigration lawyer to help you through the necessary applications. After two years as an *inversionista* you can apply for permanent residency. A second method to obtain residency as an *inversionista* is to buy a home worth over US $40,000 and loan the government another US $10,000. They use the money to fund low-income housing and, after two years, you get paid back (in *colones* not dollars), with interest.

Worker (A5 or A6 Visa)

Upper management executives and ESL teachers are eligible for A-6 visas; high-level technicians and scientists are issued A-5 visas. These employees fill jobs that are considered necessary to Costa Rica's economy and development.

Family Relation (A4 Visa)

Marry a Tica or Tico, and you can be eligible for residency with an A-4 visa. If you have immediate Costa Rican relations, they can sponsor you. You will first be granted a Conditional Permanent Residency Permit (CPRP), then a Lawful Permanent Residency (LPR).

The Association of Residents of Costa Rica *(Casa Canadá, Av 4 & Calle 40, ☎ 506/233-8068, fax 233-1152, www.casacanada.net/arcr)* is an excellent source for information and advice. This non-profit organization will also handle visa applications for a reasonable fee. Advertisements in the *Tico Times* can also point you to legal and residency advice.

■ Information Sources

Books to consult for more information include:

- ***The New Golden Door to Retirement and Living in Costa Rica***, by Christopher Howard (www.liveincostarica.com).
- ***The Official Guide to Living and Making Money in Costa Rica***, by Christine Pratt (a *Tico Times* reporter)
- ***The Legal Guide to Costa Rica***, by Roger Petersen
- ***Choose Costa Rica for Retirement***, by John Howells

Maybe we'll write one someday.

For more about Costa Rican laws for real estate and residency information, visit **www.costaricalaw.com/legalnet/realty.html**.

Gay & Lesbian Travel

Vivamus, mea Lesbia, atque amenus.
My sweetest Lesbia, let us live and love.
~ Catullus, 87-54 B.C.

Despite Costa Rica's conservative Catholic sensibilities, Ticos' reputation for tolerance and acceptance has made it a popular destination for gay and lesbian travelers. But it's a little like the "don't-ask, don't-tell" policies of the US military – gays and lesbians are welcomed as long as they do not blatantly advertise their sexuality. Overt public displays of affection by same-sex partners make many Ticos uncomfortable, especially in rural areas. To a lesser degree, the same can be said of public displays of affection between heterosexual couples. Discretion is the key to having a good time.

Despite turning away a busload of gay partygoers a few years ago, the Manuel Antonio area attracts a large number of both foreign and local homosexuals. Nearly all the hotels welcome same-sex partners, we mention exclusively gay accommodations throughout the text.

For tips and complete information about gay travel, get in touch with **La Asociación Triángulo Rosa** *(☎ 506/258-0214, atrirosa@racsa.co.cr)*, a gay rights organization, or check out the English- and Spanish-language website *(www.gaycostarica.com)*. A full-service travel agency that specializes in gay and lesbian travel in Costa Rica is **Tiquicia Travel** in Barrio Amón (*Calle 3 between Avenidas 7-9, ☎ 506/256-6429, www.tiquiciatravel.com*).

In San José, several hotels go out of their way to welcome gay and lesbian travelers, like **Hotel Kekoldi** *(Av 9 & Calle 3, ☎ 506/223-3244, www.kekoldi.com)* a gay-friendly hotel that features wonderful painted murals on the walls and continental breakfast. This big green Caribbean-style Victorian property sits on a busy corner in a centrally located neighborhood. Kekoldi is a Bribri word that means "tree of water."

Colours *(150 meters west of Farmacia Rohrmoser; US ☎ 800/277-4825, CR ☎ 506/296-1880, www.colours.net; pool, TV, restaurant & bar, breakfast included)* is a gay guest home in the Rohrmoser neighborhood of San José. Popular center of social and group tourism activities.

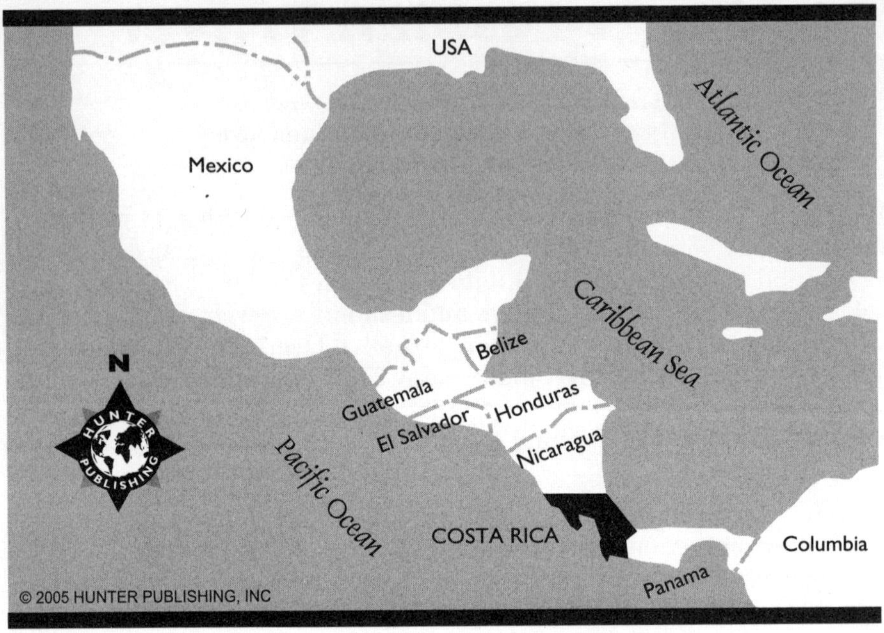

Travel Essentials

There is a third dimension to traveling,
The longing for what is beyond.
~ *The Silk Road*, Jan Myrdal

So you're off to the Costa Rica for a vacation adventure. Good choice. You'll be welcomed by friendly people and offered a chance to get to know them and their beautiful country – Ticos are justifiably proud of it. While you are there, remember to respect both them and the environment. After all, you may be on vacation, but they live there.

INFORMATION SOURCES

To get general information and brochures about Costa Rica, call the **Tourism Hotline** (*☎ 800/343-6332*). On-line information is available from the Tourist Board (ICT, Instituto Costarricense de Turismo) at **www.visitcostarica.com**, and a host of commercial sites (see page 377).

■ Entry Requirements

For stays of up to 90 days, Americans and Canadians need proof of citizenship, such as a passport or certified birth certificate and a photo ID. Europeans and residents of South American countries must have a current passport. No visas or inoculations are required. Not only is a passport better for airport security reasons, but it also carries more authority with bureaucratic Costa Rican officials. Avoid the hassle and get one in advance. If your passport expires within six months of your departure date, it's better to get a new as some airlines may prevent you from boarding due to immigration laws.

AUTHOR TIP: When you're in Costa Rica, lock your passport in the hotel safe and carry a photocopy (made before you leave home) when you go out. It's a safe way to have an ID.

If you know you'll need an extended visa before you go, or need more specific information, contact the appropriate Costa Rica embassy: in **Canada**, 135 York Street, Suite 208, Ottawa, Ontario K1N 5T4, ☎ 613/562-2855; in the **United Kingdom**, 14 Lancaster Gate, London W2 3L, ☎ 0171-706-8844; and in the **United States** at 2112 S Street NW, Washington, DC 20008, ☎ 202/328-6628. Consulates are in **New York** (☎ 212/509-3066), **Los Angeles** (☎ 213/380-6031), **Chicago** (☎ 312/263-2772), **Houston** (☎ 713/266-0484), **Miami** (☎ 305/871-7485), **Atlanta** (☎ 770/951-7025), **New Orleans** (☎ 504/581-6800) and **Denver** (☎ 303/696-8211).

Passports

US passport application forms can be obtained at major post offices, courthouses or federal passport offices in larger metropolitan areas. You can also download an application from the State Department at **http://travel.state.gov**. Apply two or three months before your trip. The site above also offers tons of information, including a very useful **Consular Information Sheet** that contains special safety or health alerts.

Canadians can apply at the **Central Passport Office** in Ottawa, ☎ 800/567-6868. For **Brits** it's the **London Passport Office**, ☎ 0171-271-3000.

I was well acquainted with the gag that if you look like your passport picture you need a vacation. But I was unprepared for the preponderance of thug-like pictures I found in the course of processing them.

~ Frances G. Knight, Director US Passport Division

US Passport Agencies

If you didn't leave it until the last minute, apply for an American passport through your local Clerk of Court or post office. Federal Passport Agencies tend to have long lines during the busiest months.

■ BOSTON

Thomas P. O'Neill Federal Building
10 Causeway Street, Suite 247
Boston, MA 02222-1094
Open 9 am-4 pm
Region: Maine, Massachusetts, New Hampshire, Rhode Island, upstate New York and Vermont

■ CHICAGO

Kluczynski Federal Building
230 S. Dearborn Street, Suite 380
Chicago, IL 60604-1564
Open 9 am-4 pm
Region: Illinois, Indiana, Michigan and Wisconsin

■ HONOLULU

First Hawaiian Tower
1132 Bishop Street, Suite 500
Honolulu, HI 96813-2809
Open 9 am-4 pm
☎ (808) 522-8283
Region: American Samoa, Federated States of Micronesia, Guam, Hawaii and Northern Mariana Islands

■ HOUSTON

Mickey Leland Federal Building
1919 Smith Street, Suite 1100
Houston, TX 77002-8049
Open 8 am-3 pm
Region: Kansas, Oklahoma, New Mexico and Texas

■ LOS ANGELES

Federal Building
11000 Wilshire Blvd., Suite 13100
Los Angeles, CA 90024-3615
Open 8 am-3 pm
Region: California (all counties south of and including San Luis Obispo, Kern and San Bernardino) and Nevada (Clark County only)

■ **MIAMI**
Claude Pepper Federal Office Building
51 SW First Avenue, 3rd Floor
Miami, FL 33120-1680
Open 9 am-4 pm
Region: Florida, Georgia, Puerto Rico, South Carolina and US Virgin Islands

■ **NATIONWIDE**
31 Rochester Avenue
Portsmouth, NH 03801-2900
Open 9 am-4 pm
Handles applications for passports by mail (Form DSP-82).

■ **NEW ORLEANS**
Postal Services Building
701 Loyola Avenue, Suite T-12005
New Orleans, LA 70113-1931
Open 9 am-4 pm
Region: Alabama, Arkansas, Iowa, Kentucky, North Carolina, Louisiana, Mississippi, Missouri, Ohio, Tennessee and Virginia (except DC suburbs)

■ **NEW YORK**
Rockefeller Center
630 Fifth Avenue, Suite 270
New York, NY 10111-0031
Open 7:30 am-3 pm
☎ (212) 399-5290
Region: New York City and Long Island
Note: New York Passport Agency accepts *only emergency* applications from those leaving within two weeks.

■ **PHILADELPHIA**
US Custom House
200 Chestnut Street, Room 103
Philadelphia, PA 19106-2970
Open 9 am-4 pm
Region: Delaware, New Jersey, Pennsylvania and West Virginia

■ **SAN FRANCISCO**
95 Hawthorne Street, 5th Floor
San Francisco, CA 94105-3901
Open 9 am-4 pm
Region: Arizona, California (all counties north of and including Monterey, Kings, Oulare and Inyo), Nevada (except Clark Co.) and Utah

■ SEATTLE
Henry Jackson Federal Building
915 Second Avenue, Suite 992
Seattle, WA 98174-1091
Open 8 am-3 pm
Region: Alaska, Colorado, Idaho, Minnesota, Montana, Nebraska, North Dakota, Oregon, South Dakota, Washington and Wyoming

■ STAMFORD
One Landmark Square
Broad and Atlantic Streets
Stamford, CT 06901-2667
Open 9 am-4 pm
Region: Connecticut and Westchester County (New York)

■ WASHINGTON
1111 19th Street, N.W., Room 300
Washington, DC 20522-1705
Open 8 am-3 pm
Region: Maryland, northern Virginia (including Alexandria, and Arlington and Fairfax counties) and the District of Columbia

■ Measurements

Measure twice, cut once.
~ carpenter's rule

Costa Rica remains on **Central Standard Time** year-round. Weights, measures and temperatures are **metric**. Gasoline is sold by the liter and distances are measured by kilometers. **Dates** are expressed with the day first, followed by the month. For example, Christmas Day is written 25/12. Often, hours are posted in military time, a 24-hour format. Two in the afternoon is 14:00, 10 at night is 22:00. If you're confused about it, simply subtract 12 hours from the time given after noon until midnight.

■ What to Take

Everyone does it. It starts with "what if?" – "What if it rains?" "What if it's cold?" "What if we go out to a fancy place?" "What if you spill ketchup on your shirt and my dress gets caught in a taxi door?" Then we wind up packing too much.

So here's the scoop – pack light, then take all your stuff back out and cut the quantity in half. The idea is to bring only enough clothes to last until you get to a laundromat, washing machine or store. Try to take no more than four to five days' worth of clothes, no matter how long your trip. In

Going Metric?

To make your travels a little easier, we have provided the following charts that show metric equivalents for the measurements you are familiar with.

GENERAL MEASUREMENTS

1 kilometer	=	.6124 miles
1 mile	=	1.6093 kilometers
1 foot	=	.304 meters
1 inch	=	2.54 centimeters
1 square mile	=	2.59 square kilometers
1 pound	=	.4536 kilograms
1 ounce	=	28.35 grams
1 imperial gallon	=	4.5459 liters
1 US gallon	=	3.7854 liters
1 quart	=	.94635 liters

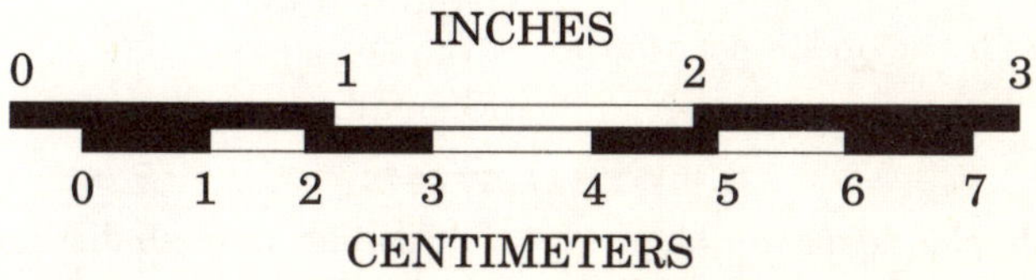

TEMPERATURES

For Fahrenheit: Multiply Centigrade figure by 1.8 and add 32.

For Centigrade: Subtract 32 from Fahrenheit figure and divide by 1.8.

CENTIGRADE		FAHRENHEIT
40°	=	104°
35°	=	95°
30°	=	86°
25°	=	77°
20°	=	64°
15°	=	59°
10°	=	50°

San José you'll find a number of self-service and drop-off laundromats (*lavanderías*) that will take care of your clothing for you.

PACKING LIST

CLOTHING	TOILETRIES
❒ comfortable walking shoes with non-skid soles	❒ shampoo
❒ several pairs of cotton socks	❒ toothbrush and toothpaste
❒ underwear	❒ prescription medicine
❒ sandals	❒ comb/brush
❒ one/two pair no-iron slacks (great when jeans are too heavy).	❒ women: make-up as desired
❒ one/two pair shorts	❒ personal hygiene products
❒ women: two modest lightweight dresses or skirts	❒ sunscreen, UVA/UVB protection SPF 30+ (sunscreen takes 30 minutes to become effective after application)
❒ two shirts or blouses	❒ good insect repellent, a must for back country travel
❒ three T-shirts	❒ aloe gel for sunburn
❒ swimsuit (carry it with you everywhere)	❒ Imodium AD and aspirin
❒ sleepwear	❒ triple antibiotic cream
❒ light cotton sweater or sport jacket	❒ Band-Aids
❒ water-resistant windbreaker	❒ facecloth
❒ hat and sunglasses	*If you're roughing it, also bring*
❒ fanny pack and money belt	❒ bar soap
❒ umbrella	❒ hand towel
❒ flip flops for use on slippery bathroom floors	❒ toilet paper
	❒ small size sink stop

OTHER ITEMS

❒ resealable plastic freezer bags (multiple uses)	❒ Spanish-English dictionary or phrasebook
❒ antiseptic towelettes	❒ sports watch
❒ handheld calculator	❒ earplugs
❒ small flashlight	❒ passport, ID, driver's license
❒ sharp penknife (be sure to pack this in luggage you are checking)	❒ paperback novel (exchangeable at many hotels)
❒ notebook and two pens	❒ sewing kit with small scissors (packed in baggage being checked)
❒ sun shield (stick-on dark plastic or round pop-open type), if renting a car	❒ laundry bag
❒ camera, film and extra camera batteries	❒ address book
	❒ credit cards and an ATM card
	❒ phone card
	❒ this book (very important)

> **AUTHOR TIP:** Take older clothes that you won't miss if you stain them or the laundry fades or shrinks them. You can always buy more there.

Toiletry needs depend on what you're doing and where. The better hotels have shampoo and soap – just about everything else is readily available in stores and supermarkets, so don't bring a gallon. Leave any open toiletries and unwanted clothes in Costa Rica.

> **SHUTTER BUGS:** Get your camera fixed or buy a new one at **Cameras Equipos** *(☎ 506/224-4186, Nikon Pentax, Mamiya),* **Taller de Equipos fotograficos** *(☎ 506/233-0176, Canon)* or **Tecfot SA** *(☎ 506/222-0996, Minolta, Olympus, Fuji).* All are located in San José.

■ Customs & Immigration

Rules on the **possession of drugs or firearms** are fairly simple – don't have any, period. Drug possession penalties in Costa Rica include prison terms and the US embassy can do little to help. It's just not worth it. You may find marijuana treated lightly, especially in beach towns along both coasts, but don't be convinced that if you're caught nothing will happen.

When you arrive at the newly remodeled Juan Santamaría airport, the Customs check requires you to push a button in a red light/green light system. If the light goes red, they look in your luggage. Airport security has tightened noticeably after September, 2001, and delays are more common.

On your return to the United States you can bring in goods, valued at up to US $400 per person, duty free. Canadians are allowed CAN $500. Fruit and uncanned foodstuffs are forbidden. Thankfully, coffee is allowed. Costa Rican coffee is heavenly.

US residents may not bring Cuban cigars back into the States Residents of other countries should check with their Customs office. We sincerely hope that archaic embargo is ended by the time you read this. Archeological artifacts or items made from endangered species are strictly forbidden.

■ Way to Go

Flying to Costa Rica

In 1999 for the first time, Costa Rica welcomed over one million visitors, a majority from North America. Nearly all arrived by plane into San José's **Juan Santamaría International Airport**, 20 minutes from downtown San José. A whole flock of airlines from around the world service Costa Rica, including one of our favorites **US Airways** *(☎ 800/622-1015, www.usairways.com)*; **Continental** *(☎ 800/525-0280, www.continental.com)*, with twice daily flights from Houston and one from Newark; **American** *(☎ 800/433-7300, www.aa.com)* from Miami and Dallas; **Delta** *(☎ 800/221-1212, www.delta-air.com)* with links from Atlanta; **United** *(☎ 800/241-6522, www.united.com)* connects in Mexico City or Guatemala from Chicago, LA and DC; and **LACSA** *(☎ 800/535-8780, www.grupotaca.com)*. LASCA was formerly the national airline of Costa Rica and is now part of a Central American conglomerate with many direct flights and connections. It takes somewhere between three and seven hours to fly in from most major US cities to San José. Some charters are now flying direct to Liberia's new international airport, making it convenient if you're going to Guanacaste or Nicoya.

Students and faculty might do well phoning **Council Travel** *(☎ 800-226-8624, www.counciltravel.com)*.

Regional airlines connect to various small airports around the country from either Juan Santamaría, the main airport in San José, or Tobias Bolano in the Pavas suburb. See the *Cross-Country Flights* section on page 98.

Recommended Operators

Any good US travel agent can make advance arrangements with a number of reputable tour wholesalers such as Tico-owned **Nature Tours** *(☎ 800/444-3990)*, based in Louisiana, who have many years experience in Costa Rica. You might even try contacting them directly – ask for Carlos or Marinela. They're very knowledgeable. In addition, several good tour companies in the States can make all or many of your arrangements for you. Expert Costa Rica tours can be made with **Holbrook Travel** *(☎ 800/451-7111, www.holbrooktravel.com)*, who owns several properties in Costa Rica so it claims a vested interest in tourism. They are an excellent agency with a good track record. Other operators are **Costa Rica Experts** *(☎ 800/827-9046, www.costaricaexperts.com)*; **Costa Rica Connection** *(☎ 800/345-7422, www.crconnect.com)*; **Tropical Travel** *(☎ 800/451-8017, www.tropicaltravel.com)*; and **Rico Tours** *(☎ 800/280-7426, www.ricotours.com)*. All specialize in Costa Rica. One of the better eco-adventure vacation agencies, especially for Costa Rica's Pacific coast destinations, is San Diego-based **Pacific Conservation Services** *(US ☎ 619/835-9705, www.conservationvacation.com).*

Costa Rica Trekking *(☎ 506/771-4582, www.chirripo.com)* has professional guides that run three organized bike treks in the Talamance Range, including one in Chirripó.

Also see the section on *Local Adventures, Tours & Day-Trips* in San José (page 129) for local Costa Rican-based tour operators and agencies who can arrange your entire vacation or just day-trips and activities.

■ Coming & Going

If you don't know where you are going,
you will probably end up somewhere else.
~ Lawrence Johnston Peter, 1919-1990

Arriving

Unless your hotel is picking you up at the airport, the main choice of transportation upon arrival in Costa Rica is by taxi. One-way taxi tickets are sold at the airport at a fixed price of around US $15.

The less common choice of transport directly to downtown is by public bus, which you can catch across from the airport entrance. Ask the driver to be sure the bus you're on is heading for San José. It drops you downtown across from the Parque Merced, Av 2 and Calle 12. The bus isn't a good choice if you're carrying a lot of luggage.

There are numerous car rental desks at the airport; their shuttle will take you to your auto. If you plan to explore only around San José, consider that a car is a liability in the traffic and parking is hard to find. Public buses or taxis are the way to go.

Leaving

If airport transfers aren't included in your hotel price, it's most convenient to take a taxi to the airport for your departure. Or perhaps you can make a deal with a van driver at one of the hotels who pick up for the package vacation crowd. Agree on the price before you leave.

From downtown, a public bus takes passengers with luggage to the airport. The bus terminal is the one for Alajuela, Av 2 and Calle 12, across from Parque Merced. Service is very frequent and takes a half-hour or less.

Don't forget to leave yourself enough money to pay the **departure tax**, about US $26. You can pay it at any San José travel agent's office or at the airport. If you pay at the airport, avoid the men outside the terminal who claim to be legit and will happily collect your money and even provide a receipt. It's not legal. Flight reconfirmation and reservation phone numbers in San José are: **Continental** *(☎ 506/296-4911)*, **American** *(☎ 506/257-1266)*, **Delta** *(☎ 506/257-3346)*, **LACSA**, *(☎ 506/ 296-0909)*, **United** *(☎ 506/220-4844)* and **US Airways** *(☎ 506/520-0507)*.

The Land Between the Oceans

Little drops of water,
Little grains of sand,
Make the mighty ocean
And the beauteous land.
~ *Little Things*, Julia A. Carney, 1823-1908

IN THIS CHAPTER

The country known as Costa Rica – Spanish for "Rich Coast" – is part of a thin isthmus land bridge that joins the large continents of North and South America. Although part of what we know as Central America, Costa Rica is a geographic component of continental North America. It has two distinct coastlines: the Atlantic (or Caribbean) coast, and the long Pacific shore. Its diverse beaches attract a multiplicity of wildlife and its waters are home to gigantic whales as well as tiny mollusks in delicate circular shells.

The Pacific coastline wiggles and stretches 1,016 km (631 miles) around three peninsulas – **Buriya**, **Nicoya** and **Osa** – and in the process forms two significant gulfs, **Golfo de Nicoya** and **Golfo Dulce**. Although it has a narrow land mass – little more than 300 km (200 miles) at its widest – a series of rugged *cordilleras* (mountain ranges) part its length and form what's also known as the great Continental Divide. It has been reported that from atop Costa Rica's highest point, Cerro de Chirripó (Mount Chirripó) at 3,820 meters (12,530 feet), you can see both oceans. The same has been reported at Irazú Volcano (3,432 meters/11,257 feet), the highest volcano, but you would have to be very lucky because clouds and haze obscure the view. High mountain ranges separate the climate between east and west coasts. Occasionally, when one half of the country is being drenched by rain, the other may be having a sunfest.

Geography

The Art of Biography, Is different from Geography.
Geography is about Maps, But Biography is about Chaps.
~ Edmund Clerihew Bentley, 1875-1956

Bordered to the north by Nicaragua and to the south by Panama, Costa Rica is squeezed into 51,100 square km (19,730 square miles) between the Caribbean/Atlantic Ocean to the east and the mighty Pacific to the west. This figure includes the small but important offshore island, **Isla del Coco** (25 square km), made famous as the story location for the *Jurassic Park* movies. The protected island national park is a favorite dive and nature tour spot, but only for the dedicated – it lies at about 5°30′ latitude north and 87°05′ longitude west, 535 km (335 miles) from Cabo Blanco in the Pacific Ocean.

Costa Rica is a little smaller than the state of West Virginia, and in Central America only El Salvador and Belize are smaller. More than half of the country's's 3.6 million inhabitants live in the Central Valley, or Meseta Central. **San José** is the capital. It's lively, colonial, disarming, crowded, charming, shabby, demanding, easy-going, difficult, parochial, modern and fast-paced – all at one time! It's also home to the government and a fairly large percentage of the population. San José lies nestled in a 1,178-meter/3,875-foot hollow surrounded by mountain peaks.

Costa Rica is divided into seven geo-political provinces: **Guanacaste** and **Alajuela** make up the country's northwest corner; **Puntarenas** hugs most of the Pacific shore; **San José** and **Cartago** form the heartland; **Heredia** runs north from San José to the Nicaraguan border; and **Limón** occupies the entire Atlantic coastline, sharing the southern half of the country with Puntarenas.

Although not a large country, Costa Rica's topography provides a divergence of air currents, affecting precipitation. A dry tropical clime distinguishes its northwest corner – in the province of Guanacaste and on the Nicoya Peninsula – and a tropical rainforest covers the south. The northwest attracts visitors drawn to the sunny Pacific beaches. Along the southern Pacific and all of the Caribbean coastline the climate is humid and tropical, with many areas measuring their annual rainfall in meters. Inland and down to the shore itself, the rainforest reigns supreme.

■ Mountains

Of the four major mountain ranges that form Costa Rica's spine, the oldest and tallest is the **Cordillera de Talamanca**, in the southern half of the country. Because of its non-volcanic origin, it boasts 16 distinct summits that exceed 3,000 meters (9,868 feet), including **Chirripó**, the tallest. The northern mountain ranges are the **Cordillera de Guanacaste** and the smaller **Cordillera de Tilarán**. The Tilarán range abuts the **Cordillera Volcánica Central** that surrounds San José in the middle of the country.

The broad depression between the Central mountains forms the **Meseta Central**, a valley at a height of between 1,000 and 1,500 meters (3,289 and 4,934 feet) above sea level. The volcanic soil and temperate climate make this region the country's agricultural breadbasket.

There are two very different lowlands to either side of these four mountain ranges. Low-lying mangrove marshes, rivers, natural canals and soft sandy shore mark the 212-km (133-mile) Atlantic coastline. The uneven Pacific coastline features a rocky base, with dramatic bluffs and hidden beaches.

■ Rivers

The reflections of the moon on one thousand rivers are from the same moon: The mind must be full of light.
~ Hung Tzu-ch'eng

When clouds from the onshore winds reach the cooler elevations of the highlands, they release their moisture as rain – lots of rain. Numerous rivers, large and small, drain the water either east or west, back to the sea. From early times until today, many of these rivers have served as a means of transportation in the mountainous countryside and low-lying mangrove flatlands.

Rancheros *row cattle across a river.*

Major transportation arteries of interest to visitors include the following:

- **Tortuguero** and the **Barra del Colorado** canals in the northern Caribbean, **Chirripó Caribe, Río Estrella** and the **Matina River** in the central and southern Caribbean;

- **Río San Juan**, which borders Nicaragua, and the **Sixaola**, on the Panama border;
- **Bebedero** and **Tárcoles** in the north Pacific, the **Río Frio** and the **Caño Negro** lakes near Los Chiles in the north;
- **Sierpe** and **Río Grande de Térraba** in the southern region;
- **Tempisque** on the Gulf of Nicoya.

And of course the ones that attract the most attention are the rafting rivers that have made Costa Rica famous in adventure travel. There are Class III-IV+ rapids on the these rivers: **Sarapiqui**, **Parrita**, **Pascua** (V)**, Peñas Blancas**, **Toro**, **Reventazón** and, our all-time favorite, the **Pacuare**. The large **Lake Arenal** offers world-class wind surfing, and **Río Corobici** features an easy raft float.

■ Beaches

The coconut trees, lithe and graceful, crowd the beach...
like a minuet of slender elderly virgins, adopting flippant poses.
~ William Manchester

The beaches (*playas*, pronounced PLY-yahs) of Costa Rica are as diverse as the country itself – you'll find a mixed bag of soft white sand, black volcanic sand, pebble beaches, brown or gray gritty sand, muddy mangroves, rocky outcroppings, soft lapping waves, roaring surf, hidden reefs, spectacular dawns and breathtaking sunsets.

If you are looking for a beach vacation with lots of sun, sand and surf, the beaches of **Guanacaste**, on the Nicoya Peninsula, offer some of the best. Stiff competition comes from the Caribbean beaches south of Limón down to the border with Panama – specifically **Cahuita** and **Puerto Viejo**. Surfers should check out our admittedly amateur evaluation of the best beaches to catch a wave on page 16.

■ Volcanoes, Nature's Hot Spots

We are dancing on a volcano.
~ Narcisse Achille, 1795-1856

The most spectacular and exciting feature of Costa Rica's central and northern mountain ranges are the active volcanoes that form their high summits. Active not just a millennium or two ago; several of your friendly neighborhood volcanoes are still going strong. These are **Poás**, **Irazú**, **Barva**, **Orosí**, **Tenorio**, **Rincón de la Vieja**, **Santa Maria**, **Miravalles** and, the most active volcano of southern Central America, **Arenal**.

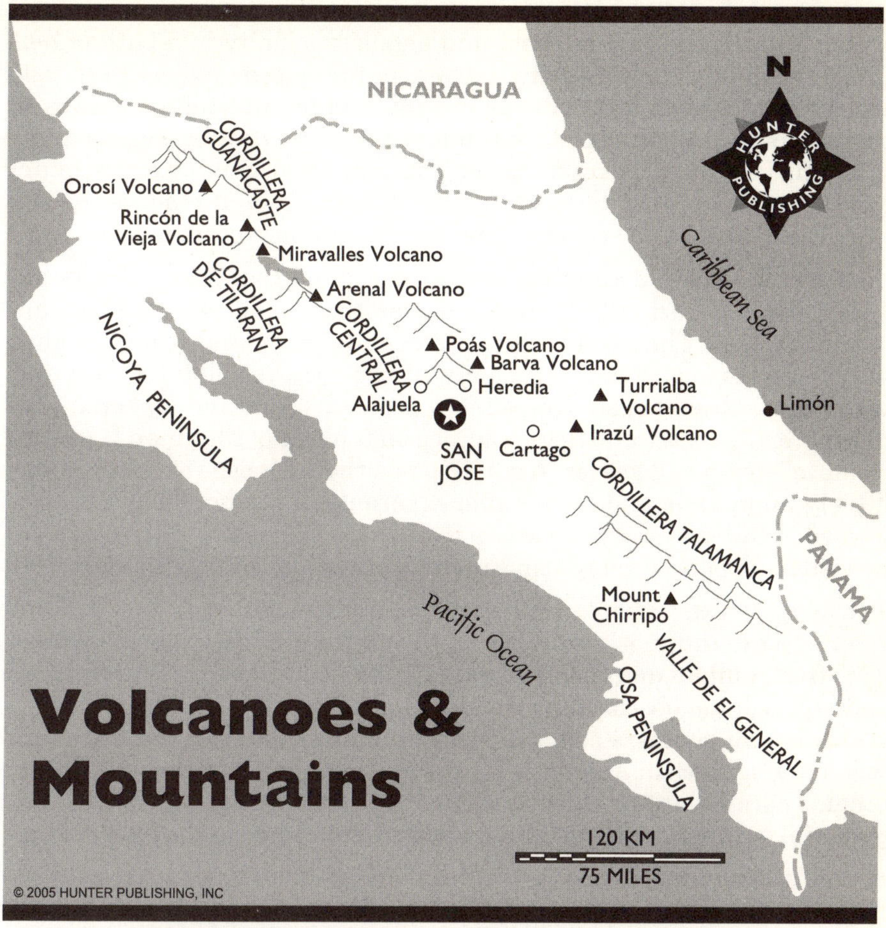

FIERY PHENOMENON

The word "volcano" comes from the ancient Romans, who believed the tiny island of Vulcano, off northern Sicily, was the location of the fiery forge where the god Vulcan tempered Jupiter's thunderbolts. The myth reflects modern scientific findings closely, both for the tremendous hot fire of eruptions as well as the fact that volcanoes attract lightening strikes around their peaks.

The country's quieter mountainsides, fertile with the soil of eruptions eons ago, make ideal land for farmers. Most of these volcanic peaks are classified as extinct or dormant. But you can never be sure, as the small farming village of Tabacón, on the slopes of Arenal, found out in 1968. The entire village disappeared under a lava flow.

Only a few kilometers below our feet, a sea of hot molten magma waits to break through the earth's crust and engulf humanity in a bath of hellfire. Fortunately, the only place it manages to succeed is where the crust's tectonic plates meet and grind against each other. In addition to causing earthquakes, these cracks form vents, which allow the hot magma to rise to the surface where it spews up, only to cool down, once again becoming solid rock. Lava is the term for molten magma when it reaches the surface. Rock upon rock, eruption after eruption, a mountain is formed.

Visiting the easily accessible volcanoes around San José – **Poás** and **Irazú** – is not particularly dangerous as these now vent only gas and steam, not lava. Their water-filled craters and lunar landscapes are absolutely fascinating. From the safety of the viewing platform, we once watched a volcanologist, dressed in a protective suit and air tank, walk down into the crater and take readings on the side of the green lake amid clouds of steam and gasses. A day-trip to either of these volcanoes makes a truly impressive and memorable experience – certainly one not to be missed. Other active volcanoes up the *cordillera* require a drive and can be visited as part of a day-trip itinerary or with an overnight stay.

The biggest attraction – and growing even bigger – is the dynamic **Arenal Volcano**, near La Fortuna. Arenal is a huge, cone-shaped volcano that rumbles and shoots rocks, lava, smoke and ash out its two main vents. When not obscured by clouds, it presents itself as a massive dark triangle with boulders as big as cars tumbling down its sides. At night, when the sky is clear and the lava flowing, it displays a light show like no other. There's a pretty real sense of danger from being so close to such a powerful giant, but this thrill only adds a great deal to the experience.

Arenal's appealing bonus is the hot springs that bubble up from the ground to feed warm-water streams – popular as simple swimming holes or incorporated into a spa. Due to the volcano's activity in the last few years, tourism has increased dramatically, with many new hotels and restaurants opened in the area.

■ Earthquakes

But did thee feel the earth move?
~ *For Whom the Bell Tolls* (1940), Ernest Hemingway, 1898-1961

Costa Rica welcomed us on our very first night in San José with a *temblor*, a minor earthquake. Centered outside Cartago, 35 km/22 miles away, it measured about 4.0 on the Richter scale. People from California may hardly notice these kinds of rumbles, but it woke us East Coasters from a sound sleep.

An earthquake, called *terremoto* (moving earth) in Spanish, is caused by the shifting of sections of the earth's outer shell (tectonic plates). A severe earthquake can release as much as 10,000 times the energy of the first atomic bomb. Major earthquakes are relatively rare in Costa Rica, although the country has very active underground faults. In fact, Costa Rica is second only to Guatemala in Central American seismic activity.

THE RICHTER SCALE

The scale that measures earthquakes is named after North American seismologist Charles F. Richter, who developed the magnitude numbering system in 1935. Each number of the scale represents a tenfold increase in the amplitude of ground movements as measured by a seismograph. An earthquake measuring 6.0 is 10 times as severe as one of 5.0.

Costa Rica's worst earthquake in recent times occurred on April 22, 1991, when a 7.1 *terremoto* struck the Caribbean side of the island, causing severe damage to Limón and destroying rail and many road connections between there and San José.

Environment

The rain it raineth on the just
And also on the unjust fella:
But chiefly on the just, because
The unjust steals the just's umbrella.
~ *Sands of Time*, Lord Bowden, 1835-1894

■ Climate

As in most Central American countries, you can choose your climate in Costa Rica not so much by moving north or south, but by changing your altitude. You could also change your attitude, but that's a different story. The beach communities at sea level on both coasts can be very hot and humid, while the climate of the Central Valley, for example, at 1,500 meters/4,934 feet, is spring-like year-round.

Unlike most of North America, Costa Rica has only two seasons: the rainy season, which runs from May to November, and the dry season, from December through April. During the rainy season, rain generally falls in the afternoon and evenings; mornings are frequently sunny and bright. And, depending on your location, it doesn't necessarily rain every day. Then again, we once sat through a solid three-day downpour deep in the rainforest. The sound that heavy raindrops make on the thick foliage of the forest is hypnotic and reassuring that life goes on. In the dry season clouds can also roll in and precipitate, especially in the afternoons.

The dry season is more popular time with tourists. The rainy season has its advantages, though there is little difference between the summer and winter temperatures. The intense rainfalls you may encounter are known as *temporales*, while the soft rains of March are known as *aguaceros de cafeleros*, or coffee grower's showers.

Overall, Costa Rica's climate is a very pleasant "tropical," with an average temperature of 22°C (72°F).

■ Ecology

To have arrived on this earth as the product
of a biological accident, only to depart through human
arrogance, would be the ultimate irony.
~ Richard Leakey, anthropologist

Its reputation as an ecological paradise is what first drew us to visit Costa Rica – and it keeps us coming back. Costa Rica has been described as the "poster child" for ecological protection for the world, and with that label comes a great deal of responsibility – and public scrutiny. Costa Rica boasts 25% of its land area dedicated as "wildlife protected," and is rightfully proud of its remarkable national park system. Its conservation effort is important because of its unique position in the north-south corridor between the two larger American landmasses. A primary stop on the evolutionary highway, it has an incredible multiplicity of individual animal, plant, insect and bird species. Tiny Costa Rica may cover only 0.03% of the planet's surface, but is the natural habitat for as much as 5-6% of the world's bio-diversity.

It leads the world, including the United States, in meaningful ecological preservation efforts. Each year thousands of students research and study in its natural outdoor laboratories and thousands more tourists enjoy eco-touristic diversions that help make the forests more valuable in their natural state than either logged or cleared for farming. Unique "biological corridors" of undeveloped land have recently been created that, at least on paper, will allow animals to move unhindered between protected areas, instead of trapping them in pockets of parks.

The de-centralized overseer of forests, wildlife and protected areas is the Sistema Nacional de Areas de Conservacion (**SINAC**), governed in turn by the Ministry of Environment and Energy (**MINAE**). They pursue the goal of reaching sustainability in safeguarding the country's natural resources.

INBIO PARQUE

InBio is the **National Biodiversity Institute** *(www.inbioparque.org)*, a private, non-profit organization whose mission is "to promote a greater awareness of the value of biodiversity and thereby achieve its conservation and improve the quality of life for society." This ambitious biological and educational project is actually attempting to record and sample every species of plant and insect life in Costa Rica. North American universities and drug giants such as Merck are helping fund the project in the hopes of finding new medicines and cures from the rainforests' plants. Species collections and park trails are open to the public. Guided tours available. See page 133.

But despite Costa Rica's successes in land management, it would be Pollyannaish to ignore the half of the glass that's empty. Costa Rica lost almost half of its forest cover between 1950 and 1990. In 2000 the government sold oil-drilling rights all over the country, including the ecologically sensitive offshore in the southern Caribbean, despite heavy environmental opposition. Fortunately, that decision may be held up in court for many years. Illegal miners, cattle ranchers and loggers still operate, at times openly, in protected areas. And big business and government still sometimes think with the old "bigger-is-better" tourism theories, approving mega-developments such as the attempted imitation of Cancún on Papaguayo Bay and the infamous resort hotel projects by the Spanish hotel chain, Barceló. (We go out of our way in this guide not to publicize any properties we believe have caused an inordinate amount of ecological damage.)

Earth has its boundaries but human stupidity is limitless.
~ Gustave Flaubert, French novelist, 1821-1880

A close look at the 25% preserved land figure reveals that only two reserves (a relatively minuscule 3,285 acres) are "Absolute," which means they have absolutely no human interference. Property that belongs to the government, wetlands, "protected areas" and "other protected wildlife areas" make up 6% of the 25%. These are called "protected" on paper, but often lack oversight against poachers, loggers, miners and squatters. The sheer size of some of these properties makes them hard to manage, and there's also a lack of money to pay rangers and park workers to patrol and maintain. Ironically, with the success of conservation efforts comes a perception by other countries and outside environmental groups – who have helped fund these projects in the past – that Costa Rica is in good shape. Funds from such agencies are now being directed to other places around the world in dire need of protection.

The remaining preserved land is designated as Biological Reserves (0.4%), National Wildlife Refuges (3.4%), Forest Reserves (5.5%) and the well-known National Parks system (11%). The phenomenon of "private reserves" – acreage bought and protected as natural areas by individuals or organizations – has supplemented the national effort and contributed in no small measure to the patchwork of protected areas.

Fortunately for everyone concerned, two or three generations have matured since local and foreign environmentalists first raised the public's ecological consciousness to the level that helped create the national park system. Today's *Ticos*, as Costa Ricans call themselves (short for *hermaniticos*, little brothers), tend to support the preservation of their natural heritage, even when it comes out of their pocket. So far, continued funding from world environmental organizations and the financial success of local eco-tourism programs in poorer sections of the country, helps keep pressure on the federal government. Environmental regulations are enforced and, in some instances, the government has even created new reserves and is expanding the acreage of existing protected areas.

All of this is good news for the world, the environment and for all of us who enjoy eco-tourism as a great way to travel.

■ Eco-tourism

We abuse land because we regard it
as a commodity belonging to us.
When we see land as a community to which we belong,
we may begin to use it with love and respect.
~ Aldo Leopold, American ecologist, 1886-1948

The first tourism for pleasure was confined to the rich and powerful, an indulgence for the Greek and Roman civilizations. In the 18th and 19th centuries, wealthy Europeans and Americans toured the cultural centers of Europe or vacationed in country summer mansions. But everyday people stayed home and worked. Thomas Cook organized English rail excursions as early as 1840, when a middle class emerged from the industrial revolution. But mass tourism only got a start 100 years later, when air travel began to shrink the world. Inevitably, that innovation brought developing "Third World" countries within reach as leisure destinations.

In an effort to attract tourists to sun, sand and surf, countries such as Mexico and Jamaica licensed builders to change or exploit the natural environment to create resort areas. Slowly it became apparent to many that the costs of such developments – both environmental and social – could be higher than the benefits. Eco or "green" tourism developed with the heightened environmental awareness of the 1960s and 70s. It is generally defined as low-impact activities in nature that help preserve and sustain the environment and benefit native people. The hope is that this type of tourism can prevent and even reverse environmental destruction. The tourist trade in Costa Rica, in a country described by one naturalist as "one big safari park," is a test of that hope.

It's Not Easy Being Green

Unfortunately, the designation "eco-tourism" is much like the nomenclature "organic" – it means different things to different people. Everyone at least agrees on the objective: a "win-win" situation for the environment, the tourist, the travel industry and the local people. But whether it actually achieves its promise is another issue.

> For an insightful and in-depth look at the problems and solutions of green tourism, read ***Ecotourism and Sustainable Development***, by Martha Honey, published by Island Press.

Various environmental organizations have similar definitions, but it is generally agreed that true eco-tourism projects should meet these seven criteria:

1. Covers tourism only to natural areas
2. Minimizes ecological impact
3. Builds environmental awareness
4. Provides direct financial benefits for conservation
5. Provides financial benefits and empowerment for local communities
6. Respects local culture
7. Supports human rights

When traveling, it is very easy to slip into enjoying "eco-tourism lite," a phrase coined by Ms Honey. In this mindset, we whiz through the forest canopy on high wires, neither appreciating the forest nor edifying ourselves – but having tons of fun. On a higher scale, hotel chains and tourism operations can claim they are green and benefit the environment because they use recycled toilet paper or biodegradable soap. That is not wrong in itself – every little bit helps – but it is not a particularly big part of the solution and shouldn't be advertised as such. The "lite" experience tends to enjoy nature without being overly concerned about its preservation. So remember, if you are concerned, when accommodations are chosen, small is beautiful.

The Certificate of Sustainable Tourism

In response to the demand for more green tourism, Costa Rica's government authorities developed a Certificate of Sustainable Tourism (CST) that they hoped would root out the "green washers" (businesses that abuse the concept of eco- or sustainable tourism). Costa Rica's tourist board, ICT, awards Blue Flags to clean beach destinations and lists certified "green leaf" eco-friendly properties on their CST website *(www.turismo-sostenible.co.cr)*. A hotel can receive a maximum of five green leaves after completing an extensive questionnaire. Unfortunately, their website's arrangement is a little confusing. Select a list of hotels geographically and those with only one or two leaves are not distinguished from more ecologically sensitive properties with three or four. Only when you select a list by "green levels" does it distinguish the number of leaves.

A big flaw in the designation process, argues fellow author and environmental proponent, Beatrice Blake, is that the survey favors large businesses and fails to credit smaller, more environmentally conscious hotels with lower consumption per guest. This is also a common complaint by small hotels against the tourism board. Also, the survey does not give special consideration for those with private nature reserves. Many of the more rustic and ecologically friendly lodges don't even make the list.

However, CST's green leaf approval does help travelers choose hotels that are at least attempting to follow ecological practices, especially those located in urban areas. Let's hope that, as the process is fine-tuned, the certification process will get better and even more hotels will respond to the demand for environmentally responsible practices.

> **AUTHOR NOTE:** We note in our reviews the tours, lodges and hotels we think have a positive environmental awareness and leave out those we suspect of "green washing."

■ National Parks, Reserves, Preserves & Refuges

Only in one spot is there today some of the wildlife that was formerly everywhere in the northwest.... Here are the puma and manigordo, deer, peccary, tepiscuintle, pizote, kinkajou, chulumuco, kongo, carablanca, and miriki. The jaguar and tapir are already extinct.... Two years more and the mountain will be dead. Who is going to save it?

~ Olof Nicolas Wessberg, in a passionate letter about Cabo Blanco to World Wildlife Organizations in 1960.

The "golden toad" (*bufo perigienes*) is a beautiful animal. A brilliant orange-colored male, it's the kind of toad even a princess would kiss – if she could find one. Originally discovered in 1964 by biologists Jay Savage, Norm Scott amd the late Jerry James, it was known to exist only in its breeding ground, the Monteverde Cloud Forest. In the early 1980s, biologists counted more than 1,500 adult golden toads (*sapo dorado* in Spanish) in an annual survey. The next year the researchers found only 11. By 1987, they could find only 1. Since that day, golden toads have never been seen again. As a group, frogs were on the planet long before dinosaurs, so why then are they disappearing all over the world? Especially from such a protected area such as Monteverde? No one is sure, although it seems to be related to environmental pressures. Whatever the reason, scientists agree that the plight of the frogs is a biological warning sign for our Earth.

Researchers are looking closely at the correlation of the frog's disappearance with El Niño and global warming, but another theory speculates that an alien organism or microscopic pest may have been carried in by an unsuspecting scientist – or visiting eco-tourist – and caused a plague. That would be ironic. After all, the national parks, along with private reserves such as Monteverde, were begun specifically to preserve and protect Costa Rica's biological diversity.

The story of the creation of Costa Rica's remarkable national park system is one of hard work, imagination, dedication and a tremendous amount of luck.

> The story also makes for a great read. Pick up a copy of ***The Quetzal and the Macaw***, by David Rains Wallace, a Sierra Club book.

Ecological awareness has always been a theme in Costa Rica's political history, perhaps because of the country's natural beauty. In 1775 Spanish colonial governor Don Juan Fernandez de Bobadilla issued a proclamation against slash-and-burn farming, "since the practice is followed by sterility of the soil." In 1833, green belts of permanent farmland and forests around cities were mandated, and in 1846 the government set aside forested watershed areas. By 1859, Costa Rica had set aside for preservation all uncultivated lands in a 15-km/nine-mile zone on either side of the main rivers in the Central Valley.

The first attempt at establishing a true national park system came in 1939, when a law created a two-km/1.25-mile protected zone around the two local volcanoes, Irazú and Poás, overlooking the Meseta Central. A second zone was created on either side of the then newly built Inter-American Highway that bisected an old-growth oak forest. Despite the law, the sides of the volcanoes were cleared for pasture, and the oak trees were cut and shipped to Spain to be made into wine barrels. By the early 1960s, several foreigners (*extranjeros*) had taken preservation into their own hands, buying land to prevent its degradation. Archie Carr, a herpetologist, worked with local people to save the turtles at **Tortuguero**. Ornithologist Alexander Skutch and botanist Leslie Holdridge bought farmland to preserve part of a rainforest. Their holding became the famous **La Selva** private preserve. Way back in 1951, a group of expatriate Quaker dairy farmers purchased a cloud forest in the mountains of Tilarán that would become **Monteverde Cloud Forest Reserve**.

So it may be fitting that the "god-parents" of the first large Costa Rican park were foreigners as well. Aspiring fruit farmers Karen and Olof Wessberg landed in the Nicoya Peninsula on Costa Rica's northwest Pacific coast in 1955. They had bounced around the world from Denmark and Sweden, their respective home countries, to Ecuador, Guatemala, the United States and Mexico. They finally settled in Montezuma, a small town on the southern Nicoya coast, because they enjoyed its quiet natural beauty. They especially loved living next to one of the last wilderness areas on the peninsula, **Cabo Blanco** (White Cape). In 1960 they woke up to find someone cutting trees and clearing a small *finca* (farm) in the primary forest. Olof became determined to buy the entire tract (at $10 an acre) in order to have the government put it off-limits to development. His unique appeal yielded offers from the British World League Against Vivisection, Sierra Club, Philadelphia Conservation League, Nature Conservancy and the Friends of Nature. However, Costa Rican bureaucracy had no experience in managing such an ambitious endeavor (or the large Santa Rosa tract they had acquired in 1966). In 1969, in typical Latin American make-do fashion, they hired two graduate students – **Mario Boza**, 27, and **Alvaro Ugalde**, 24 – to run a new national park system. They could not have made a better choice. On a trip to the US in 1960, Mario Boza had been so impressed with the Great Smoky Mountains Park that upon his return he drew up a master plan for Poás Volcano as if it were a national park. His master's thesis would become the blueprint for other Costa Rican parks. Alvaro Ugalde was a biology student at UCR who had taken a national parks management course at the Grand Canyon in the US. He worked as a volunteer, along with a Peace Corps member, physically managing the Santa Rosa tract before anyone was hired to do so. His first task after he was hired? Running the new Santa Rosa Park.

If the Wessbergs' efforts gave physical birth to the park system idea, these two uncommonly gifted environmentalists raised it to adolescence. Boza and Ugalde's foresight and hard work gave Costa Rica the foundation required to build upon a relatively modest start and create a viable national park system that is now the envy of the world.

Park fees are about US $8 for all foreign tourists, but cheaper for local residents. It's a way of funding the lesser-used parks and, although many foreign visitors feel it's unfair, we believe it's a good way to insure the park system's survival (despite the fact that the fees go into general reve-

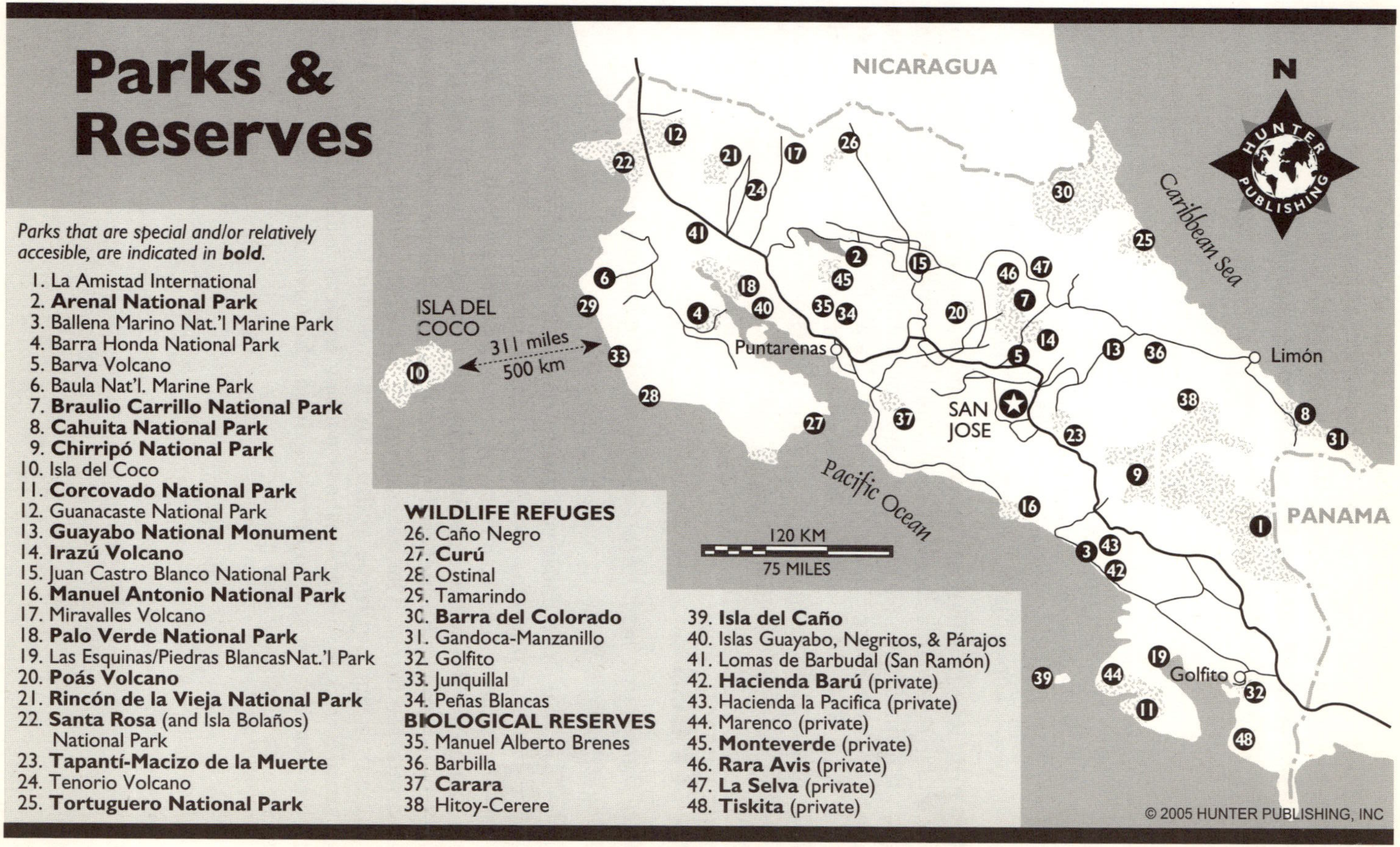

Land Between the Oceans

nues and are not park-dedicated). Though the government still struggles with funding (the land in several parks has not been paid for, many years after its expropriation) Costa Rica has found that protecting its natural heritage has paid unexpected benefits well beyond everyone's expectations.

Flora

Flowers changed the face of the planet.
Without them, the world we know – even man himself – would never have existed.
~ Loren Eiseley, American anthropologist, 1907-1997

With over 15 different ecosystems and transitional zones, it's no wonder Costa Rica is a botanist's and gardener's delight. Even if you can't tell the difference between a heliconia and a jacaranda, you can appreciate the verdancy and color of Costa Rica's flora. It's everywhere you look. And so is beautiful **impatiens**, an imported flower that blankets the floor of shady forests, especially along roadsides. They are wonderfully colorful but are pushing out native plants.

■ Trees

Keep a green tree in your heart and perhaps the singing bird will come.
~ Chinese proverb

Costa Rica chose the **guanacaste** (*Enterolobium cyclocarpum*) as its national tree, although it had an abundance of other choices. The diversity of trees concentrated in Costa Rica is amazing. You'll recognize **coconut palms** on the beach and **pine** trees at higher elevations, but not perhaps *gmelina*, or **teak**, which are grown on lowland tree farms. The hearts of palm tree (from which palm oil is extracted) is the **pejibaye palm**. An interesting tall tree is the **Indio desnudo** (naked Indian), whose peeling bark is a deep brick-red. The tree is also known as "sunburned gringo," for obvious reasons. Its fruit is a favorite of white-faced monkeys.

A farmer works on his "living fence."

LIVING FENCEPOSTS

As we traveled the country, we couldn't get over "living fence-posts," a phenomenon attributed to the combination of climate and the properties of native *arboles* (trees). As they do all over the world, ranchers and farmers here use cut tree trunks or branches to make inexpensive fences. But once these twigs are stuck in the rich Costa Rican soil, the nearly-dead branches take root and grow, creating a row of sapling trees that divide fields.

Several tree species perform this amazing task, including Indio desnudo, erythrina (used to shade coffee), fig, madre maton and madre de cacao. Spiny **cactus** is also used as fencing in Guanacaste and Nicoya.

The lopped tree, in time, may grow again.
~ *Time Goes By Turns*, Southwell, c.1595

The giant **ceiba** is a sacred tree to indigenous people whose beliefs have passed down from the ancient religion of the Maya. Their influence extended this far south. The **milk** tree in Corcovado has a drinkable milky latex sap and its sweet fruit is edible. **Almond** trees are commonly used for shade. They are not the same as those in the US, but the nut tastes similar.

Jacaranda are large open-crowned trees with deep purple-blue blossoms in the dry season. Another colorful tree is the **African tulip tree**, which has deep green leaves and brilliant red-orange blossoms tinged with yellow.

The dangerous **manchineel tree** (*manzanillo*) grows on and near beaches. The fruit is highly poisonous and, although the tree's shiny green leaves and spreading branches give good shade, don't sit under it.

One of the most important trees in Costa Rica's history is the **cacao**, the seed of which is used to make chocolate. The country was once a leading grower and still has a sizeable commercial crop.

Mangroves

Mangroves look more like shrubs than trees, but coastal mangroves (*mangle*) areas are a critical part of the ecosystem. Their many functions include protecting fish, sponges, coral and marine life under their intertwined roots, and sheltering bird nests in their brushy foliage above ground. Along the sea, mangroves dampen winds, high waves and flooding when storms hit. They trap soil in their mangled root system that eventually fills in to become land. Their importance as a wetland anchor and natural wildlife haven cannot be exaggerated.

Strangler Fig

A reed before the wind lives on,
while mighty oaks do fall.
~ English proverb

Strangler figs (*ficus*) begin their life innocently enough, one among many epiphytes living on the branches of a rainforest tree. But soon their long woody roots wrap around the host tree on their way to the forest floor. After many years, competition with the fig kills the host tree. It is not uncommon to come across a hollow, lattice work-like *ficus* formed in the shape of the tree that rotted out within it. One canopy tour in Monteverde features a climb inside the fig tree's tube-like structure up to the platform.

Strangler figs vines climbing a tree.

■ Rainforests

I am for the woods against the world,
But are the woods for me?
~ *The Kiss*, Edmund Blunden, 1896-1974

One of the biggest misconceptions about the rainforest environment is that it rains all the time. Of course, when it does rain, it rains a lot. The designation "rainforest" has several variations, usually distinguished by vegetation and other characteristics. Rainforests are found in regions of low seasonal variations, with at least 1,800 mm (70 inches) of annual precipitation. Yet one type of rainforest doesn't get much rain at all. A famous example is Monteverde, which is a cloud forest. Cloud forests are frequently smothered in moisture-laden clouds, where most of the water comes from the constant drenching of dew.

Deforestation is an ongoing concern.

Another typical characteristic of rainforests is the poor soil in which they thrive. Nutrients that feed the trees come from a sophisticated composting cycle that provides a thin but potent layer of good soil. Once cleared for cattle or farming, the thin soil quickly washes away or is used up, leaving barren ground. What this means is that rainforests cannot be re-generated – they are gone but not forgotten. Only 6% of the world's rainforests remain and they are disappearing at the rate of over 100,000 acres per day. That's an area the size of New York State every year!

The tragedy affects us all, even when we live far from the rain-forest. They are the lungs of the earth, where oxygen is exchanged for carbon dioxide, and they clean and recycle water. Tropical rainforests support at least half of the world's plant and animal species.

Apart from being important watersheds, rainforests contain medical compounds that can benefit mankind. For example, more than 70% of the plants known to produce drugs with anti-cancerous properties are tropical. Cures for malaria (quinine) and amoebic dysentery (ipecac) are only two of the hundreds of medicines derived from botanical sources. Cortisone is another, and so is diosgenin – the active agent in birth control pills – which is derived from wild yams. Since time began, man has used plants as medicine. Indigenous peoples around the world, including the few left in Costa Rica, still use plants and home remedies medicinally with great effect. The rainforest also provides commercial and edible products, such as rubber, vanilla, coconuts, resins, starch, thatch, dyes and bananas. Undoubtedly, there are thousands of new medicines waiting to be found in the rainforest, yet man is cutting them down at a staggering rate. What a shortsighted legacy for our children.

Recently, some agronomists have been theorizing that, because of its topography, volcanic soil and climate, Costa Rica's lost forests may have a greater ability than other cleared rainforests to respond to managed regrowth. Only time will tell.

You'll probably hear mentioned "primary" forest and "secondary" growth when talking about *la selva* (the forest). The thick, heavily vegetated woods you see, which may seem as if it should be called virgin forest, has actually been cut and re-grown in the last couple of decades. Primary or virgin forest is rather thin along ground level, its lower level growth restricted by the thick canopy where most of the rainforest life resides.

■ Flowers, Fruit & Fragrance

Avocado growers denied publicly and indignantly
the insidious, slanderous rumors that avocados were aphrodisiacs.
Sales immediately mounted.
~ *Food*, Simon & Schuster, 1980

It has always intrigued us to find plants that barely survive on our kitchen windowsill at home growing wild and abundant in the jungle. **Philodendron** and **dracaena** become huge in the jungle and are grown as export crops. **Hibiscus** thrives here, as do exotic herbs and the stunning ornamental **heliconia** flowers. The ubiquitous large red, orange and pink blossoms of the **bougainvillea** – yet to become one of our favorite semi-tropical flowers – appear in every garden, framing colonial courtyard arches or growing along a wall. They also grow wild along the road, as does the **mimosa**, a low-growing plant with lavender puffball flowers. The gigantic (up to two meters/6.6 feet wide) leaves you may see crossing Braulio Carrillo park are **poor man's umbrellas** (yes, this is the proper name), and really are used for as umbrellas by local people caught in a shower. (That is, except those in parks, where it is illegal to cut a plant.)

Oranges, **coconuts, guayaba**, **papaya** and **mango** are the homegrown fruits of the countryside. You'll see vibrant green **rice** paddies in the lowlands of both coasts. **Sugar cane** is an introduced, grass-like plant with feathery flowers, which can grow to five meters (15 ft) in height. But coffee and bananas are king. They are the two main export crops and both have a long, close history with Costa Rica. See page 71 through 76 for their story.

Coffee is a tree-like shrub with glossy dark leaves and white flowers. It grows best at a mid-level altitude under shade trees. Each red berry contains two coffee beans, which are harvested from October through January. Costa Rican coffee is among the best in the world.

Bananas were first introduced in the New World in 1516. They are a squat, palm-like tree whose delicious soft fruit grows in bunches on a drooping stalk. The tropical plants thrive here. Bananas like a lot of moisture but cannot thrive with wet roots. On the banana plantations, trenches are dug on either side of the crop rows to drain the water. Blue insecticide-impregnated bags cover the bunches until maturity.

Bananas are harvested green and shipped around the world. Another popular variety with locals is the **plantain**, served cooked as a side dish at breakfast, lunch and dinner. You'll love it.

■ Orchids

By happy alchemy of mind,
They turn to pleasure all they find.
~ Matthew Greene, 1696-1737

Costa Rica is renowned throughout the world for its beautiful orchids. Exotic blooming orchids are the largest family of flowering plants in the world and they thrive in warm, humid climates. The many varieties sport hundreds of flower styles, from tiny delicate petals running along the stem to bold blossoms and big thick green leaves. Costa Rica claims up to 1,500 different varieties, 75% of which are epiphytes, and provides much of the world's commercial supply. Epiphytes (from the Greek for "upon plants") attach to host trees and gain their nourishment from airborne dust and rain. These are not parasitic relationships because they do not feed on their hosts. In the amazing world of nature, many tree orchids make use of variety-specific pollinators – bees, ants, hummingbirds, wasps and moths – for fertilization. The **purple guaria orchid** (*Cattleya skinneri*) is the national flower.

> **DID YOU KNOW?** Orchids were named by Dioscorides, a Greek physician who likened the species tubers to male testicles – "orches."

Ground-dwelling epiphytes (Costa Rica boasts some 170 species) are known as bromeliads, named for Swedish botanist, Olaf Bromel. The most famous bromeliad is the pineapple. Bromeliads provide a tiny, self-contained eco-system in their protected core where water collects.

A great place to see and learn about all these magnificent flowers is **Lankester Botanical Gardens** outside Cartago on the way into the Orosí Valley. See page 170.

Ferns are orchid relatives, and they are grown as cash crops in Costa Rica. Those large black mesh tents that cover the hillsides at cooler elevations are usually protecting ferns grown for export.

Fauna

In the parched path I have seen the good lizard,
the one-drop crocodile, meditating.
~ *El Lagarto Viejo*, Frederico García Lorca, 1898-1936

The animal kingdom in Costa Rica is larger than life. The country has many microclimates and it teems with a staggering range of wildlife, partly because of its location on a land bridge between continents.

If your object is to see any of the more shy creatures in remote locales (take your pick from 850 species of birds, 208 mammals, 220 reptiles, 330 species of hummingbirds, 34,000 insects, 130 freshwater fish and 160 species of amphibians), your best chance is to hire a local naturalist guide.

Some of the common mammals include the **coati** (*pizote*), a dusk and dawn hunter related to the raccoon (*mapache*). The **collared peccary** (*saíno*) resembles a pig and lives in large groups in the forest. The rodent-like **agouti** (*guatusa*) can be found foraging on the forest floor near rivers and streams.

Tapirs (*macho de monte*), huge 250-kilo (550-lb) mammals, are endangered, partly because they're prized as delicacies on the dining table, and partly because of shrinking habitat. The **vampire bat** (*vampiro*) attacks cattle in the northwest, but are only one of over 100 species of bats in Costa Rica.

Famous frogs are, of course, the **golden toad** (*sapo dorado*), now feared to be extinct, the **red-eyed tree frog** and the **poison dart frog** (*dendrobates pumilio*). The latter is a tiny frog that advertises its toxicity with its bright color. They are less than an inch long and can be found under low plant leaves.

The leathery-scaled, olive-gray **crocodile** (*crocodilo*) has beady eyes that stare blankly from its head as it skims the surface of the water hunting for frogs, fish, birds and small mammals. They and their slightly smaller cousin, the dark brown **caiman**, hunt mainly at night. During the day they sun themselves along the banks of rivers and mangrove swamps.

But it's the **snake** (*serpiente*) that most people worry about when trekking around the forest. And Costa Rica has 162 species. But take heart, only 22 are poisonous. So, it's a mixed blessing if you encounter one. The chances are that you'll be with a naturalist guide, who generally carries anti-venom as a precaution. Snake-bitten tourists are very rare. We have only seen one snake in the wild, a small but venomous **yellow eyelash viper**, sunning itself in the crook of a tree when we were safely riding an aerial tram.

Snakes generally slip away when humans approach, but the one that is responsible for the most bites is the aggressive **fer-de-lance** (known by its Spanish name, *terciopelo*) It has an olive-brown to dark-brown skin with light color "X" markings along its back and sides. If you encounter one, stand very still until it calms down, then try to get back out of range. If you do get bitten, seek help immediately. The fer-de-lance sometimes strikes first and asks questions later. The king of snakes is the **boa constrictor**, which kills its prey by crushing it in a tight coil. When you're hiking, stay on the trail.

You can admire these snakes safely in the *serpentarias* in San José, Grecia, or Parque Viborana near Turrialba.

■ Primates

There are 193 species of monkeys and apes,
192 of them covered with hair.
The exception is a naked ape, self-named homo sapiens.
~ *The Naked Ape*, Desmond Morris, 1968

Monkeys are a favorite with tourists and always cause a stir when they pass overhead in trees. They are intelligent, forest-dwelling social animals that travel in extended family groups, called troupes. The three main species you'll encounter in Costa Rica include the most common, the large **howler monkey** (*mono congo*), a black, relatively slow-moving vegetarian primate. The alpha male, with his troupe of up to 20, is usually the biggest of the bunch and if he's annoyed he'll let out the growl (a guttural who-who-who) that can be heard for long distances. Be careful standing underneath howlers, they'll sometimes throw fruit or, worse, try to pee on your head.

The **white-faced capuchin** (*mono cara blanca*) is a smaller, more rapid, treetop-dwelling insect eater. Their moniker comes from the hood of white fur on their shoulders, chest and face. They can be found on the Caribbean lowlands, and in Osa, Manuel Antonio, Monteverde and Guanacaste.

The blond-chested, black-handed **spider monkey** (*mono araña*) is famous for its long prehensile tail, which acts as a third hand. These agile communal monkeys can leap an incredible 10 meters/33 feet from branch to branch.

A fourth type of monkey is much less visible than those mentioned above. The **squirrel monkey** (*mono titi*) can be found only along the lowland

Pacific coast. Its black head, olive-green shoulders and orange hands, feet, back and calves, make it easily distinguishable. See the section on Manuel Antonio, page 293, for information about a concentrated effort to save these scampering tree dwellers.

■ Cats

Cats and monkeys, monkeys and cats – all human life is there.
~ *The Madonna of the Future*, Henry James, 1843-1916

The king of the jungle is the **jaguar** (*tigre*), the largest of the New World cats. It holds a special place in indigenous culture: it is the form taken by the sun when it descends into the underworld at nightfall. Its image symbolizes power and strength. A male jaguar may reach over six feet in length and can weigh in at 136 kilograms (300 lbs). Its short coat, spotted much like a leopard, ranges from grayish-gold to reddish-tan, with spots grouped in small circles known as rosettes. Unlike the leopard, however, the jaguar's rosettes surround solid spots. Occasionally, jaguars are black all over. Jaguars feed on tapirs, peccaries, foxes, turtle eggs, rodents, even deer – but rarely man. Because of the dwindling habitat – each big cat requires a forested area of 100 square miles – they are vulnerable to extinction in Central America.

The **jaguarundi** (*león breñero*) is the smallest cat, with a low-slung long body resembling that of a weasel. Jaguarundi range in color from brown to gray and are slightly larger than a household cat. They stand 14 inches at the shoulders and weigh as much as nine kilograms (20 lbs). A sinuous tail takes up nearly half of the cat's 35- to 55-inch length. Already a rare animal, it's becoming rarer as its natural habitat in wild thickets and lowland forests is cut and burned for ranching.

The **ocelot** (*manigordo*) is one of Latin America's most beautiful and rare cats, noted for its creamy tan fur and dark spots with open centers. Ocelots usually weigh from 20 to 32 lbs and grow to 33-40 inches in length. Like most wild cats, they maintain territories marked by their scents. They're solitary ground hunters, but are agile enough to climb trees if threatened. Their main predator is man, who values their fur for coats. They are a protected endangered species – so poachers do the only hunting.

The **puma**, which is also native to the United States and Canada, is otherwise known as the cougar or mountain lion. A full-grown male puma may be nearly as big as a jaguar and weigh 91 kilograms (200 lbs). Its soft fur coat runs from reddish to gray to brown. This big cat is an amazingly agile climber, able to leap 13 meters (40 feet) in length and an astounding five meters (15 feet) high. The puma can successfully drop from a height of 18 meters (56 feet).

You'll have to be very lucky to see any of these felines in the wild; we were thrilled to have a jaguarundi cross our path when we were on the Nicoya Peninsula.

■ Sloths

The haste of a fool is the slowest thing in the world.
~ Thomas Shadwell, 1642-1692

The family of *edentata*, indigenous to the Americas, includes **anteaters**, **armadillos** and **sloths**, a favorite of guides who seem to know all their favorite hang-outs in trees (usually *guarumos*) along your route. Costa Rica is home to two types of sloths, the often-viewed three-toed sloth, a diurnal animal, and the seldom-seen nocturnal, two-toed sloth. The brown three-toed and rare two-toed sloths are tree-dwelling leaf eaters with an incredibly sluggish, fermentation-based digestive system. They move very slowly to conserve energy, so slowly that their Spanish name is *perezoso*, which means "lazy." Anteaters and armadillos are lower to the ground, but harder to find because they're nocturnal.

A sloth surveys the rainforest scene.

■ Turtles

The turtle lives 'twixt plated decks
Which practically conceal its sex,
I think it clever of the turtle
In such a fix to be so fertile.
~ *The Turtle*, Ogden Nash, 1902-1971

There are five major species of large sea turtles that nest on Costa Rica's shores, and their mostly nocturnal egg-laying is a wonderful thing to see. But be careful not to disturb them and always go with a licensed, experienced guide. Six of the seven turtle species worldwide are endangered.

The **green turtle** (*Tortuga verde*) mates and lays its eggs on the beach several times a year. Green turtles are especially common at Tortuguero, where the Caribbean Conservation Corps was begun. They measure about a meter (3.3 feet) in length and weigh 75-200 kilos (165-440 lbs).

Loggerheads (*cabezona*), with their massive bird-jawed skulls, have short fins and grow to a little over one meter (three feet) in length. They nest on other beaches but seem to gather in larger numbers at Playa Grande near Tamarindo.

The black, narrow-finned **leatherback turtle** (*baula*) is named because of its leathery hide in place of a shell. Leatherbacks grow as large as two meters (six feet) and weigh up to 680 kilos (1,500 lbs). That's living large! They come ashore on both coasts but especially at Playa Grande near Tamarindo, Tortuguero and the Gandoca Manzanillo Refuge.

On the other side of the coin, the **hawksbill** (*carey*) is one of the smallest marine turtles at about one meter (three feet) or less and only 91 kilos (200 lbs). Because of its highly valued spindle-shaped tortoise shell, it has been hunted to near-extinction.

The **Olive Ridley**, also called the Pacific Ridley (*lora*), nests at Ostinal near Playa Nosara and at Playa Nancite in Santa Rosa Park. There is no other sight in the world like a beach full of Ridley turtles storming ashore to nest. To huddle on a deserted beach late at night with only the brush stroke of the Milky Way to illuminate your world is quite an experience. Turtles return to the same beach each year and lay their precious eggs by digging a shallow hole in the sand with their flippers. Any type of unnatural light or noise will disturb the giant lumbering females and can cause them to abort their nest. Once covered over, the hatchlings emerge about 60 days later and crawl toward the surf. If they're lucky. Between wrong turns and predators – sea gulls, large fish, raccoons, foxes and human poachers – rarely do more than 4-5% grow to maturity. Costa Rican laws severely restrict the harvesting of sea turtles; so if turtle is ever on a restaurant menu, please don't order it.

■ Birds & Butterflies

If I had to choose, I would rather have birds than airplanes.
~ Charles A. Lindbergh, quote recalled on his death, 26 Aug. 1974

When birdwatchers die, they want to go to heaven in Costa Rica. Over 850 species of birds make their appearance in the diverse ecosystems here, many on migratory vacation from colder climes. The national bird of Costa Rica is the **yigüirro** (*Turdus grayii*), pronounced *yi-GWE-ro*, is a robin without the red breast, and is renowned for its melodious song. But the most spectacular native bird is the **quetzal** (pronounced KATE-zal), a brilliant green cloud forest dweller with a wispy, 60-cm/24-inch plumed tail. These large birds (up to 35 cm/14 inches tall) are found only at altitudes between 1,200 and 3,000 meters (4-10,000 feet). They are most commonly seen at mating time between February and April, feeding in fruit trees, notably in Monteverde and Tapantí-Macizo de la Muerte parks. In the belief that they could not survive in captivity, they became a symbol of freedom in Central America.

Oropéndolas are the large black birds with yellow tails that build those fascinating pendulum-shaped nests you see hanging from branches. The nests look like woven Christmas tree decorations. And where would the tropics be without the symbols of tropical climates, the **toucan** and the **scarlet macaw** (*lapa roja*). The macaw's vibrant colors begin with red and orange and then add yellow, gold, blue and green. Because of its beauty, and its apparently monogamous mating characteristic, it is prized as a pet. Export is illegal, but still this Pacific coast dweller – and its cousin, the **green macaw** in the Caribbean – are in great danger of extinction in the wild. The toucan's big multi-colored beak is equally distinctive and a thrill to see.

The rainbow-billed toucan is a symbol of the tropics.

Dedicated birdwatchers should pick up ***Birds of Costa Rica***, by Gary Stiles and Alexander Skutch (Cornell University Press) or ***Travel & Site Guide to Birds of Costa Rica***, by Aaron Sekerak (Lone Pine Press).

To make birdwatching easier, many forest hotels and lodges put fruit out on stands to attract a large variety of birds. We also enjoy watching the many different colors of **hummingbirds** Of the 330 known hummingbird species, about 65 are native to Costa Rica.

VOICES OF THE FOREST

You can listen at home to the sounds of Costa Rican nature with a stereo CD called ***Voices of the Cloud Forest***, depicting a day in Monteverde, and the ***Costa Rican Bird Song Sampler***, an audio guide to recognizing forest bird songs. Both are produced by David Ross at Cornell's Laboratory of Ornithology (www.birds.cornell.edu/lab_cds.html).

The large **blue morpho butterfly** is quite a stunning sight against the deep green of the forest. Brilliant and plain butterflies abound in gardens, as well as in the warm rainforest.

If you're a serious fan of these fascinating, delicate insects, pick up ***Butterflies of Costa Rica***, by Phillip de Vries, a huge paperback for US $37.50.

DON'T BUG ME

My favorite rejection letter came from an airline magazine to which I'd pitched an article about Mexico's Monarch butterflies. 'No thanks,' they wrote back, 'we featured Costa Rica last year.'

~ Ron Mader, editor, www.planeta.com

Dr. Richard Whitten has assembled a world-class collection of weird and wonderful insects in his **Jewels of the Rainforest Museum** at Selvatura Park in Monteverde (see page 219). His colorful butterfly and bug displays are as much works of art as they are educational tools. Get him talking about his work and you'll never get away. But what a genuine pleasure to meet someone who really loves his work.

The People, Their History

Pura Vida

~ unofficial national slogan of Costa Rica that translates as "pure life."

IN THIS CHAPTER

You won't see a "Yankee Go Home" banner at a demonstration in Costa Rica. In fact, you'll see many signs that read, *Bienvenidos*, "Welcome." **Ticos** (TEA-coes), or Ticas (TEA-caz) are a warm and welcoming people that really like North Americans – and they show it. The gringo community of full- and part-time residents is quite large, with many Canadian and American retirees drawn by the climate, social benefits and lower cost of living.

TICOS & TICAS

Costa Rican people call themselves Ticos or Ticas (female). These words stem from *hermaniticos*, and *hermaniticas*, meaning little brothers and little sisters.

The Costa Rican culture is typical of Latin America in that it is predominantly Catholic and conservative – but not stridently so. The people here, mostly descended from the Spanish with a mix of indigenous and African, are more racially homogenous than in other Central American countries. The population on the Caribbean coast around Limón is more predominately African, a result of the importation of Afro-Caribbean workers for the banana fields. The official language is Spanish, though many college educated Ticos and those in the tourist trade speak English. On the Caribbean side, most people speak both languages.

Ticos are generally well educated, with a 93% literacy rate – higher than the United States. An old Costa Rican saying claimed, "We have more teachers than soldiers," and that is still true today, some 60 years after Costa Rica abolished its army. The country has a history of peace and stability unmatched among its neighbors and is known for its tolerance. This status has made it a natural asylum for penniless refugees as well as wealthy deposed dictators.

The lack of armed conflict reflects the political and social characteristics – compromise and non-confrontation are important social tools. Sometimes infuriatingly so. "People in other countries can be categorical – but not Ticos," explains playwright Melvin Méndez. "We beat around the bush to avoid saying 'No,' a syllable that seems almost rude to us. Rather than hurt someone, we say one thing and do another."

If you stay long enough you'll find dealing with the layers of bureaucracy frustrating, with long lines at banks and offices. Life runs on Tico time, which means that a 2 pm appointment may be 2:30 or even 3:30. On the other hand, all these things pale in comparison to the genuinely pleasant nature of *los Ticos*.

History

■ People Before Time

The past is the only dead thing that smells sweet.
~ *Early One Morning*, Edward Thomas, 1878-1917

Long before Christopher Columbus first sailed from the Old World into the New, Meso-America had an "old" world already. People inhabited this part of the isthmus that is present-day Costa Rica for at least 11,000 years. Some of the relics left by the earliest Stone Age settlers show both North and South American influences – a sign of Costa Rica's importance as part of the land bridge between continents.

Archeologists believe the Maya influence from the north extended into the northwestern corner and Nicoya Peninsula of Costa Rica. The people that lived there, principally **Chorotegas**, had an organized and structured civilization with rigid class lines. A Cacique – a warrior chief and high priest – led them. The Chorotega was the largest of Costa Rica's many tribes. They left no written records, only highly stylized art and pottery. Their craftsmen worked in jade, gold and stone, and created the functional but artistic three-legged stone *metates* used for grinding corn. Some are still in use today, a thousand years later, by rural people. Chorotega pottery is glazed and is most often a light beige color with black markings. Local artists today have revived the lost indigenous methods and produce some unique and appealing works.

Semi-nomadic peoples lived in the eastern and southern tropical forests, raising corn and cassava. These **Carib** tribesmen chewed coca, a habit of the Andean and Inca cultures. The fierce **Boruca** inhabited the high Talamanca region in the south and Pacific zones. They lived in huge cone-shaped communal huts that could hold nearly 100 people. The **Corobicis**, another tribe, were thought to have a matriarchal culture.

In the drier regions of the Central Valley and highlands, indigenous people built stone foundations and large, stockade buildings that held extended family groups. They cobbled their pathways and created aqueducts and drainage systems in the style of their southern cousins.

Stone balls at the Children's Museum.

But the most fascinating remnants here came from the **Diquis**, a lost native civilization who left behind thousands of near-perfect **spherical stone balls**. These remarkable balls are found only in Costa Rica's Southern Pacific zone. Some are as small as oranges and some are huge – as big as two meters/6.6 feet in diameter and weighing over 14,500 kilos (16 tons). You'll find some specimens in the Gold Museum, the National Museum and the Children's Museum in San José. In many private yards throughout the country, the balls are used as decorative garden ornaments. A great place to see and touch the balls in an undisturbed setting is on Isla Caño, off the Osa Peninsula.

But who made these granite, andesite and sedimentary stone balls and why? They have been found scattered all over the countryside. Were they markers or religious artifacts? No one knows, as the people who created them have long since disappeared.

The country's most important archeological site is **Guayabo**, an excavated city on the slopes of Turrialba Volcano. For an unknown reason it was abandoned about 100 years before the Spanish arrival.

Recently, researchers have postulated that instead of the long accepted belief that Costa Rica had a small indigenous population, 400,000 to 500,000 people may have called Costa Rica home at the time of Columbus. The conquest quickly reduced that figure by 95%. Today, 40,000 native people, divided into eight cultural groups, live on 22 reserves, most in the remote south. The **Chorotegas** on the Matambú reserve in Nicoya and the **Guatusu** near Arenal are the only Maya descendants left. The other remaining tribes – **Boruca**, **Bribri**, **Cabécares**, **Térrabas**, **Teribes**, **Guaymis** and **Malekus** – are related to the South American **Chibcha** civilization of Columbia.

■ The Conquest

Happy is the country that requires no heroes.
~ Bertold Brecht, 1898-1956

Christopher Columbus gazed from his ship at the rich green vegetation of the shore near the present-day Costa Rican city of Limón, which he called "Cariari." It was September 8, 1502. Still convinced he would find a passage to India, he was on his fourth and final trip from Spain to the Americas, a decade after his first voyage opened the New World to Europeans. Following his discoveries, Spanish conquistadors and carpetbaggers got rich exploiting the islands of Hispañola and Cuba. This trip was Columbus's last attempt to discover a route to the Orient – plus finally to acquire a little something for himself.

The Carib natives he encountered on the shore of this verdant land wore gold pendants around their necks and seemed friendly enough. They approached his ship with goods to trade and stories of the wealth and gold of the lands farther south. Immediately, he decided to petition the Spanish Court to govern this rich coast of **Veragua**. Fortunately, Columbus could sail better than he could name a land because the Veragua moniker was soon dropped in favor of his descriptive adjectives, rich coast (*costa rica*).

Traditionally, hollowed logs served as boats.

When he finally made it back to Spain, however, Isabella had died and the King refused to see him. Columbus spent his final few years in failing health and would never return to the New World. Ironically, the joke would be on those that cheated him out of Costa Rica, for the rich land he expected turned out to be one of the poorest of Spain's American colonies. Impassable terrain, huge mountains, raging rivers, floods, heat, swamps, at least 19 separate hostile tribes, plus a lack of mineral wealth would keep the tiny colony out of the spotlight right into the 1800s and independence.

Those factors made the eventual "conquest" of Costa Rica more like a stern test of survival than a military victory. For years, would-be conquerors had to content themselves with excursions close to its Pacific shores, reached by sea from Panama after Vasco Nuñez de Balboa discovered the Pacific Ocean in 1513.

Costa Rica was considered an inhospitable country due to earthquakes and volcanoes. This historic photo shows Cartago after a flood in 1963.

Captain Gil González organized the first major invasion in 1522. He and his men acquired enough gold to make the "rich coast" name stick for good, but they failed to establish a permanent settlement. González's remarkable exploits include marching his men up the coast to Nicaragua. According to his own accounts, he baptized 32,000 Indians along the way. But the treasury of gold nearly cost him his life when he returned. Threat came not from the natives, but from jealous Governor Pedrarias of Panama, a ruthless tyrant who had even beheaded his own brother-in-law, Balboa, discoverer of the Pacific.

The grandson of Columbus, Luis, mounted an expedition in 1546, after the King of Spain at last granted his family's long-sought title, Duke of Veragua. Most of his 130 men were lost and the New World's last direct link with its European discoverer ended.

Finally, in 1563, **Juan Vásquez de Coronado** founded the settlement of Cartago, Costa Rica's first capitol and its first real city. Coronado could be considered the true "conqueror" of Costa Rica. During his tenure he organized expeditions and explorations around the country and made alliances with warring indigenous tribes. By the late 1560s, after Coronado was lost at sea on his way back to Spain, the native inhabitants of Costa Rica were either in slavery, dead from the many diseases that decimated the population, or living in remote, inaccessible areas.

■ Colonial Times

Gaily bednight, A gallant knight
In sunshine and in shadow, Had journeyed long,
Singing a song, In search of Eldorado
~ Edgar Allan Poe, 1809-1849

Colonial times were hard for all. Costa Rica, rich in flora and fauna, did not have the easily accessible gold that spurred the Spanish to conquer and settle the New World. And its native population, scattered and decimated, made the *encomienda* system – where local natives became the slaves of landowners – less than successful. For lack of manpower, most farms became family farms, and the national myth holds that because everyone had to work for survival, no class system developed here as it did in Mexico or Guatemala. The legacy of hard-working, independent-minded farmers is the basis for Ticos' love of democracy. Of course, that version glosses over the maltreatment of natives and Caribbean-Africans – yet one cannot deny that Costa Rica's long-time democratic leanings ultimately avoided the worst of the social turmoil that plagued its Latin neighbors.

But the economy was another story. Although the land was relatively fertile, the rough terrain hampered exports. Things got so bad that in 1709, cacao beans (the sole export) became the official currency. A present-day historical quip about colonial times is: "What was well distributed in Costa Rica was not wealth but poverty." When Irazú Volcano erupted in 1723 its ashes blanketed Cartago. This was the country's largest city, yet it consisted of only 70 adobe and thatch houses and two churches.

For years, Costa Rica was the colonial backwater of the Kingdom of Guatemala, of which it was a part. By 1821 its 65,000 inhabitants were all but forgotten – so they were surprised to hear that Guatemala had declared independence from Spain on behalf of all Central American countries on September 15. The four largest Meseta towns of Cartago, San José, Heredia and Alajuela each met separately to declare their own independence. Jealous and suspicious of one another, the towns agreed to remain neutral toward the future of Costa Rica until the "clouds of the day disappear" – a decision often cited as a classic example of the Tico tendency to procrastinate.

Conservative forces in Cartago and Heredia wanted to align Costa Rica with Mexico, while republican leaders in San José and Alajuela wanted complete independence. And, like city states of ancient Greece, each of the four cities insisted on being the new capital. In March 1823, a quick battle in the Ochomongo hills (a hilltop monument commemorates the fight along the Cartago-San José Highway) resulted in a republican victory, independence and, later, the designation of San José as capital.

■ Democracy & Coffee

Liberty is always unfinished business.
~ American Civil Liberties Union slogan

Wealthy landowners and aristocracy met in San José and elected **Juan Mora Fernández** as the first chief of state. As a leader, Mora encouraged coffee growing and began modest exports of the bean through Chile to Europe, where the dark brew was becoming a fashionable drink. Mora's second successor, a domineering San José lawyer, **Braulio Carrillo**, came to power in 1835. Carrillo imposed liberal reforms and revised anachronistic civil and penal codes. He withdrew from the Central American Federation, which Costa Rica had joined in 1824, and also paid off the "English debt," Costa Rica's share of a debt incurred by the failed Federation. His greatest legacy, however, was his strong promotion of coffee production throughout the Central Valley. He gave free trees to the populace to plant in their yard, and offered free land to anyone who would grow coffee on it. Carrillo returned as the country's first dictator when a successor tried to roll back his reforms, but his despotic ways did not sit well with the now democratic country. Exiled to El Salvador in 1842, he was assassinated there three years later.

Fortunately, Carrillo's agrarian efforts had insured Costa Rica was well positioned on Christmas Day, 1843, when the English captain, William Le Lacheur, sailed into Puntarenas looking for cargo. Growers in San José, with plenty of coffee to sell, trusted him with their goods on consignment. Two years later Le Lacheur returned with plenty of pounds sterling – the beginning of direct trade, an economic relationship with England, and the first good times for Costa Rica.

Coffee, one critic observed, "became a religion instead of a mere crop." If there was no big class difference before, the rise of an *aristocacia cafetalera* changed all that. Coffee gained the nickname *grano de oro*, or "grain of gold." The coffee boom brought coffee barons, wealth and development to ports such as Puntarenas and Puerto Limón, roads and railways, and new hospitals and schools. But a monocrop and consequent monoculture breeds its own set of financial and social problems. For the first time a privileged class system, based on coffee profits, emerged.

In 1847, Congress named as the first president **José María Castro Madriz**, a 29-year-old editor and publisher. His wife, Doña Pacifica, designed the national flag. But his reforms met opposition from the powerful coffee barons and he was replaced. When the country's second president, **Juan Rafael Mora Porras**, tried to open a state bank to provide favorable credit to small farmers, the coffee elite greedily squelched his reforms as well.

For many years, brightly painted oxcarts carried the precious bean down from the mountains to the Pacific port of Puntarenas. However, the coffee oligarchy realized that in order to stay competitive the country needed better access to the Atlantic. In 1871, to finance a new rail route to Limón, a deep-water shipping port on the Atlantic side, the government borrowed $8 million dollars from England. But when coffee prices hit bottom in 1900 it caused a severe food shortage and famine and the unfavorable financial terms of the loan hobbled the country's economy for 40 years.

COFFEE TIME

You get a proper cup of coffee in a copper coffee pot.
~ popular English tongue-twister

The origin of coffee lies in the legends and myths of Africa and the Middle East. One story tells of Kaldi, an Ethiopian goatherd who found his animals eating at a dark-leaved shrub bearing red berries. Another legend attributes the discovery of coffee to Omar, an Arabian dervish exiled to the African wilderness. He survived by brewing the berries he picked from coffee bushes. Whoever discovered it, coffee is considered native to Ethiopia.

By the early 1500s, coffee had made its way around the Middle East, and Arab patrons of coffeehouses lingered over the sweetened black brew. These early coffeehouses introduced the drink to European traders, who recognized it as a potential crop for their various tropical colonies. But the Arabs prevented the Europeans from taking live bushes in order to keep their monopoly. The Dutch finally obtained a coffee plant from Yemen and began cultivating coffee commercially in 1616. Sacks of beans labeled from plantations in one of their East Indian colonies, gave rise to one of coffee's best-known nicknames, "Java."

The credit for introducing coffee to the New World goes to Gabriel Mathieu de Clieu, a French naval officer. In 1720, he sailed for the French colony of Martinique with three coffee seedlings, obtained under highly questionable circumstances. Becalmed en route, de Clieu shared his water ration with the seedlings. His sacrifice paid off. Once planted on his estate in Martinique, the bushes flourished. From there, coffee cultivation spread to other countries in the New World. Costa Rica ranks 11th in world production and exports 280 million pounds of top-quality beans.

■ Bananas & the Jungle Train

I think I can, I think I can, I know I can.
~ *The Little Engine That Could*, Watty Piper

The Costa Rican equivalent of the Panama Canal was the **Atlantic Railroad**, described by one engineer as "the most difficult damned piece of railroading in the world."

By the mid-1800s, coffee had become very big business for small Costa Rica and its growers needed to ship more competitively to their European markets. They wanted a rail line to run 194 km (120 miles) east from Alajuela to Limón, the deep-water Atlantic port where Columbus first landed. The challenge was how to build it over impossibly high mountains, through formidable jungle and over swampy lowlands.

The railroad project, which started optimistically in 1871, soon ran into trouble after nearly 4,000 workers died from disease and accidents in laying the first 20 miles of track. **Minor Cooper Keith**, a brash, young, charismatic North American with an adventurer's spirit, soon talked his way into directorship of the project. With construction in disarray, his bulldog determination pushed it forward. When Costa Rican workers refused to work outside of the Central Valley, Keith recruited first Chinese and Italian laborers, then Jamaicans and other West Indian workers. In 1884, he renegotiated the British loans for more favorable terms. As part of his compensation, he was granted the concession to operate the railroad and a lease on 323,887 hectares/800,000 acres – nearly 7% of the country – adjoining the rail line. Keith determined to cultivate bananas in the tropical lowlands to raise more funds for the construction. This proved to be either lucky or a stroke of genius.

Orange you glad I didn't say banana?
~ punch line of an old "knock knock" joke

The "coffee" railroad was finished by 1890, although the first freight was actually bananas. This side-venture for Minor Keith proved so successful that he merged his plantations with Boston Fruit to found the infamous **United Fruit Company**. Vilified as the worst example of foreign exploitation and economic domination of Central America, United Fruit proved to be the modern equivalent of the Spanish carpetbaggers of years before. Plantation workers often endured harsh conditions for little pay and were totally dependent on the company for housing, medical care and supplies from company stores. When Panama Disease infested bananas on the Atlantic side in 1934, black workers were not allowed to follow the relocation to Golfito in the Pacific lowlands. After a 72-day strike in 1985, the company pulled up stakes in Golfito and returned to Limón. Whenever the UFC left for greener plantations, its former workers and their families were devastated.

United Fruit's domination of the banana trade lasted until the late 1950s, when **Standard Fruit** (Dole) broke the monopoly, but even now the various banana companies' influence and economic power are clearly visible. As late as 1994 the police near Sarapiquí fired upon striking union workers. Since the strike in 1985, most of the independent agricultural unions have been broken and replaced by management-friendly "workers' associations," which every worker must join. Wages hover around US $70 a week for back-breaking, and sometimes dangerous, manual labor.

Part of the danger to workers is from agro-chemicals used to fertilize and protect the banana trees. Herbicides and fungicides are used liberally and the blue bags you see covering the banana bunches are impregnated with pesticides to protect the fruit. Although a new plastic recycling plant has been built near Siquirres to stop ground disposal of the used blue bags, banana plantations are a major source of pesticide pollution as well as deforestation. In the last 15 years, production has grown 50% and 45,000 hectares/111,150 acres – five times as much land as 1967 – are now cleared of jungle and planted with bananas.

During earlier days of United Fruit, black Jamaican workers wanted nothing to do with the *mestizo* Spanish workers – and vice versa. The company took advantage of that to divide the workers when it came to rights and wages. In 1909, Marcus Garvey, the Jamaican-born North American black separatist, visited Costa Rica and was appalled by the deplorable working and living conditions. Even though the *mestizo* workers were not in much better shape, it was mostly blacks who stood up against the company in bloody and deadly strikes for better pay and conditions. Their justifiable complaints have lasted into modern times. Labor laws and more sympathetic governments have at least relieved some of the most egregious of workers' problems.

Culturally, Keith's importation of English-speaking workers of African descent also had a lasting impact on Costa Rica. Isolated from the Central Valley Spanish-speaking government, black residents kept English as a first language, although it can be somewhat hard to understand if spoken in dialect. And governing the province from the highlands has sometimes presented problems for both cultures. Over the years, Ticos have had to face up to their own institutional racism, despite their reputation for tolerance and inclusion. However, these days, blacks have integrated successfully into Costa Rica's Spanish-centric society, while most white Costa Ricans have done the same in the Afro-Caribbean culture of the Atlantic side. As bilingual speakers in San José, Limón residents sometimes even have a workplace advantage over their Spanish-speaking countrymen. Economically, however, the standard of living in Limón and along the Caribbean side in general, where most of the country's black population resides, is visibly lower than in the Central Valley.

Despite a rich income from bananas, the Caribbean side of Costa Rica was often treated as a poor cousin. Until 1970, when a highway was finally built, the old Atlantic Railroad was still the principal means of transportation between the capital and Limón. The city's most common complaint – one that has been made for many years – is that the government ignores the Atlantic side in favor of other regions.

But the eastern side of Costa Rica, once dependent almost exclusively on banana production, now struggles to attract its fair share of tourism. Bananas provide Costa Rica with millions of dollars in annual revenue and jobs for thousands of workers in economically depressed areas. We always look for the "Grown in Costa Rica" labels when we shop for our daily dose of potassium-rich bananas.

On the roads you'll see many Dole, Chiquita and Del Monte brand trucks carrying precious bananas for export to the United States and Europe – Costa Rica is second only to Ecuador in world banana production – but the heralded Atlantic Railroad is no more. Financial losses caused the suspension of passenger traffic on November 20, 1990, just days short of the railroad's 100th birthday. This piece of history, nicknamed the "Jungle Train," became a popular tourist excursion, taking visitors through spectacularly beautiful countryside from San José to Limón. The train's *coup de grace* came on April 22, 1991 when a powerful earthquake caused landslides that swept away large parts of the line. A small portion of the line is still running and reincarnate as a tourist attraction.

Walker's War

It's a rough trade – war's sweet to them that never tried it.
~ *The Antiquary*, Walter Scott, 1816

One of the most bizarre incidents in the bloody history of Central America occurred near the end of Mora Porras' presidency. In 1855, **William Walker**, a Tennessee native, conquered neighboring Nicaragua with a ragtag mercenary army of Confederate sympathizers and carpetbaggers. Walker isn't mentioned much in North American history books but he is well remembered in Central America.

A boy genius, Walker graduated from the University of Nashville at age 14 and received law and medical degrees from the University of Pennsylvania by age 19. However brilliant he was as a student, he failed as a doctor, lawyer and journalist. In 1849 he tried his luck, also unsuccessfully, in the California Gold Rush. His mind increasingly unbalanced, Walker came to believe that his true calling in life was to be a soldier of fortune. Not just a solider, but a leader of soldiers. To that end he joined a "liberation" expedition into Baja Mexico, sponsored by a pro-slavery group, the Knights of the Golden Circle. Before being driven out of Mexico, he egotistically declared himself, "President of Sonora and Baja California."

After taking over Nicaragua's government, Walker immediately legalized slavery. But his grandiose schemes ran afoul of Cornelius Vanderbilt, the powerful North American millionaire who owned large business interests there and hoped to build a trans-ocean canal through Lake Nicaragua. Vanderbilt encouraged Costa Rica to go to war and Walker conveniently supplied an excuse. With pro-slavery interests backing him (they hoped to expand the pool of potential slaves), Walker and a few thousand men (known as "filibusters") invaded Costa Rica in March 1856. President Mora raised an army of 9,000 Ticos and marched on Walker's encampment, headquartered in a large farmhouse in Guanacaste province. On April 11, Juan Santamaría, a Costa Rican drummer boy, torched Walker's farmhouse roof before dying in a hail of bullets. Santamaría became a national hero and the old farmhouse is now a national monument in the middle of Santa Rosa Park. The international airport in San José is named Juan Santamaría in his honor.

Tragically, after driving Walker out of Guanacaste and then Nicaragua, returning Tico soldiers brought back cholera and an ensuing epidemic killed 20% of Costa Rica's civilian population. Walker later met his own fate in a way that showed how demented he had become. In 1857 he tried to conquer Nicaragua again, but was taken prisoner. Paroled in 1860 he sailed to Honduras, seized a port customs house, and immediately declared himself "president." Flushed out by Honduran soldiers, he took refuge on a British man-of-war but, after insisting he was the rightful

President of Honduras, they put him back ashore. An army firing squad promptly executed him.

Modern Times

The experience of democracy is like the experience of life itself – always changing, infinite in its variety, sometimes turbulent, and all the more valuable for having been tested by adversity.
~ Jimmy Carter, 39th US President

Costa Rica went through growing pains typical of all nations after independence. For example, the efficacy of its free compulsory educational system, instituted in 1885, raised enough political consciousness among the common people to spark a rebellion against a corrupt government in 1919. Federico Tinoco, Costa Rica's last dictator, was finally brought down – not by soldiers, but by teachers and students after their protest demonstration was fired on by his supporters.

Bullet holes on the turret of Fort Bellavista in San José are a sign of Costa Rica's less peaceful times.

In 1941, the day after Pearl Harbor, Costa Rica declared war on Germany and Japan, a largely symbolic gesture since they had no navy or troops of any quantity to send overseas. President Rafael Angel Calderón, a liberal who had previously instituted a social security system, labor code and other social guarantees, confiscated the property of Costa Rican families of German ancestry. He cited the 1944 U-Boat torpedoing of a United Fruit merchant ship (the *San Pablo*) in Limón as his reason. It was a serious political blunder because many of these families had been living there for generations and were part of the financially powerful coffee elite. The action set the stage for a Civil War four years later, a war that would claim the life of one in every 300 Ticos.

■ Civil War

Give peace a chance.
~ John Lennon, 1940-1980

Like the War Between the States in America, Costa Rica's **War of National Liberation** in 1948 defined the nation. No single, simple cause made the country's men and women take up arms, but in Costa Rica's war, one man, **Don Pepe (José) Figueres**, became the symbol of the conflagration in much the same way Abraham Lincoln did in the US.

José Figueres at a press conference, 1948.

Figueres, a true compassionate conservative, is a legendary figure in Costa Rican history and is termed the "Grandfather of Costa Rica." Self-educated in Boston, he returned to Costa Rica in the 1920s with an idealistic, utopian vision for his country's future. The profits from his farm, La Lucha Sin Fin (Endless Struggle), were used to benefit the local community. But he stayed out of politics until 4th of July riots denouncing the San Pablo attack erupted in San José. On the radio, Figueres criticized Calderón as unable to govern and insure public order. He was promptly arrested and exiled to Mexico, where he used his time to plan a revolution against what he and others considered a corrupt regime.

Despite opposition from the coffee oligarchy and a shaky alliance with the Communist Party, the 1944 election of Calderón's hand-picked successor, Teodoro Picado, was particularly violent and fraudulent. Things got worse in 1948 when Picado's term ended and new elections were announced. Calderón himself ran against the opposition candidate, Otilio Ulate. Although Ulate apparently won handily, out-going President Picado negated the results because of charges and counter-charges of fraud. Ulate was arrested and one of his advisors assassinated.

Figueres marched from his farm toward San José with 600 volunteers to unseat a government he felt had lost all credibility. But the army and police forces prevented a repeat of Costa Rica's quick, one-day inter-city war after independence. This modern conflict would take 44 days and exact a much higher loss of life. Despite an invasion in the north by Nicaragua to support Calderón, Figueres' forces captured Cartago and Limón and forced the government to surrender. Bullet holes around the turrets of the National Museum in San José are visible reminders of the 2,000 deaths in this short, but sad, civil war.

Besides the abolition of the army, a source of continual pride today in Costa Rica, Figueres' temporary dictatorship also granted suffrage to women and extended citizenship to all people born in Costa Rica. This was particularly beneficial to the people of the Atlantic region, many of whom had previously been denied the rights of citizenship. Don Pepe Figueres was twice elected president, the last time from 1970-1974. He died a national hero in 1990.

Recent Memory

I have measured out my life with coffee spoons.
~ *The Love Song of J. Alfred Prufrock*, TS Eliot, 1888-1965

The 1980s were especially turbulent times in Central America. In 1979, communist Sandinista guerillas overthrew the oppressive, dictatorial government of Anastasio Somoza in neighboring Nicaragua. At first, Costa Rica supported the efforts of the Sandinistas, but it later served as a haven for Contra rebels, who fought against them in a long, bloody, US-backed Nicaraguan war. America gave large amounts of aid to Costa Rica in exchange for allowing the Contras to operate along their northern border, yet it only postponed the dire effects of an inflation-ravaged economy. Meanwhile, Costa Rica's foreign debt swelled to $3.8 billion and unemployment rose to over 15%.

In the first serious unrest in modern times, a US embassy van was firebombed in San José in March 1981. A chauffeur and three marines were injured. The terrorists, only some of whom were Ticos, were captured. In 1993 a group that called itself the Death Commando took over the Supreme Court building and held justices hostage inside. The bad guys were at first thought to be Columbian radicals, but it turned out they were led by a pair of Tico brothers who wanted money. They were disarmed and arrested.

In 1986, when Ticos elected **Oscar Arias Sánchez** president, it proved to be a providential choice. He adopted an ambitious policy of federal government reduction and restructuring. With painful belt tightening, the economy slowly recovered. However, President Arias is most remembered for his Central American Peace Plan, which helped end the war in Nicaragua. He was awarded the **Nobel Peace Prize** in 1987. The award reinforced Costa Rica's world standing as a tolerant, pluralistic, peace-loving country.

Costa Rica's reputation as a sanctuary has been strained more recently by the huge influx of economic refugees from Nicaragua. Because of a lack of adequate health care in their native country, many of the refugees have inundated Costa Rica's social welfare system and its free health care system has serious financial woes. Both petty and violent crimes have risen dramatically in the last few years, which any Tico taxi driver will tell you is the fault of the "Nicos." Whether that's true or just racism (there is a racial difference most North Americans can't distinguish), Nico refugees have altered Costa Rican culture – both for the better and for the worse.

Being There

Americans have always been eager for travel,
that being how they got to the New World in the first place.
~ Otto Freidrich, *Time* magazine, 1985

IN THIS CHAPTER

Bienvenido a Costa Rica – welcome to Costa Rica! You shouldn't encounter too much culture shock upon arrival. Even if you're unused to the ways of Latin America, Ticos are used to you. Unlike other poorer, less-visited countries of the region, Costa Rica is essentially a modern nation with a cosmopolitan outlook toward the world. Plus, there are so many North Americans living and visiting here, the society cannot help but be accommodating. Costa Rica is often called a paradise. It boasts a beautiful country, modern infrastructure, tolerant society, religious freedom and a stable democratic government. Although it's not quite perfect, most tourists find Costa Rica offers them an ideal vacation that leaves nothing but fond memories.

A young Costa Rican boy grins as he eats corn at a festival.

Culture & Customs

Men of culture are the true apostles of equality.
~ Matthew Arnold, 1822-1888

Like other Latin cultures, Costa Ricans are very family-oriented and Sunday is the big day for family outings in local parks. The park at La Sabana, for example, is filled with picnickers and joggers or groups playing soccer or basketball, and there are even free exercise and aerobics classes.

COSTA RICA STYLE

We were impatient to get a bus back from Poás Volcano one Sunday and tried our hand at hitchhiking. A charter bus stopped for us and from the windows a score of friendly faces beckoned us inside. It was a neighborhood outing organized by the local people. Two other "guests" slept in the back while we chatted with anyone and everyone. It was a priceless opportunity to get to know Costa Ricans personally. To this day we remain friends with one family who invited us home for dinner.

Because it is not tropical with scorching hot afternoons, Costa Rica has not developed the culture of *siestas*, afternoon naps, that you find in countries like Mexico.

Costa Ricans are very polite and non-confrontational; they tend to be late for appointments, dance very well and are very friendly. Both Ticos and Ticas find blondes with light eyes attractive, while some North American men are attracted to the reputation Ticas have of being old-fashioned and devoted. While they are that, some men think that Ticas will willingly put up with the bad behavior that other women reject. Guess what? You're in trouble if you think so.

Blonde is *rubia* (female) and *rubio* (male); girlfriends are *novias* and boyfriends are *novios*.

Como los globos que flotan en las fiestas,
tengo, para no desinflame, un nudo en el estómica.
Like the balloons that float at parties,
I have a knot in my belly so I won't go flat.
~ Alberto Forcada, Mexican poet and author

■ Holidays & Fiestas

Ticos know how to put on a party – big time. If you are lucky enough to share a holiday or fiesta with the local people, it will add a whole new dimension to your vacation.

We had a marvelous time at a Mother's Day celebration in the small park opposite the metal school building in San José. We were surrounded by friendly smiling families. Ladies danced in colorful native costumes, food vendors sold myriad tempting treats, and the children rode ponies and competed in games (when they weren't being chased by the fantastic *payasos*, gigantic papier-mâché clown heads). It offered us the chance to be part of the local culture. We had a great time.

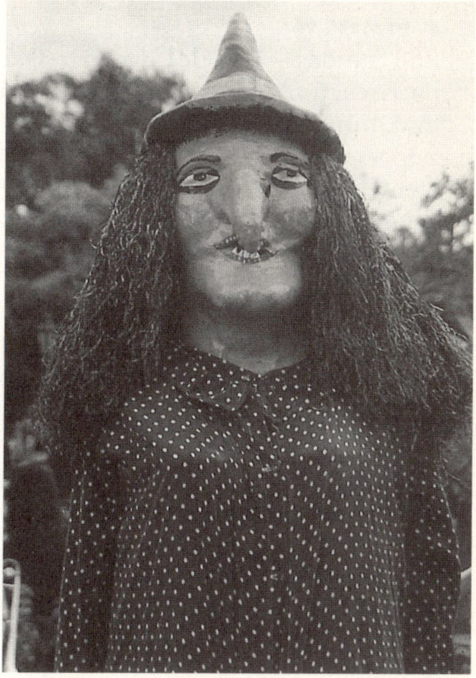

Fiesta mask.

Some festival dates vary each year. Check with the **Tourist Board** *(US & Canada, ☎ 866/267-8274, www.visitcostarica.com)* for exact details.

■ JANUARY

January 1 – **New Year's Day**.

First two weeks – **Fiesta de Palmares** in Palmares, a quiet village 56 km/35 miles west of San José. Carnival rides, bullfighting, music and folk dancing.

Week of Jan. 15th – **Fiestas de Alajuelitas**. Oxcart parade to an iron cross overlooking town, honoring the Black Christ of Esquipulas. Also, **Fiestas de Santa Cruz** in Guanacaste, with marimba music, folk dancing and more.

■ FEBRUARY

First week – **San Isidro de General Fair** has livestock shows, industrial fairs, bull teasing and an agricultural and flower exhibition.

Last week – **Sun Festival** is an annual gathering for a fire ceremony to celebrate the Maya New Year on Feb. 25th. Look for info in

San José. Same week is the **Puntarenas Carnival**, offering a week of fun in the sun in a working town that knows how to party.

■ MARCH

2nd Sunday – **Día del Boyero** (Oxcart Driver's Day), San Antonio de Escazú. A parade of colorful oxcarts, along with competitions and animal blessings.

2nd week – **International Arts Festival** throughout towns in the Central Valley. One of the best of its kind. Most cultural events take place in San José (check dates at www.festivalcostarica.org).

March 16-26 – **Fruit Festival**, Orotina. Fruits and vegetables from all across the country are exhibited and sold. Rides, food, lectures and concerts.

Mid-month – **Pilgrimage**. A religious procession beginning in Cartago and ending in Ujarrás at the ruins of the first church in Costa Rica.

March 19th – **Saint Joseph's Day**. St. Joe's namesake neighborhoods celebrate with special masses and fairs. People from San José picnic at Poás Volcano.

March or April – **Holy Week**. Religious processions depict crucifixion. Holy Week is especially popular in San José, Cartago and Heredia. **Easter Sunday** features a joyous procession of Resurrection. Popular, but many city dwellers head for the beach for a long weekend.

■ APRIL

April 11th – **Juan Santamaría Day** commemorates Costa Rica's national hero of battle with William Walker (see page 233). Celebrations all week, with parades, bands and dances. Especially big in Alajuela.

Last week in April – **University Week**. Concerts, exhibitions and parades at the University of Costa Rica in San Pedro.

Last week of April through first week of May, **Artisan's Fair**. San José hosts a popular craft fair.

■ MAY

May 1st – **Labor Day**. The President gives his annual State of the Nation address and Congress elects new leaders. There are many marches. City of Puerto Limón celebrates with picnics, dances and cricket matches.

May 15th – **San Isidro Labrador's Day**. Namesake towns honor the Patron Saint of farmers and farm animals. There are parades and fairs, and a priest blesses crops and animals.

May 17th – **Carrera de San Juan** is a big cross-country race challenging runners over a tough 22.5-km/14-mile course.

May 29th – **Corpus Christi Day**. Religious celebration and national holiday.

■ **JUNE**

Third Sunday – **Father's Day**. Dad's special day. Ask him for more money.

June 29th – **St. Peter** and **St. Paul Day**. Popular religious celebrations for namesake towns.

Three residents celebrate at a fiesta.

■ **JULY**

Saturday closest to July 16th –**Virgin of the Sea**. In salute to Puntarenas' Patron Saint, Virgin of Mt. Carmel, there is a regatta of decorated fishing boats and yachts in the Gulf of Nicoya. Parades, sports events, firework displays and religious masses are held.

July 25th – **Guanacaste Day**. Celebrates Guanacaste's 1824 decision to become a province of Costa Rica (instead of Nicaragua). Liberia holds fiestas, parades, folk dances, bullfight and concerts.

■ **AUGUST**

August 1 & 2nd – **Virgin of Los Angeles Day**. Honors Costa Rica's Patron Saint, La Negrita, with a nationwide pilgrimage to Cartago. Worshippers crawl on their knees in a procession. Also that day is **Our Lady of Angels** in Pardos, near Cartago, where figures topped by huge papier-mâché heads, called *payasos*, reenact a battle between the Moors and the Spanish.

August 15th – **Mother's Day**, a national holiday. Mothers are treated to special meals, candy and flowers. Call home.

August 30th – **San Ramóns Day**. Neighboring towns parade 30 saints through the streets to the San Ramón church.

■ **SEPTEMBER**

September 15th – **Independence Day**. All of Central America celebrates their mutual Independence Day. In Costa Rica, student runners carry a "Freedom Torch" from Guatemala to Cartago, timed to arrive at precisely 6 pm on the 14th, when everyone in the country stops and sings the national anthem. Parades on the 15th.

■ OCTOBER

Early to mid-October – **Carnaval** in Puerto Limón. Mardi Gras-style parades, floats and dancing in the streets. This town knows how to party.

October 12th – **Dia de la del Pilar**. The San José district of Tres Rios celebrates its Patron Saint.

October 12th – **Fiesta del Maíz**. Corn is the focus in Upala with parades and costumes made entirely of corn husks, grains and silks.

■ NOVEMBER

November 2nd – **All Soul's Day**. Day of the Dead, which begins on the 1st, is observed by family visits to graveyards to leave flowers for departed loved ones.

End of November – **Oxcart Parade** down the Paseo de Colon, San José. Begun in 1997 to honor the oxcart heritage. Entries come from all over the country.

■ DECEMBER

All month – The **Lights Festival** in San José features homes and businesses decorated with lights. Parades, concerts and nightly firework displays.

Week of the 8th – **Fiesta de los Negritos**. Indian rituals combine with Catholic concepts to honor the Virgin of the Immaculate Conception. Costumes, drums, flute music and dance in the indigenous village of Boruca.

Week of 12th – **Fiesta de la Yegúitta** (Little Mare) in Nicoya. Virgin of Guadalupe is honored with ancient Indian rituals and special foods, processions, fireworks and concerts.

Mid-December – **Posada** season begins. Carolers go from house to house (many collecting for religious donations) and Tico friends, co-workers and families get together in homes and restaurants for long joyful meals.

Mid-December to end of month – **Festejos Populares** (Popular Festivals). South San José fairgrounds at Zapote put on the country's largest and most unusual year-end bash with rides, food, bull teasing, music and fireworks. As many as 200 people cram into the bullring and a bull is released into the crowd. Reminiscent of the "Running of the Bulls" in Pamplona, Spain, except there is no place to run. If Hemingway were still alive, he'd be dying to go.

December 25th – **Christmas Day**. Traditional dinner includes *tamales*, corn meal pastry stuffed with meat and wrapped and cooked in corn husks or banana leaves. Christmas Eve mass is the Mass of the Rooster, *Misa del Gallo*.

December 26th – **Tope**, the daddy of all horse parades, downtown San José.

December 27th – **Carnival**. A huge parade with floats and music takes place in downtown San José.

December 31st - January 2nd – **Fiesta de los Diablitos**. Indians of the southern Boruca region near Golfito enact a fight/dance between Indians, *diablitos*, and Spaniards. Indians dress in burlap sacks with elaborately painted masks; the Spaniards are two athletic young men in a bull costume. Village flute and drum music.

Food & Drink

Never eat more than you can lift.
~ Miss Piggy

You won't find native Costa Rican food listed high on the world's culinary scale. Perhaps it's too plebeian for some. Unlike Mexico's, it is neither spicy nor complex. But it is high on our list of comfort cuisine.

■ Local Foods

The most typical dish in all Costa Rica is ***gallo pinto***, "red rooster," a rice and bean dish served with breakfast and sometimes lunch and dinner. Different recipes include herbs or garlic, but the basic ingredients are always black beans and rice, bell peppers and onions. Don't pass up an opportunity to try **lizano**, the Costa Rican imitation of English Worcestershire sauce. This piquant blend of vegetables, chiles and sugar is used on almost everything, like ketchup. If you get addicted, it's now available in the US on your grocer's Spanish foods shelf.

Fried **plantains** (a very large, firm variety of banana, also referred to as cooking bananas) are a common, sweet side dish to many breakfast and lunch meals. On the Limón side, rice and beans are flavored with coconut and Caribbean spices. Hot peppers are also more popular there. Side salads are generally made with cabbage slaw, red and green, flavored with oil and vinegar or mayonnaise. In a warm climate they prove very refreshing. The daily specials and cheapest dishes in most smaller restaurants are called ***casados***, which means "married." Your choice of fish, meat or chicken is served side-by-side (that's where the "married" part comes in) with a scoop of rice and slaw salad. This is the lunch of choice, especially in *sodas*, small mom and pop eateries.

The most typical dish for lunch and dinner in the home (and popular in restaurants too) is ***arroz con pollo***, chicken strips mixed in rice with vegetables. Fresh fish is also common and wonderful.

Some other meals and snacks you may come across include ***olla de carne***, a heavy meat and vegetable soup; ***picadillos***, a hash of potatoes, plantains and veggies; and ***empanadas***, corn flour dough filled with meats, chicken or fruit and fried. Empanadas are the preferred bite at street carts or doorway snack bars.

Costa Rica does top Mexico in desserts, especially for those with North American tastes. The national dessert is ***tres leches***, three milks, our favorite. Unfortunately, this cake can vary in quality. Of course, there's **flan**, custard with caramel or coconut. It's ubiquitous in Central American cuisine. Be sure to check out the offerings of bakeries, ***panaderías***, where you select your treats with a pair of tongs, place them on a baking tray, then take them to the counter and pay.

LEARN THE LINGO

To drink in Spanish is *beber*, but a waiter will ask *para tomar?* – what will you have?

■ Coffee, Beer & Batidos

Don't go to Costa Rica without having **coffee**. It's a higher quality than typical North American everyday fare, and better than Starbucks. *Café con leche* is coffee served in hot milk, like a cappuccino, only less expensive. **Café Britt** is usually considered the country's best brand, but it is also a bit more expensive than others. It's available in dark and light roasts. Buy your coffee in a supermarket and save over gift store prices. Many other local brands are worth trying too. Buy bags marked *puro*, not *traditional*, unless you like your sugar premixed in with the grounds. Ground coffee is marked *molido*, and whole bean is *grano entero*. You can also order Café Britt in the US *(☎ 800/462-7488, www.cafebritt.com)*.

With a proliferation of fresh fruit, there are no national refreshments as delightful as *batidos* or *jugos*. ***Jugos*** are fresh juices (*jugo de naranja* is orange juice), while ***batidos***, or ***refrescos***, are juice shakes made with water or milk. We are absolutely addicted to *batidos* and love to try different exotic flavors, such as mango, blackberries (*mora*), tamarindo, pineapple (*piña*), guava, papaya, banana and – our latest addiction – strawberry (*fresca*) in milk (*leche*). In August when the volcanic soil of Poás yields bumper crops of red ripe strawberries, there is nothing so wonderful as a lunch in the cool fresh mountain air accompanied by a strawberry milkshake in a tall glass. Yum.

BATIDOS

Batidos and their counterparts, *liquados* in Mexico and *smoothies* in the USA, are truly delicious – and they are drunk at any time of day. Made by blending fruit, crushed ice, water or milk together into a smooth liquid, their variations are as endless as one's imagination and Costa Rica's bountiful selection of fruit. We tried countless concoctions using one, two or three fruits and enjoyed them all. But our favorites are fresh strawberry or luscious blackberry – Costa Rica has the largest blackberries we've ever seen – both made with milk. Another similar treat is *horchata*, a rice-milk drink. All these drinks are a delicious, healthy, addiction. Ask them to use purified water, *agua purificada*, when you order. Or try them with milk.

Not as well known outside of Costa Rica is the wonderful **beer** they brew. Our favorite is **Imperial**, a lager, but you should decide on yours by trying them all! **Pilsen** is a light gold pilsner, while smooth **Bavaria** is a bit heavier and darker than Imperial. **Tropical** is a lightweight but a good brew in the hot lowlands. Be aware that many Ticos drink their beer in a glass with ice, so if that's sacrilegious to you, better make sure you say, *sin hielo*, (SIN YEL-low). The only brewpub we know of in Costa Rica, **K&S**, at Plaza Cristal, 600 meters south of La Pops in Curridibat, makes a very tasty beer that they serve along with darn decent mid-priced meals. Local bands play the weekends and the pub offers expansive views.

The national drink of the nation's alcoholics is **Guaro**, a sugar cane-based hard liquor that tastes like rubbing alcohol.

Since the first German beer brewers came to Costa Rica it has been the custom in *cantinas* and restaurants to serve a little snack with your drink. ***Bocas***, which translates to "mouths" or "mouthfuls," is the name of the appetizer that nowadays may or may not come with your drink. Alas, the economics of scale have restricted the freebies to smaller bars in the countryside, or friendly places in town. The many kinds of *bocas* offered vary from ***ceviche*** (raw fish in lime juice) to ***chicharrones***, heart-clogging fried pork rinds, and everything in between – including rice and beans. Ask for free *bocas*, but remember the standards of cleanliness in your local watering hole may not meet yours.

Alternative Accommodations

If a man does not keep pace with his companions,
perhaps it is because he hears a different drummer.
Let him step to the music which he hears,
however measured or far away.
~ *Conclusion*, Henry David Thoreau, 1817-1862

If you are looking for lodgings more intimate than a hotel, Costa Rica offers an ever-growing number of alternatives. Perhaps a home stay with a Costa Rican family or in-the-rough camping sounds good. But don't overlook some of the exquisite small hotels and inns we mention throughout the book – many of them boast tranquility, intimacy and charm, without the impersonal nature of a large hotel. In fact, the lack of large, cookie-cutter hotel chains insures most Costa Rican accommodations have distinct personalities. And that appeals to us.

■ Down on the Farm

Farming looks mighty easy when your plow is a pencil
and you're a thousand miles from the cornfield.
~ Dwight D. Eisenhower, 34th US President

COOPRENA *(☎/fax 506/259-3605, www.turismoruralcr.com).* Eight agricultural cooperatives, ranging from Monteverde in the north to the southern Osa Peninsula, offer guests a rare opportunity to tour their farms and interact with resident families. Participants stay in rustic lodges, the proceeds from which are used for communal projects, such as fixing up roads, improving schools and cleaning the local soccer field. Good family-style food and fresh milk. They have their own tour company called Simbiosis Tours, specializing in local initiatives in ecological and land tourism. It's an admirable effort to convert tourist dollars directly into grass-roots conservation projects, encourage local people to follow sustainable practices and improve the quality of life.

Corcovado Agroecoturistic Association *(Puerto Jimenez, Osa, ☎ 506/735-5440, www.ecoturism.co.cr/osanatural)* offers accommodations, meals, tours and logistical support for individuals or scientists in and around the Corcovado National Park on the remote Osa Peninsula.

If you've come to Costa Rica to get an education, think about the tropical agro-ecological farm **La Flor de Paraíso Environmental School** *(☎ 506/534-8003, www.la-flor.org)*, near Paraíso, outside Cartago. It offers Spanish language and culture courses, organic farming, artisan workshops, tropical rainforest regeneration projects and a medicinal plant garden. Paying volunteers are needed for the experience of a lifetime.

■ Bed & Breakfasts

It used to mean a suspicious-eyed old woman with knitting needles who hustled you upstairs to sleep on itchy sheets and had fried eggs ready for you in the morning – ready since the night before.
~ On improvements in Britain's B&B hotels, *Esquire*, 1986

The allure of bed and breakfasts, whether a spare room in a private home or an establishment with a number of rooms for rent, is their homey atmosphere and the opportunity to share close contact with others. Breakfasts are included, obviously, but often dinners can be added to the package for a small charge. Most per-night prices are well below hotel rates, but not always.

To make arrangements in advance, you can phone any of the following booking agents, who do a good job matching clients to compatible Costa Rican families.

B&B BOOKING AGENTS

Bell's Home Hospitality *(☎ 506/225-4752, fax 224/5884, www.homestay.thebells.org)* organizes stays in the San José area. Many are spare bedrooms in 50+ private homes. Bell's offers explicit directions and, for a fee, will arrange airport pick-up. Write them at Dept 1432, PO Box 025216, Miami, FL 33102.

Bed and Breakfast Group *(☎ 506/289-8638, crnow@amnet.co.cr, www.costaricainns.com).* Founded in 1991 by a gringa, this group has links to 200+ B&Bs, rooms in private homes, and small hotels across the country. Many good contacts in rural areas near eco-attractions.

TurCasa *(☎ 506/258-3381, www.turcasa.com)* is a women's homemaker cooperative that offers "homestays" with their own families. Airport pick-up possible.

Besides B&Bs, another interesting way to stay in a homey atmosphere is to sign up for study at one of the many Spanish-language schools. They will make your residential arrangements in a participating local middle-class family's home.

Hostels

Youth must have its fling.
~ *Pirates of Penzance*, Gilbert & Sullivan, 1879

These dormitory-style accommodations, segregated by sex, with shared baths, are a low-cost alternative to private rooms in hotels or B&Bs and a great way to network with fellow travelers. Not just for "youth," these inexpensive hostels offer card-carrying travelers of any age, secure, clean and relatively comfortable lodgings. You'll find a number of senior travelers mixed with baby-boomers and college-aged backpackers from around the globe. Costa Rica boasts some 14 hostels from San José to Limón and places in between. Visit the Hostelling Costa Rica website for more information, www.hostelling-costarica.com.

HOSTELRY RESOURCES

Hostelling International *(in US, ☎ 202/783-6161, in Canada, ☎ 613/748-5638, R.E.C.A.F. in Costa Rica, ☎ 506/244-4085, www.hostels.com)* offers information on membership cards, reservations and details on each hostel.

Camping

And the night shall be filled with music,
and the cares that infest the day,
shall fold their tents like the Arabs,
and as silently steal away.
~ *The Day is Done*, Henry Wadsworth Longfellow, 1807-1882

If you like to commune with nature up close and personal, try a stay at one of Costa Rica's small selection of private camp sites. Also, a fair number of the national parks provide rustic campgrounds; their fee is around US $2 per night, plus the $8 park entrance fee.

Most campsites provide potable water; we recommend taking your own bottled water just to make sure. Showers and restrooms usually are included, and occasionally you'll have access to a public phone. Some sites, however, can be less than basic and you may have to scrounge around for facilities. Remember not to leave valuables in your tent unless there is someone around to protect them. Lastly, you may find it more enjoyable to choose the dry season for your trip because camping during the rainy season can be a muddy affair.

CAMPGROUNDS

PRIVATE CAMPGROUNDS NEAR THE BEACH:

Montezuma Beach - Camping Los Monos *(☎ 506/642-0048)*. Open Dec.-May; call before arriving.

Tamarindo - Bagatsí *(☎ 506/659-9039)*.

Playa Junquilla - *(☎ 506/666-5051)*. Three km/1.8 miles before Cuajiniquil.

Pochote Beach - Camping Tino Zeledón, no phone.

Tambor Beach - Camping Los Malinches *(☎ 506/683-0264)*.

Potrero Beach - Mayraís Camping & Cabinas *(☎ 506/654-4213)*.

Jacó Beach - Camping Madrigal *(☎ 506/643-3329)* and Camping El Hicaco (no phone). Unless you're a party animal, it's best to avoid both camps on holiday weekends.

IN THE CENTRAL VALLEY NEAR SAN JOSÉ:

Alajuela Area

Laguna Fraijanes Recreational Park *(☎ 506/442-2166)* offers huts without lights for up to four people for about US $11. A full-service hut runs $20 for two. Chalets sleeping eight, $25. US $2 per person with own tent, or $3 if you need to rent a tent. Bring linens. A restaurant is nearby.

Heredia Area

Bélen Trailer Park *(☎ 506/239-0421)*.

Roble Alto Camp *(☎ 506/237-1453)* has rustic cabins that sleep eight. Bring your own food and blankets. Laundry, kitchen, sports courts and pool. Costs around US $30 per *cabina*.

Getting Around

I'd rather go by bus.
~ Prince Charles, age six, when asked if he was excited about sailing to Tobruk on the Royal Yacht.

First, the good news: Costa Rica is a small country that allows you to travel from the Central Valley down to either coast in a short amount of time. That puts many eco-adventure destinations within easy reach, even for day-trips. The bad news is that some roads, especially in more rural areas, can be a trial. More good news is that Costa Rica has an excellent bus service as well as inexpensive domestic flights to remote corners of the country. Consequently, the best mode of transportation depends on where you're going, when, and for how long.

■ Travel by Bus

If you have a single destination in mind, such as Arenal, Montezuma, Cahuita or Tamarindo, **first-class bus service** is the way to go. Vehicles are modern, travel almost non-stop, and are driven by someone who knows the roads from experience. There is no central terminal in San José for direct buses to other cities, so city planners have spread out the bus stops around town to lessen congestion. See page 387 in the *Appendix* for a schedule and San José bus stop/terminal locations. You can also check for changes online at www.costaricabybus.com.

San José & Vicinity

If you're staying in San José, then local buses are the best way to get around. Traffic is typical for big cities – a pain in the butt – so having a car in town is often more of a liability than it's worth. Not to mention the parking, or the lack thereof.

During our first stay in Costa Rica, we commuted daily into San José downtown from the eastern suburb (*barrio*) of San Pedro. Buses run frequently on popular routes and are very inexpensive, perhaps 90¢ to the suburbs, 50¢ around town. Look for the name of your destination painted on the front windshield (not so easy to spot at night). Few bus drivers speak English, but many passengers do, so ask for help if you're unsure. Many of the local buses are old "Bluebird" converted school buses made in the US, and few have buzzers to alert the driver to your stop. A whistle or a yell, usually works. If you don't know the exact stop, ask the driver when you first get on to let you know.

Downtown is a good place to walk. Most everything you're there for is concentrated in a 12-block area. Taxis are available and inexpensive. Remember, they're *very* scarce on rainy rush hour evenings! Rates go up a bit later at night and if you travel to out-of-the-way neighborhoods.

> **WARNING:** A place to avoid at night is the Coca-Cola market area in the northwestern end of town. If you're returning to downtown from points north or west at night, get off a little early on Paseo Colon and take a taxi or the Cementerio bus from there.

Around the Country

Don't be afraid to travel anywhere by bus, even if you get lost and wind up somewhere other than where you planned. Chances are it will be a pleasant adventure. It always was for us.

For excursions from San José into the countryside, we took either direct, first-class buses to destinations such as Jacó, Quepos or Puerto Viejo (maximum fare is under US $20), or semi-local buses to Cartago, Alajuela, Orosí and Heredia (about US $2).

When we vacation in Costa Rica, we usually ratchet down our lifestyle and travel by bus. It's more ecological and there is no better place to be environmentally aware than Costa Rica. On the other hand, driving around Costa Rica in a rental car is a popular and convenient way to see the country. You can stop to take pictures, go down side roads or change your mind about your destination en route. See page 97 for some driving tips.

■ Car Rentals

Have you ever noticed....
Anybody going slower than you is an idiot,
and anybody going faster is a maniac?
~ George Carlin

Renting a car is a good way to get around Costa Rica. It's a small country; consequently many parts of it are within easy driving distance. Roads are generally well marked, although rarely with the route numbers you see on maps. Instead, occasional signs along the road tell you the number of kilometers to a larger city. Major routes are well paved and easy to follow.

Insurance

Before you decide if you want to get around by car, there are some caveats to renting. On top of the daily or weekly rental cost, you need to buy Costa Rican auto insurance as yours is not valid here. We recommend getting as much insurance as possible, including no deductible. It may cost more, but remember these are foreign cars on foreign roads in a foreign land. Many of the secondary roads in the mountains and along the shore are unpaved, unless you count the dirt between the yawning potholes. Due to heavy rains, hot sun, *temblores* and mountainous terrain, even paved roads are susceptible to deterioration and the occasional mudslide.

Rugged Terrain

Now I know how many holes it takes
to fill the Albert Hall.
~ The Beatles

Unless you stick to the main tourism areas in the Central Valley, you'll need to rent a four-wheel-drive vehicle. A few backwoods roads require you to cross small rivers without the aid of a bridge. Keep in mind that car rental agencies do not want you to cross the rivers in your car, even if it's a four-wheel-drive vehicle. We got out, rolled up our pants legs and checked the depth first. It's also wise to follow the tire tracks of previous vehicles.

Be cautious too when driving in the mountains, where sudden fog or rain can envelope the road, reducing visibility to near zero. Another consideration is the lack of guardrails, even when there is a 1,000-meter drop on the side of serpentine, hairpin curves. What a vista! In beach areas to the west, count on a lot of dust, especially in the dry season, and plenty of heat. Get a car with air conditioning if you're off for Guanacaste or Limón.

If you're basing yourself out of San José for the first few days or last night of your stay, you don't need a car in town. Some larger provincial cities offer car rentals, so you can always fly or take a bus to say, Liberia, and then rent a car. To get around the choking downtown traffic, San José built a *Periférico*, a bypass road to the south of the city that begins near the suburb of Escazú and ends just north of the San Pedro suburb. Along it are some hard-to-see traffic lights and absolutely insane traffic circles. Be brave. This road works best when it's not rush hour. Speaking of rush hour, the Paseo Colon becomes one-way during that time.

You'll soon see that Costa Rican drivers are *mucho loco*, and roads are in unpredictable condition. Our best advice is to slow down. Sooner or later, you'll be stuck behind a stinky, slow-moving truck and anxious to get where you're going – but don't worry, be happy. This is your vacation, not a commute to work.

See next page for instructions on what to do and who to call if you're involved in a car accident. Call your rental company first; most have 24-hour help numbers.

Car Rental Companies

We were very happy with the rates and service at **Poas Car Rental**, a Costa Rican-owned company with pick-up service from the airport. Their airport office is at the Hampton Inn *(US ☎ 888/607-7627; Costa Rica ☎ 506/442-6178; www.carentals.com).*

Other local renters are **Tricolor** *(☎ 506/440-3333, www.tricolorcar-rental.com)*, **Toyota Rent a Car** *(☎ 506/223-8979, www.toyotarent.com)*, **Europcar** *(☎ 506/257-1158, www.pregorentacar.com)*, **Elegante** *(in US, ☎ 800/582-7432, in Canada, 800-445-6499, ☎ 506/257-0026, www.elegante rentacar.com)*, and **Tropical** *(☎ 506/442-8000)*.

Siesta Campers rents fully equipped VW Westfalia camper vans *(☎ 506/289-3898, www.edenia.com/campers)*, excellent for their high clearance. Want to rent a **Harley Davidson motorcycle**? Ask **Maria Alexander Tours** in Escazú *(☎ 506/289-5552, www.mariaalexandra.com)*. Rates include helmet, rain gear and lock.

International automobile agencies with toll-free numbers in the US are **Avis** *(in US, ☎ 800/331-1212, 506/442-1321)*, **Budget** *(in US, ☎ 800/527-0700, 506/441-4444)*, **Hertz** *(in US, ☎ 800/654-3131, 506/441-0097)*, **National** *(in US, ☎ 800-328-4567, 506/441-6533)*, and **Thrifty** *(in US, ☎ 800-376-227, 506/442-8585)*.

RULES OF THE ROAD

1. Unless posted otherwise, the urban speed limit is 40 kmh (25 mph); on highways it's a crawling 60 kmh (37 mph). Around schools or hospitals, 25 kmh (15 mph).

2. Ceda el Paso means "Yield, Right of Way," which means you give way to oncoming traffic. It's a very common traffic sign at the many single-lane bridges.

3. Seatbelt use is required.

4. Driving on beaches is strictly prohibited.

5. If you are in accident, do not move your car until the police *(☎ 911 or 117)* tell you to do so.

6. Drive defensively!

You'll get a slew of free maps in Costa Rica, many of which are barely adequate at best. The best all-around map for the country is produced by **Toucan Guides** *(http://costa-rica-guide.com, $9.95)*, not only because it is accurate, well indexed, and loaded with handy information, but it features a soft, flexible coating that resists tears and ubiquitous rain from Spain. Not to mention sticky fingers. Get it before you go or download the PDF file. If you're a cheapskate, there's a free map – filled with ads but very useable – produced by www.costaricamap.com. It is available at many hotels and tourist information booths. **International Travel Map**, out of Vancouver, BC, Canada, produces a detailed country map. We bought one in San José 7th Ave. bookstore.

■ Hitchhiking

Increasingly, hitchhiking is a thing of the past around the world, with most "rides" offered only in more remote rural areas. We sometimes offer rides to walkers on the side of the road, but haven't encountered any true hitchhikers. If you're thinking of hitching, take all the precautions you would at home, such as not accepting a ride with someone who has been drinking or, if you're female, not accepting a ride from a carload of guys. It's always best to travel in pairs, but if you're hitchhiking that may mean a longer wait. Gringos are in the habit of passing you up and most Costa Ricans might think you're rich enough to take the bus which, given the low cost and safety, is probably a much better idea.

Picking up locals is another thing, especially in remote areas. We sometimes offer rides to people walking alongside the road, usually uphill, in order to practice our Spanish until the next village. When you let them off, it's a custom – at least among older folk, who may be as poor as church mice – to dig their hand into a well-worn pocket and ask, *Cuanto cuesta*? How much?

Simply smile and reply, *Por nada* (for nothing).

■ Cross-Country Flights

A good way to cover the distances to the corners of Costa Rica is by air. To spend a weekend in Quepos or Tortuguero, for example, you can easily and relatively cheaply fly in puddle jumpers (twin or large single engine planes). Most flights last a half-hour or less, depending on stops.

Two main domestic airlines are **Nature Air** *(Bolaño airport, west of **Pavas**, ☎ 506/220-3054, fax 220-0413, www.natureair.com)* and **Sansa** *(Grupo Taca office building, near Sabana Park, ☎ 506/221-9414, fax 255-2176 or for advance in the US ☎ 800/535-8780)*, which flies out of Juan Santamaría airport.

During the high season, flights are often full so you need to book in advance or at least try to get reservations as soon as you arrive in Costa Rica. Both airlines sell tickets through local travel agents – a good way to assure service. Nature Air flights are slightly more expensive but more dependable with their schedules. We favor them. Surfboards add another US $15-20 surcharge. Don't bring a lot of luggage – there is a strict one-bag limit Cost is $60 and up, depending on your destination.

Nearly all flights take off and return in the mornings as Costa Rica's weather tends to deteriorate in the afternoons, especially in the rainy season. All flights are VFR, Visual Flight Rules, because clouds and very tall mountains don't mix.

> **AUTHOR TIP:** You stand a better chance of getting a flight without a reservation when you're coming back into San José than if you're flying out. Keep calling to ask about cancellations, or have a travel agent call their contacts.

Learn Spanish in Costa Rica

I must learn Spanish one of these days,
Only for that slow sweet name's sake.
~ *The Flower's Name*, Robert Browning, 1812-1889

The best way to learn Spanish is to take advantage of the many immersion courses offered at numerous language-study institutions around the country. On our first trip to Costa Rica we registered for a six-week course at the Forester Instituto Internacional in Los Yoses, located just a 20-minute walk from downtown San José.

Most schools offer two-week schedules, small groups, half- or full-day courses, plans with tour options, academic credit and area lodging. Economical homestays, with a Spanish-speaking local family, provide a golden opportunity to practice your classroom learning. Homestays usually include breakfast and dinner with the family – total immersion in the culture as well as the language.

We spent a half-day in class and used the other half to explore the countryside on our own, getting on buses and going where they went. At the end of our stay our educational results were mixed – one of us has a fairly good handle on the language, the other, hmmm – but in terms of getting to know Ticos and their country, we did very well!

■ Language Schools

Below are listed some of the many schools currently offering courses in Costa Rica. Prices start as low as $200 per week, including home stays.

In San José

Forester Instituto Internacional, Los Yoses San José (*☎ 506/225-3155, fax 225-9236, www.fores.com*). Professional school in a trendy neighborhood that still houses several embassies.

Intensa, Los Yoses San José *(☎ 506/225-5009, fax 253-4337, www.intensa.com)*. As the name suggests, it offers intense study as well as one-on-one classes. Optional full day.

Instituto de Español, Guadalupe, San José *(☎/fax 506/283-4733, www.intensivespanish.com)*. Ask for their 2-for-1 special.

Ilisa, San Pedro *(in US, ☎ 800/ILISA4U; in CR, ☎ 506/280-0700, fax 225-4665, www.ilisa.com)*. Highly regarded.

Costa Rican-North American Cultural Center, Barrio Dent and also in Sabana Norte *(☎ 506/207-5000, fax 224-1480, www.cccncr.com)*.

Instituto Británico, Los Yoses *(☎ 506/225-0256, fax 253-1894, instbrit@sol.racsa.co.cr)*.

Costa Rican International Language Academy & Latin Dance School, Barrio California, San José *(☎ 506/233-8938, fax 233-8670, www.telecomcr.com)*. What could be better than learning to speak with your dance partner? They offer cooking, too!

Centro Linguistico CONVERSA, San José *(in US, ☎ 800/354-5036, in CR, ☎ 506/256-3069, fax 233-2418, www.coversa.co.cr)*. Also a campus in Santa Ana in the foothills.

University of Costa Rica, San Pedro *(☎ 506/207-5634, fax 207-5089, www.cariari.ucr.ac.cr/~filo/espanol.htm)*. Eighty-hour course taught by the professional staff at the School of Philology and Literature. University ID, library privileges and cultural activities.

In Escazú

Language and International Relations Institute (**ILERI**), Escazú *(☎ 506/289-4396, fax 228-1687, ilerist@sol.racsa.co.cr)*.

Lisa Tec B&B Language School *(near Cariari Golf Course, ☎ 506/239-2894, fax 293-2894, www.ltspanish.com)*. Not exactly in Escazú, but a tranquil language school and B&B.

In Heredia

Intercultura, Heredia, Av 4, Calle 10, *(☎ 506/260-8480, fax 260-9243, www.spanisintercultura.com)*. Complete offerings including Latin dance, activities and volunteer opportunities. Once-a-month classes at Jacó beach.

Pura Vida Institute, Heredia *(☎ 506/260-6269, fax 237-0387, www.costaricaspanish.com)*.

Instituto Profesional de Educación DAZA, Heredia, also in Liberia & Guanacaste *(☎ 506/238-3608, fax 238-0621, www.learnspanish-costarica.com)*.

Rural & Beach Schools

Centro Panamericano de Idiomas, 125 meters/411 feet east of cemetery in San Joaquín de Flores, outside of San José *(☎ 506/265-6866, fax 265-6213, www.cpi-edu.com)*. Homestays and volunteer programs. Also in Monteverde and Flamingo Beach.

Rancho de Español, Alajuela *(☎/fax 506/438-0017, www.ranchodeespanol.com)*. Quiet and small.

Spanish Language and Environmental Protection Center (SEPA), San Isidro del General *(☎ 506/770-1457, fax 771-1903, www.online.co.cr/sepa)*. It's located in a non-tourist area, so it's a real immersion experience.

Montaña Linda, Orosi *(☎ 506/553-3640, fax 533-2153, www.geocities.com/thetropics/paradise/6728)*. School and hostel.

Escuela D'Amore, Manuel Antonio, Quepos *(☎/fax 506/777-1143, www.escueladamore.com)*. The school of love? Hmm.

Horizontes de Montezuma, Montezuma Beach *(☎/fax 506/642-0534, horizontes@mail.ticonet.co.cr)*.

Sports

Bullfight critics row on row
Crowd the vast arena full,
But only one man's there who knows
And he's the one who fights the bull.
~ Federico Garcia Lorca, 1899-1936 (a favorite quote of JFK)

Costa Ricans enjoy sports in their daily lives: kids and adults play basketball, soccer, a little baseball, and swim. Maria del Milagro París was the first Tica to qualify for the swim finals at the 1980 Moscow Olympics. Costa Rican sisters Sylvia and Claudia Poll, world-class swim competitors, train in San José. Sylvia won the country's first medal, silver, in 1990. Then in Atlanta, her younger sister Claudia won a gold medal at the Olympics. The country went crazy with pride!

Bullfighting is pretty much confined to holiday celebrations except in Guanacaste, which is cattle country.

■ Soccer

But *every* Tico is a soccer – *fútbol* – fan. Soccer is the national sport, with 12 teams in the Primera Divisíón that play every Sunday and Wednesday night during the season from December

Soccer match at Saprissa Stadium, San José.

to May/June. Lately, Costa Rica has done very well in international matches. In 1990 the national team advanced to the second round of the World Cup in Italy by beating Scotland and Sweden.

Happily for us, the soccer rivalry between the United States and Costa Rica is heating up, especially after close and physical matches in qualifications to World Cup 2002. World Cup competition takes place every four years and Costa Rica and the US are in the same region, perennially fighting to qualify for the three limited spots against powerful Mexico, Honduras, Guatemala and several Caribbean islands, such as Jamaica and Trinidad-Tobago, who also field good teams.

FIRST DIVISION TEAMS (Equipos de Primera División)	
Alajuelense (Alajuela)	Osa
Carmelita (Barrio de Carmen, Alajuela)	Pérez Zeledón (San Isidro)
Cartaginés (Cartago)	San Carlos (Quesada)
Herediano (Heredia)	Santa Bárbara
Liberia	Santos (Guapíles)
Limón	Saprissa (Tibas)

We attended a 2002 World Cup qualifying match between the US team and the Ticos (Costa Rica won) at the Saprissa Stadium. At the end, every taxi and bus was jammed with celebrating fans, so we had to walk back to town through a gauntlet of flag-waving *futból* fanatics. Despite our American flag shirts, Tico fans proved themselves good-natured winners. If you like soccer, get thee to a stadium to experience a level of enthusiasm that goes off the scale!

Money Matters

Money Makes the World Go Round
~ Title song, *Cabaret*, 1966

Currency in Costa Rica is the **colon** (plural is *colones*), which floats against the almighty dollar. You can change dollars in banks (bring your passport) or stores, which accept them at their own exchange rate, sometimes better, sometimes worse than the official exchange. When purchases are small it hardly matters. Canadians, Europeans and Brits should convert their cash into US dollars first, as they are more common and easy to exchange while in Costa Rica. Hotel exchange rates are rarely as good as the banks or even some stores who convert money as a sideline.

Traveler's checks are falling out of favor with the proliferation of **ATMs**. Stick your card in and select the desired amount of colones; your only fee is any ATM charge that may apply. We use this method so we don't have to carry around more cash than we need. Another reason checks are being seen less frequently is because they are often changed at a lower rate than cash in Costa Rica.

If you change dollars or get an advance on your credit card in a **bank**, bring a copy of your passport for identification. For some reason, Costa Rican banks, and therefore most merchants, prefer **VISA** charge cards, although MasterCard is are accepted almost everywhere. Recently, many of the US bank charge cards, including ours, added a percentage service charge for foreign exchange purchases. In other words, the $100 you charged on a credit card for your hotel now costs you $100+. Check with your card company for their rules. If you figure out a way to get around it, let us know.

■ Tipping

Tipping is not a city in China.
~ sign on a tip can on counter

% We tip by North American standards – $2-2.50 per night for cleaners in hotels and $2 per large bag carried by bellhops. Taxi drivers get no tip as they include it in the rate. Remember to negotiate; few taxis have meters, or *marias*. (Also, seatbelt use in cars and taxis is required by law but is uncommon except for us gringos. Use them – you won't regret it.)

Guides, many of them college-educated naturalists, should be tipped as generously as you feel they're worth – they live almost exclusively on tips. Many restaurants automatically add a 10% tip, *propina*, to the bill. If we had good service, we tip another 10%.

Check our Just In Case *section, page 110, for the numbers to call if you lose your credit card.*

■ Gambling

Many of the larger hotels offer gambling casinos with slot machines and gaming tables. The Gran Hotel has a small casino and the huge pink Hotel Del Rey (Av 1 and Calle 9) has one of the more popular casinos, with a sideshow of middle-aged "fishermen" in Hawaiian shirts and young ladies of the night. The casinos in the Aurola Holiday Inn and Barceló Amon serve free buffets for gamblers, while the Gran Hotel and Del Rey will feed you at your table. On several slow evenings we've played blackjack with a one-dollar minimum, feeling rich if we won 5,000 colones and saying, well it's only $10, if we lost. Remember, the house always wins. "Rommy" is their name for blackjack, and "Tute" is a kind of poker.

Safety & Crime

See me safe up, and as for my coming down, let me fend for myself.
~ On descending a ladder, Thomas More, 1478-1535

Before we list all the precautions you can take to avoid difficulties on your vacation, we will say that Costa Rica is generally a safe destination, without the serious problems that plague poorer nations. However, it would be naive to think that crime doesn't exist. The murder in 2000 of two female American students, shot and killed in a car-jacking, brought the increase in crime to the attention of US travelers. As a result, tourism figures dropped immediately. Although two perpetrators were caught and convicted, travelers began to realize that Costa Rica has a growing crime problem. Theft is a constant concern everywhere, but especially in San José, Quepos and Limón.

We should emphasize that very few of the huge number of visitors that arrive annually ever encounter problems. We have been many times and our only complaint was being short-changed at a highway tollbooth. Annoying, but not dangerous. In fact, the most common causes of death and injury among tourists are car accidents and accidental drowning.

Common-Sense Precautions

- Never leave valuables in your car, not even in the trunk, as rental cars are obvious to thieves.
- Never leave valuables unattended on the beach.
- Sad as it is to say, you should be cautious about strangers too eager to help you find a taxi, a hotel, fix a flat tire, show you the way, or carry your bags.
- Follow common-sense rules about not carrying a lot of money around.
- Leave flashy jewelry and watches home. Lock valuables in the hotel safe and zip your camera bag up tight.
- Stay out of seedy or deserted areas, such as the Coca-Cola, at night, especially in larger cities. (The Coca-Cola area, named after a bottling plant that has long gone, is around Calles 14-18, Av 1 and 3, near the central *mercado* in San José. It has an important bus terminal. Safe during the day, pickpockets and muggers have been known to target tourists and Ticos here at night.)

Men on their own are also vulnerable. We've heard of teams of two young men and a lookout putting men in a choke hold until they pass out. We met a college-age student to whom this happened, so think with your elbows!

Inventive thieves have been known to target rental cars by giving the tires a slow leak just as the customer leaves the airport. They follow targets down the highway and, when the unsuspecting victim pulls over, rob them. If you get a flat under these circumstances, keep driving to a public area.

Despite this disheartening litany of possible crimes, we can't say enough about every day Costa Ricans – they are hardworking, sincere, helpful, caring and very honest. With a few precautions, your chances of being a crime victim are very small.

For Women Travelers

I am a woman meant for a man,
but I never found one who could compete.
~ Bette Davis, actress, 1908-1989.

Women traveling alone or in pairs are not uncommon in Costa Rica and rarely encounter extraordinary problems. Stay out of situations where you are clearly vulnerable, such as getting drunk in a bar, hitchhiking or walking home alone at night through an unfamiliar area. Sex crimes are relatively rare in Costa Rica, but a touch of machismo in men is alive and well. Don't be shocked to be called a *macha*, meaning a blond or light-skinned girl, by men on the street or ones passing in cars. It can be annoying, but don't be offended by what they think is good-natured teasing. Remember, it is also stupidly hilarious to hear lines such as *Que curvas y yo sin frenos*, which translates to "What curves and me without brakes!" You have to admit that it's more creative than "Come here often?"

Feminism is also very much a part of modern Costa Rica. Women's cooperatives are springing up around the country, such as the one in Monteverde. Contemporary, educated Ticas stand up for themselves and don't tolerate macho men very well. But *piroperos* (men who make remarks in public) are a part of the Latin culture that is dying a very slow death.

Health, Special Concerns

Al que madruga Dios le ayuda.
(God helps those who get up early.)
~ Spanish proverb

Costa Rica is the most modern and sanitary country of the Central American isthmus, so it presents few health worries. No shots are required, but if you're traveling on to more remote sections of Central America – such as Guatemala, Panama, El Salvador or Honduras – a vaccination against **hepatitis A** is strongly recommended. Contaminated water is the common source; a shot of immune globulin gives adequate temporary protection. A doctor friend of ours, who has vacationed in Central America for the past 25 years, recommends a **hepatitis vaccine** to all travelers regardless of where they go in the world – Cartago or Copenhagen.

■ The Water

Outside of San José we drink bottled water to avoid intestinal infections. We learned, however, that nothing offers fail-safe prevention – not even bottled water. Some medical sources even suggest *tourista* (gastric distress) can be caused by a combination of other factors. Its symptoms, which mimic salmonella poisoning, may include any or all the following: nausea, diarrhea, vomiting, stomach cramps and low-grade fever. Purists suggest waiting it out for three or four days, but that's hardly realistic if you've got only a week's vacation and a gazillion things to do. So here's our tried and true treatment.

If we're in a budget hotel, the first thing we do when we start feeling bad (and it comes on very quickly) is upgrade to a hotel with air-conditioning – maybe even cable TV – and a comfortable bed. A couple of aspirin and plenty of sleep are called for. If we suffer frequent diarrhea and stomach cramps, we take the recommended dose of Imodium AD. Pepto Bismol relieves the symptoms as well, but takes longer. Drink plenty of bottled water or Coca-Cola with lime or, in severe cases, hydrating fluids such as Pedialyte, available at a local drugstore. We also drink manzanilla tea (chamomile) with honey, a helpful folk remedy. Then we crank up the air-conditioner, curl up and go to sleep. We repeat the Imodium if the diarrhea returns. In about 24 hours we're usually feeling well enough to get back out and enjoy ourselves again – with some reservations.

If you're feeling really bad, have your hotel call a doctor.

If you've had a bout of *tourista*, you may still feel a little weak, so take it easy and don't over-exert yourself. For a few days you may also experience mild stomach cramps after eating. Eat light and cut out liquor and hot spices. In all our visits, we've been sick only once, but it was a memorable occasion. We learned a lesson from it. If you're really sick, go to a doctor (or pharmacist) and get an antibiotic. Don't be shy about it.

The US Public Health Service does not recommend taking any prophylactic medicines beforehand, but there are other ways to aid in prevention. Besides drinking bottled water, use it when you clean your teeth. Peel fruit before consuming. In addition, we theorize that much of the bacteria that gives problems can be eliminated with frequent hand washing. The sensory delights of Costa Rica include touching new things, so a thorough hand scrub every chance you get is a good idea. You should have a fair number of chances because many restaurants offer a sink right in the dining room and it's considered polite to wash before eating. Alternatively, use the antiseptic towelettes we recommended or consider taking along one of the new anti-bacterial sanitizing liquids, such as Purelle, available in the US.

■ Other Health Concerns

! Although getting sick is a prime concern of tourists, drowning is a major cause of death. Be extremely wary of **rip tides** when swimming on either coast. There are few lifeguards on the beaches here. A rip tide is like an underwater river pulling you out to sea. If you get caught in one, don't panic. Swim parallel to shore until out of its grip.

Other worries of new tourists are **snake bites**. Although Costa Rica has a large number of poisonous snakes, most tourists aren't in such wild areas that they're in danger. The worst offender is the fer-de-lance, or *terciopelo* in Spanish, a particularly aggressive snake with a very poisonous bite. Stay on the path, wear leather boots in the wild, and go with a guide.

If you should have a severe medical problem, most hotels will arrange a visit to the clinic or will have an English-speaking doctor make a house call. Costa Rica's doctors are well trained, so if you're sick don't wait until you get home to have someone look at you. **Malaria** and **cholera** are extremely rare, but they're not to be fooled with if you display symptoms. **Dengue fever** outbreaks (high fever and aches) have occurred in the past, spread by a daytime mosquito, and should be treated promptly. Use **mosquito repellent** in rural areas to aid in prevention. If you have any health problems after your return from Costa Rica, it may be wise to consult a physician.

You might also check with your **medical insurance** company to see if they cover expenses outside of the country. Most do, but very few will pay for emergency medical evacuations, sometimes called air ambulances. A list of companies that provide travel medical insurance can be found on the web at travel.state.gov/medical.html.

TRAVELERS WITH DISABILITIES

Unfortunately, few Costa Rican buildings, walks, curbs, buses or bathrooms are wheelchair-accessible, so travelers with disabilities have a hard time getting around. Even the terrain works against handicapped visitors – it's all up and down. **FAUNA** *(Foundación Acceso Universal a la Naturaleza, ☎ 506/771-7482, www.chabote@racsa.co.cr)* promotes tourism for people with disabilities and can help plan an itinerary. A Tico agency that specializes in day tours for those with handicaps is **Vaya con Silla de Ruedas**, in San Pedro *(☎ 506/225-8561, vayacon@racsa.co.cr)*.

Prostitution

Men will pay large sums to whores,
for telling them they are not bores.
~ WH Auden, English poet, 1907-1973.

Prostitution is legal in Costa Rica but did not get much publicity – good or bad – until recent worries surfaced about the underground growth of "sex tourism." Local and national authorities do not want the bad reputation or the social problems that go with that kind of tourism and organized trips for sex are strongly discouraged. Prosecutors crack down very hard on underage exploitation and in 2001 the first American was arrested for just that.

Most prostitutes work out of select clubs, bars or escort services and remain relatively low key unless you're looking for them. Even then, women wait to be approached and are not generally forward or aggressive. They are supposed to have a health card certifying recent medical check ups. Women from other countries have come for the money and some of the working Ticas have taken up the trade because of a lack of decent jobs available in Costa Rica's sluggish economy. Homosexual prostitutes also work out of certain bars, and many stand on street corners late at night.

Regular cautions go to anyone who gets involved. Some robberies have been associated with the trade – especially with men who drink – and some streetwalkers have tested positive for HIV. If you indulge, use condoms and care.

Telephones & the Internet

Well, if I called the wrong number, why did you pick up?
~ caption in a *New Yorker* cartoon, James Thurber, 1894-1961

The telephone service in Costa Rica is very good. Even so, cell phones are ubiquitous. You'll see them attached to the belts of men and women everywhere, probably because the phone company (ICE) takes such a long time to install land lines.

■ Phone Calls

Early in your stay it is a good idea to get a **phone card** that will enable you to make calls from public telephones. You'll find it especially handy in case of an emergency. Cards – called *tarjetas telefonicas* – are sold in units of 500, 1,000, or 3,000 colones at lots of little stores known as *pulperías*. There's a computer chip inside them that keeps track of your monetary credit. In very rural areas, where there are no public phones, you can often phone within Costa Rica from a store or hotel and they'll charge you (not a lot) by the connection time.

To reach the US or Canada directly, dial 001, then the area code and number. To charge with a calling card, dial: **AT&T**, ☎ 0800/011-4114, **MCI**, ☎ 0800/012/2222, or **Sprint**, ☎ 0800/013-0123. For international collect calls, dial 175; local directory assistance, 113; or international directory assistance, 193.

If you plan to call home a lot, consider buying a "Servicio 199" card from the local telephone company office as it has lower international rates than your American phone service. With these you can dial home directly.

To call Costa Rica from the US, dial 011, plus the 10-digit number (all Costa Rica numbers begin with 506). This is an international call.

■ Getting On-Line

Internet connections are available (but often slow) at a fair number of San José area Internet cafés and in a few hotels. Outside of San José you'll find cybercafé connections in tourist areas, college towns or larger cities. However, in every post office, large or small, there is always at least one computer hooked up to the Internet. Pay in advance for a card with a set amount of time from the postal counter and sign on using the code on the card. It's a way many students in poorer rural areas are able to be a part of the cyberspace generation.

Just in Case

All hell broke loose.
~ *Paradise Lost*, Milton, 1667

■ Credit Card Issues

VISA, the country's most popular card, can be reached at ☎ *001-800-847-2911* in the States, or try their local numbers: ☎ *506/224-2631 or 506/224-2731.* **MasterCard** and **American Express** local office is in Credomatic at ☎ *506/257-4744, 506/257-0155.* To get an English speaker, ask for their 24-hour *servicio extranjero*.

■ Emergencies

Dial 911 and you should get an English-speaking operator. If you speak Spanish, it may be faster to call police, an ambulance or the fire department directly. Dial 128 for an **ambulance**.

Crimes should be reported to the **Judicial Investigative Police** in San José, ☎ *506/295-3640*, or, if you're in the country, ask for the nearest **Guardia Rural**. To complain about corruption or government abuse, phone the Ombudsman's Office at ☎ *800-258-7474*.

In the event of a **traffic accident**, call your rental company first; most have 24-hour help numbers. Then in San José call the **Policia de**

Transito at ☎ *506/222-9330*. Do not move your car until an officer arrives, even if you are blocking traffic. You probably also have to phone the **National Insurance Institute**, ☎ *800/800-8000*, if your rental company doesn't do it for you.

To reach the **Fire Department**, dial 118.

In the event of a medical emergency, call the **Red Cross Ambulance** service by dialing 128. Private flight and ambulance service: ☎ *506/286-1818*. This is a pay-per-use emergency service.

HOSPITALS	
■ San José	
Mexico Hospital	☎ 506/232-6122
San Juan de Dios Hospital	☎ 506/257-6282
Calderon Guardia Hospital	☎ 506/257-7922
■ Metropolitan Area	
Max Peralta Hospital, Cartago	☎ 506/550-1999
San Rafael Hospital, Alajuela	☎ 506/440-1333
San Vicente de Paul Hospital, Heredia	☎ 506/261-0091
■ Private San José Hospitals & 24-Hour Pharmacies	
Clínica Católica	☎ 506/283-6616
Hospital CIMA San José	☎ 506/208-1000
(pharmacy)	☎ 506/208-1080
Clínica Bíblica	☎ 506/257-5252
	(ext. 2 for Express Service in San José)

■ Costa Rica Tourist Board

The most popular satellite office of **Costa Rica's Tourist Board**, or ICT, is under the Plaza de Cultura, Calle 5 between Av Central and Av 2, next to the Gold Museum. Call them at ☎ *506/223-1733* or try dialing 192. The out-of-the-way main office is located at the eastern end of the Juan Pablo Segundo Bridge in the former CINDE building *(☎ 506/299-5800)* in La Uruca.

■ Embassies

EMBASSY LOCATIONS		
United States	Located in Pavas, west of downtown, San José	☎ 506/519-2000, 506/220-3127 (emergency) www.usembassy.or.cr
Canada	Oficentro Ejecutivo La Sabana, Edificio #5, Sabana Sur	☎ 506/242-4400 www.dfait-maeci.gc.ca/sanjose
Great Britain	Paseo Colon between 38 and 40, San José	☎ 506/258-2025 www.embajadabritanica.com
France	Near the Indoor Club, Carretera a Curridabat,	☎ 506/234-4167 www.ambafrance-cr.org
Germany	Near Casa de Dr Oscar, Arias, Rohrmoser	☎ 506/232-5533 www.embajada-alemana-costarica.org
Netherlands	Oficentro Ejecutivo La Sabana, Edificio #3, Sabana Sur	☎ 506/296-1490 www.nethemb.or.cr
Israel	Edificio Centro Colon, 11th Floor, Paseo Colon, Calles 38 & 40, San José	☎ 506/221-6444 http://sanjose.mfa.gov.il
Italy	Calle 33, Avs 8 & 11, Los Yoses	☎ 506/234-2326 www.ambitcr.com
Spain	Calle 32, Paseo Colon & Av 2, San José	☎ 506/222-1933 embajadade.costarica@anit.es
Switzerland	Edificio Centro Colon, 11th Floor, Paseo Colon, Calles 38 & 40, San José	☎ 506/232-0052 embajada.costa.rica@thenet.ch

Handy Hints

I have found the best way to give advice to your children is to find out what they want to do and then advise them to do it.
~ Harry S. Truman, 33rd US President

This mishmash of advice should make your vacation easier and more fun. Also included are some idiosyncrasies of Ticos and gringos in Ticolandia.

- **Museums** are closed on Mondays, so plan your visits accordingly.
- Ticos don't use **street numbers**. Instead, they use directions such as "100 meters south of the Coca-Cola." It doesn't matter that the Coke plant closed decades ago and is now a bus station. Directions are getting a bit better, but there is still the occasional "20 meters west of where they burned the dead dog," whenever that was.

- Instead of replying to *gracias* (thank you) with the more common expression in Latin America, *de nada* (for nothing), Ticos acknowledge that favors aren't always worth so little; consequently they respond *con gusto* – with pleasure.
- Friends informally **greet** each other with *maje* (MA-hey), which pretty much means an affectionate, "hey, stupid."
- Without doorbells on most homes, a strange greeting to attract the house's occupants has developed – one that is purely Costa Rican. "**Upe, upe**!" is called out near the front door by vendors or anyone else who doesn't know the name of the occupant. It comes from the 1800s, when the first greeting was a religious *Ave de Guadalupe*, after the Virgin of Guadalupe in Mexico. In typical Tico slang it was quickly shortened to "Upe," (OO-pay).
- The English-language weekly newspaper, ***Tico Times***, comes out every Friday and is distributed to main tourism areas around the country. It's updated weekly online at www.ticotimes.net. A great daily website for concise news is **www.amcostarica.com**. You can sign up for with them for daily e-mail bulletins.

- **Spanish** and **English** are spoken by most of the people who live on the Caribbean side of Costa Rica.
- Ticos often drink **beer with ice**. The English really hate that.

- Although MasterCard is accepted most places, and American Express accepted in large hotels, **VISA** credit cards are preferred in Costa Rica. **ATMs** are a great way to get cash without waiting in line.
- Tiny restaurants or sandwich shops are called ***sodas***.
- We don't smoke, but plenty of people do. **Cuban cigar** smokers will love the plentiful selection here – and unlike in the US, they are legally imported. Buy from a reputable store with a humidor, but it's not legal to take them into the US.

- Male Costa Ricans are ***Ticos***, while females are ***Ticas***.
- A title of respect (and in many cases, affection) used frequently in Costa Rica is ***don*** for men and ***doña*** for women, instead of *señor* or *señorita*.
- **Latin music** radio stations pound out salsa and merengue, which is often heard playing on buses and in taxis. But you'll find an all-English station at 107.5 FM. Rock 'n roll, of course. Super Radio at 102.3 plays oldies with a daily Beatles hour at 4 pm. Top 40 hits from the 60s on up can be heard on 99.5, and 1960s and 70s rock is found at Punto Cinco, 103.5. Intellectual talk is hard to receive at 101.3. Classical stations include 96.7 and 97.1. Radio Nacional offers classical and eclectic soft music at 101.5 FM.

- Off-season is the rainy "green" season during the North American summer. **Best deals** for hotels or tours are during this time. We like it because the warm rains keep everything green and because we're cheapskates at heart.
- **Musmanni** is a country-wide chain of bakery/sandwich shop/ice cream parlors that open early, close late, and are a great source of inexpensive and filling goodies.
- **Pharmacies** are a good source for medical advice or doctor referrals, and pharmacists are very helpful in suggesting medications. Prescription drugs are generally cheaper than in the States. Look for a green cross in the window or on the sign.
- Many North Americans come down to Costa Rica for **cosmetic surgery** and **dental work**, which costs about half as much as it does in the US.
- Never leave **valuables** in your parked car, even in your trunk, if you can help it. Don't leave your purse on the seat with the window down while driving, as pickpockets have been known

to run by and grab. Thieves have been known to poke long sticks through windows of ground-floor accommodations to hook valuables left in sight. It's a drag. Be aware.

- On a hike in the rainforest or just in nature, it's important to avoid snake bite. Ask your guide if he or she carries anti-venom. While in the woods, always walk with your head down and eyes forward – especially at night. Stay on the path, wear high leather or rubber boots, and go with a guide. No sandals or canvas sneakers in the woods.
- To help keep unwelcome critters such as small scorpions at bay, do not leave your clothes or knapsack on the floor, Also, be sure to shake out your shoes before putting them on.
- Bring lots of one dollar bills to use for tips. Be generous to guides and service workers, as they depend on tips for a living.
- Be cautious about strangers too eager to help you find a taxi, a hotel, fix a flat tire, show you the way, or carry your bags, etc. If you're in a rental car and have just visited an ATM, be suspicious if you get a flat tire. Thieves have been known to puncture tires and, when they stop to help, rob you. Keep driving until you get somewhere public, and ignore anyone who stops to help (except the police).
- The expression, ***pura vida***, which translates as "pure life," is Costa Rica's unofficial motto. It is often used as a response to "what's up?" and other greetings, as a greeting of its own, and as a farewell. You might yell it with exuberance after conquering whitewater rapids or whisper it with wonder as you watch a lumbering sea turtle come ashore to lay its eggs. If you stay long enough, you may also hear it used sardonically as an adjective to describe Tico tendencies to procrastinate. Or in response to a governmental proclivity to say one thing and do another, or to express frustration at slow progress or ignorant thinking. Throw up your hands and say *pura vida*, and it means, "That's life in Costa Rica."

Prisma Cosmetic Dentistry

Your smile...

The best souvenir from

Costa Rica

Cosmetic and General Dentistry

Implants | Mercury free dentistry
Orthodontics | Nitrous oxide sedation
Laser bleaching | Porcelain inlays/onlays
100% porcelain bridges | Resin filling
100% porcelain crowns | Full gum treatment
Full on-site Laboratory | Veneers

Drs. Josef Cordero
Telma Rubinstein
Cosmetic Dentistry

Phone: (506) 291-5151, 291-5252, Fax: (506) 291-5454
E-mail: dental@racsa.co.cr
www.cosmetics-dentistry.com
300 mts East Plaza Mayor mall
Rohrmoser, San Jose, Costa Rica

San José

If you reject the food, ignore the customs,
fear the religion and avoid the people,
you might better stay home.
~ James A. Michener, author, 1907-1997

San José has been the capital of Costa Rica and the seat of its democratic government since 1823. Nestled high in the Central Valley (1,253 meters/3,770 feet), between green volcanic mountain ranges, the metropolitan city of about one million people is oriented east and west, surrounded by suburbs that cling to the gentle foothills. The nearby town of Escazú, southwest of the city off the *autopista*, is home to the largest population of North Americans living in Costa Rica, either full-time or part-time, and is consequently more upscale than San José's downtown or other suburbs.

IN THIS CHAPTER

Unlike other Central American capitals, San José is not a very colonial city, although the heart of the historic center features several beautiful old buildings. It didn't become a decent-sized city until relatively late in the 1800s, long past the colonial era, and had a modern building boom in the 1950s and 60s. Because of its economic success from coffee exports, San José became the second city in the Americas to install electricity (1884). At its heart are the National Theater, Plaza de la Cultura, Parque Central, the Cathedral and Gold Museum, which are centered in a four-block area with a long pedestrian-only shopping walkway bisecting the middle of downtown.

Despite the lack of stone and plaster colonial ambiance, many of San José's buildings boast a unique and charming architectural style known as "Caribbean Victorian." Just beyond the modern office high-rises are wood-framed clapboard houses built at the turn of the century. Graceful porticos and wrap-around verandas distinguish these appealing homes (some converted to offices or small hotels) painted in soft pastels. Many have the original metal roofs that sound rat-tat-tat in the rain.

San José is a very cosmopolitan capital with many welcoming qualities. Near-perfect weather is a big attraction to visitors as well as expatriates who call the Central Valley home. Some 250,000 foreigners, mostly North Americans, live in Costa Rica full time. The temperature is fairly consistent between 70 and 75°F. Occasional Northerlies, cold fronts, call for a jacket or sweater at night. The rainy season, when late afternoons and evenings bring showers, lasts from May to October, *mas o menos*.

As the cultural heart of the nation, San José is home to several theater groups, excellent museums, parks, a national symphony, cinemas and universities. It is also host to nightclubs, casinos, discos and fine restaurants. Residents of San José are known as *Josefinos*.

On the down side, the city is often crowded, dirty, noisy, ugly and disagreeable. It may take some getting used to – some people hate it, but the longer you stay, the more San José grows on you. We love it.

Orientation

I'm the type of guy that likes to roam around,
I'm never in one place, I roam from town to town.
And when I find myself falling for some girl,
I hop right into that car of mine, drive around the world.
Cause I'm a wanderer, yeah a wanderer,
I roam around, around, around...
~ *The Wanderer*, Dion & the Belmonts, 1961

Do you know the way to San José? Even if you do, it may be easier than finding your way around it. That's because Costa Ricans don't use street addresses. Instead, they use directions such as "100 meters south of the Coca-Cola." (The Coke plant closed and is now a bus station – you're supposed to know it was once there.) Good thing it's a small country – and a small city. We often bump into people we know or have met in our travels, and once we encountered acquaintances from the States on a San José sidewalk! So if you're walking around town, keep your eyes open for friends as well as those drivers who whip around corners without looking for pedestrians. *Muy Peligroso*!

Because the neighboring town of San Pedro, home of the **University of Costa Rica** and **Ulatina** (another university), is so much a part of San José's personality, we have included its attractions in the San José listings. It's a lively part of town, just east of the city limits along the main roadway connecting the upscale Los Yoses neighborhood, through student-friendly San Pedro, to curious Curridabat, and on to Cartago.

San José city itself is divided into various neighborhoods, *barrios*, such as Los Yoses, Amon, Otoya and Merced. Roads are in the typical Latin American grid pattern: streets, *calles*, run north and south, while avenues, *avenidas*, are oriented east and west. Bisected east and west by Av Central, *avenidas* to the north of Av Central bear uneven numbers, while those to the south are even. In the same way, Calle Central is the central north-south axis, with streets to the east using odd numbers and those to

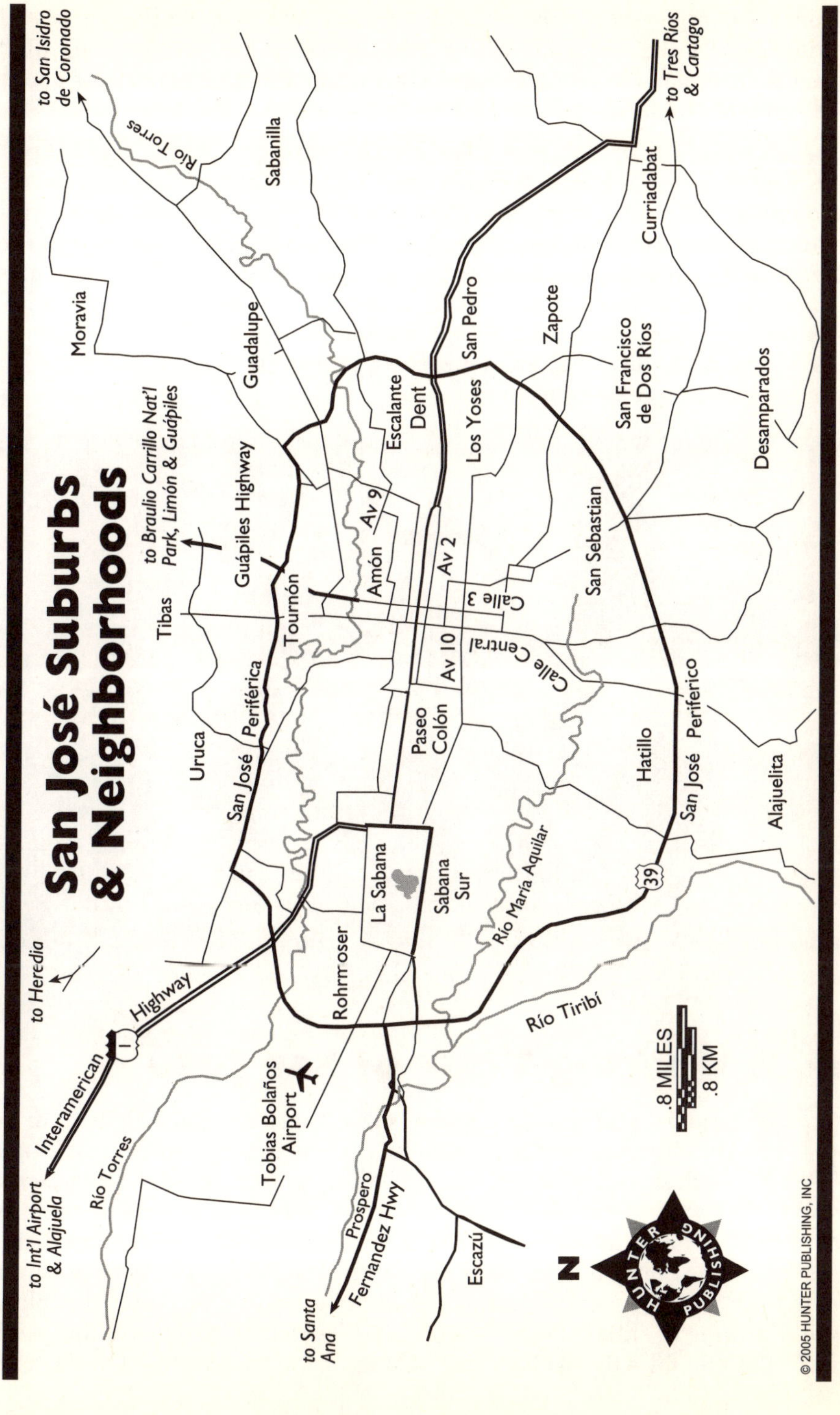
San José Suburbs & Neighborhoods
to San Isidro de Coronado
Río Torres
Sabanilla
Moravia
Guadalupe
to Tres Ríos & Cartago
Curridabat
San Pedro
Zapote
San Francisco de Dos Ríos
Desamparados
Escalante
Dent
Los Yoses
to Braulio Carrillo Nat'l Park, Limón & Guápiles
Guápiles Highway
Av 9
Amón
Av 2
Calle 3
San Sebastian
Tibas
Tournón
Av 10
Calle Central
Uruca
San José Periférica
Paseo Colón
Hatillo
San José Periferico
Alajuelita
39
La Sabana
Sabana Sur
Río María Aquilar
to Heredia
Rohrmoser
Highway
Interamerican
1
Río Tiribí
.8 MILES
.8 KM
Tobias Bolaños Airport
to Int'l Airport & Alajuela
Río Torres
Prospero Fernandez Hwy
Escazú
to Santa Ana
N
HUNTER PUBLISHING
© 2005 HUNTER PUBLISHING, INC

San José

the west using even numbers. Look for street signs up on the corners of buildings. Blocks are generally considered to be 100 meters long (329 feet) so directions to go three blocks west and two blocks north is about 500 meters (1,645 feet).

For an overview of the city, take a half-hour tour on the **Tico Tren** *(☎ 506/226-1349)*, a train engine that pulls a sightseeing car. It was brought here by ship from Key West in 1968. The family-run tourist attraction is a familiar sight in San José. You can pick it up in front of the National Theatre (Gran Hotel) or Parque Central, or just flag it down as it passes. The fare is US $3. The Tico Tren runs weekends in the rainy season and nearly daily in the high season, except when Carlos Solano, its driver/owner, is engaged as a private tour guide (see page 130). Flag him down and say hello from us.

The Tico Tren is a great way to explore the city.

■ A Walking Tour of San José

Start at **Parque Central** •7• (Av 2 between Calle Central & 2), the city's oldest park. We must confess that downtown parks are modest and generally unappealing as recreational attractions, and this park is no different. But it is a good gathering ground, especially on Sundays, and the pavilion in its center, donated by Nicaraguan dictator Anastasio Somoza, once housed a children's library. The park features popular Sunday concerts and is the hangout of the inventor of a playground game of skill called "Pica Caballo," who'll demonstrate the game and offer one for sale. Watch your belongings in the crowds. On one corner street is the **Soda Palace**, allegedly the restaurant where revolutionary plots were once hatched.

Facing the park is the most important church in Costa Rica, the **Metropolitan Cathedral** •8•. Built in 1871 after the original was destroyed in an earthquake, the interior is expansive, with elegantly painted columns made of wood. The high altar is under an ornate cupola ceiling. A side chapel, **Capilla de Santissimo Sacramento**, is decorated with carved painted flowers and leaves.

Across the street on Av 2 is the **Melico Salazar Theater** •6•, built in the 1920s and named after a famous Italian opera singer who liked Costa Rica so much he moved to San José in 1937. We attended a concert of the National Symphony Orchestra here (they now perform at Teatro Nacional, below). In addition to concerts and special events, it now hosts a folklore ballet. Stop at the box office for tickets *(☎ 506/221-4952).*

East on Av 2 is the **Gran Hotel** •9•, set back from the street in a paved plaza, Costa Rica's grande dame of accommodations. In the tradition of Central American town planning, the center of a city houses a "grand hotel," usually the oldest and most prestigious. El Gran has a small, popular casino, and its inside restaurant is quite good. But the outdoor **Café Parisien** – open 24 hours – is San José's best place to sit and eat or have a drink while the world passes by your table. That world might include European backpackers, vendors of painted feathers, Cuban cigar sellers, *ochorena* players, well-dressed theater patrons, provocative prostitutes, young language-school students, lovers young and old, camera-toting tourists, government officials, shoppers, large families, beggars – and you.

If there is one "must-see" site in San José it's the **Teatro Nacional** (National Theater, ☎ *506/257-0863, www.intnet.co.cr/tn/tn.html*) •10•, next door to the Gran Hotel. Completed in 1894 at the height of Costa Rica's coffee and banana wealth, the theater is an ornate, spectacularly beautiful testament to a bygone era – a golden age of opulence. Its neoclassical exterior is impressive, but the baroque décor is breathtaking. The entrance lobby features Italian pink marble and 22- karat gold trim.

In the horseshoe-shaped grand hall, which seats over 1,000, the floor was constructed so it could be raised to the level of the stage, creating a ballroom. Velvet-lined luxury box seats rise two levels around the auditorium; the third level is the gallery. If you have the opportunity to see a concert here, don't miss it. You don't necessarily need a suit or formal dress, but good casual clothes will suffice.

Up a Carrara marble staircase is Costa Rica's most famous painting (look up), *Una Alegoría*, by Milanese artist Aleardo Villa, commissioned in 1897. Reproduced on the colorful five-colones bank note, Villa depicts an idealized coffee harvest with sacks of the *grano de oro* being loaded onto a sailing ship. Ticos have long ago forgiven Villa for painting the coffee bushes too short and the women look like Italian grape pickers. Admission to tour the theater is about US $5 and worth every colon.

Downstairs, when you first enter the theater, is the charming **Viennese-style café** run by Café Britt. This is a wonderful place to have lunch, afternoon tea, or just coffee and dessert. Ferrario Carlo Milano painted the ceiling above your head and the café hangs changing art exhibits by local artists.

The second-floor foyer overwhelms the senses with ornate gilding, crystal, statues, paintings, columns, thick carpet, drapes, mirrors, lights and fine furniture. The floor, replaced in 1940, boasts a selection of 10 varieties of local hardwoods.

To the side of the theater is the **Plaza de la Cultura •11•**. Because of its location along Av Central's pedestrian walkway, it has become the central meeting place in the downtown, often attracting street performers.

Below the plaza, under a curving arched roof, are the **Gold Museum** *(☎ 506/243-4202, www.museodelbancocentral.org)* and the **Tourism Office** (ICT). Tourist info is available Monday through Saturday, 9 to 5. The pre-Columbian gold museum is open Tuesdays through Sundays from 10 am to 4:30 pm. Admission for the impressively rich, 2,000-piece exhibit is around US $6. It's one of Central America's largest collections. Sorry, no photos. ☎ *506/223-0528.*

Continue east another block on the pedestrian walk and turn left (north) on Calle 7 uphill toward **Parque Morazán •14•**. If you're into snakes, a quick detour on Av 1 leads to the **Serpentario •15•** indoor zoo, with our favorite creature, the Jesus Christ lizard. Head upstairs and follow your nose.

Parque Morazán features a central gazebo where Sunday concerts are often held. In addition to the tall **Aurola Holiday Inn** (with a fancy casino on the top floor, and a great view at night) on the far side of the park, this area has a couple of bars frequented by prostitutes. The most famous, **Key Largo**, is worth a look for its Caribbean Victorian mansion architecture, but you'll have to pay US $5 for a beer or a US $4 cover charge for men.

Cross Av 3 and 5 to the **Edificio Metálico** (Metal Building) **•16•**, a yellow elementary school designed by French architect Victor Baltard, who also did Les Halles in Paris. Cast of iron in Belgium in 1892, it was shipped overseas and assembled on the site. To the side is **Parque España •17•**, home to towering tropical shade trees, thick clumps of bamboo and an open-air market on Sunday. The statue of a Spanish conquistador here raised more than a few eyebrows when it was erected in 1992, indicative of the mixed feelings Latin Americans have about their history.

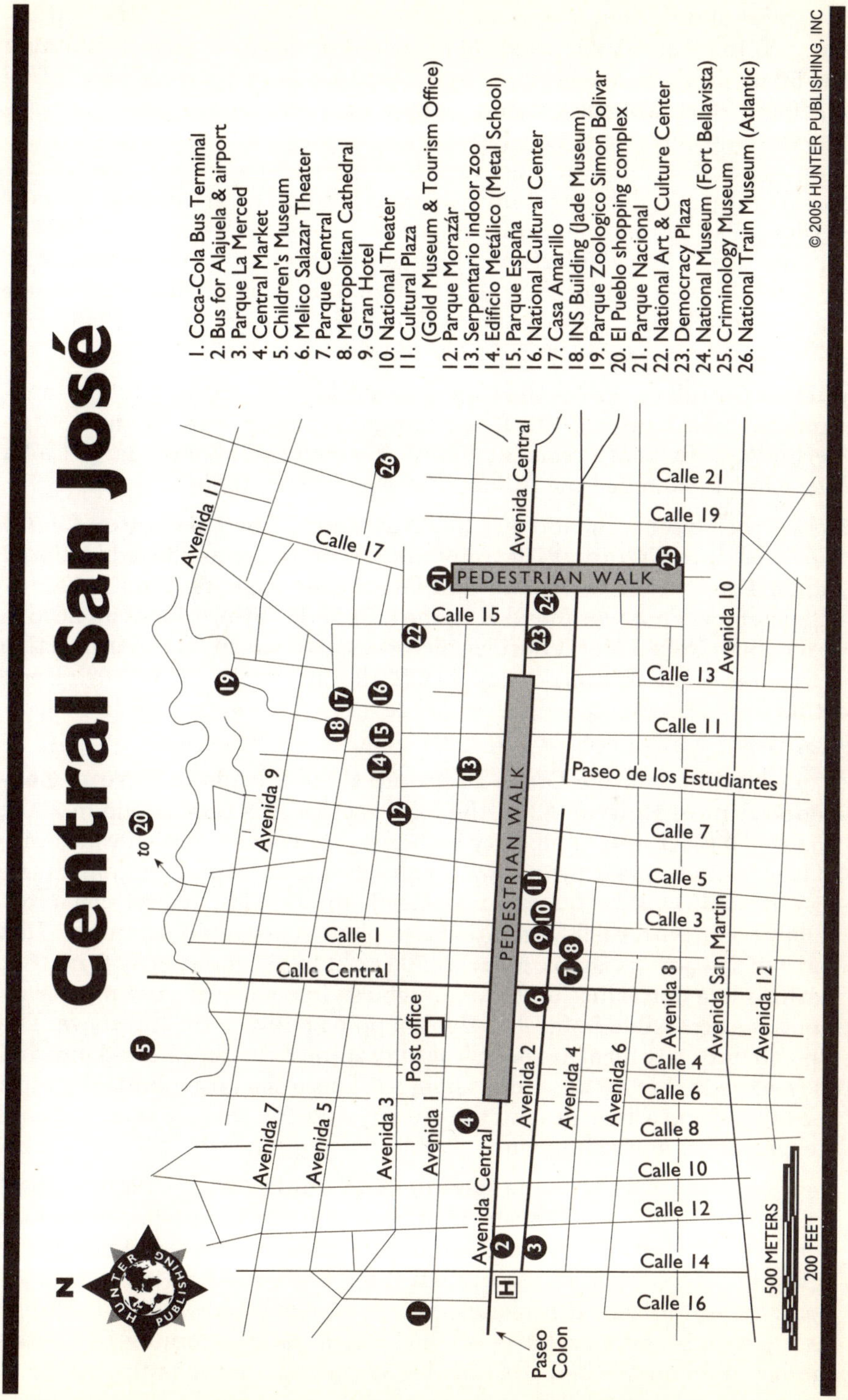
Central San José
1. Coca-Cola Bus Terminal
2. Bus for Alajuela & airport
3. Parque La Merced
4. Central Market
5. Children's Museum
6. Melico Salazar Theater
7. Parque Central
8. Metropolitan Cathedral
9. Gran Hotel
10. National Theater
11. Cultural Plaza (Gold Museum & Tourism Office)
12. Parque Morazár
13. Serpentario indoor zoo
14. Edificio Metálico (Metal School)
15. Parque España
16. National Cultural Center
17. Casa Amarillo
18. INS Building (Jade Museum)
19. Parque Zoologico Simon Bolivar
20. El Pueblo shopping complex
21. Parque Nacional
22. National Art & Culture Center
23. Democracy Plaza
24. National Museum (Fort Bellavista)
25. Criminology Museum
26. National Train Museum (Atlantic)
© 2005 HUNTER PUBLISHING, INC
N
HUNTER PUBLISHING
Avenida 11
Avenida 9
Avenida 7
Avenida 5
Avenida 3
Avenida 1
Avenida Central
Avenida 2
Avenida 4
Avenida 6
Avenida 8
Avenida San Martin
Avenida 10
Avenida 12
Calle 21
Calle 19
Calle 17
Calle 15
Calle 13
Calle 11
Paseo de los Estudiantes
Calle 7
Calle 5
Calle 3
Calle 1
Calle Central
Calle 4
Calle 6
Calle 8
Calle 10
Calle 12
Calle 14
Calle 16
PEDESTRIAN WALK
PEDESTRIAN WALK
Post office
to 20
Paseo Colon
H
500 METERS
200 FEET

The tall building that overlooks the park and school is the National Insurance Institute, **INS** (*Instituto Nacional de Seguros*), which contains the fabulous **Jade Museum** •20•, whose 11th-floor quarters were refurbished in 2001. The view from its upper floor windows is worth a photo, but you won't be able to take pics inside. Over 6,000 works of pre-Columbian art and jewelry, in jade and other precious stones, make up the world's largest collection. Many pieces came from the private collection of Carlos Balser, a renowned archeologist who came to Costa Rica in 1921 to run the Gran Hotel. Unfortunately, some of the erotic, carved stone phallic symbols were removed to allow more office space. But what is left should still bring a smile. Open Monday through Friday, 8:30 am-3:30 pm. Admission, $2. ☎ *506/287-6034.*

Just up the hill on Av 7 is the **Casa Amarillo** •19•, a grand yellow mansion that now houses Costa Rica's foreign ministry – and a piece of the Berlin Wall. In front President John F. Kennedy planted the large Ceiba tree when he founded the Alliance for Freedom in 1963.

Across the street is the fortress-like **National Cultural Center**, *Centro Nacional de la Cultura*, •18•, converted from its use as the old National Liquor Factory. Coffee liqueur and Guaro (see page 89) used to be distilled here when it was founded in the mid-1800s by president and coffee baron Juan Rafael Mora. Besides historical artifacts, it features an active art center, the **Contemporary Art Museum** *(☎ 506/257-9370).* Don't let its forbidding high walls discourage a visit. It's open Tuesday-Saturday, 10 am-5 pm. Free admission on Sunday; US $3 at other times.

A quick jaunt into the Otoya Barrio up Calle 9 leads to **Parque Zoologico Simon Bolivar** •21•, the local zoo. Until a remodeling in 2000, this was a disgraceful and dirty little zoo, but it's been transformed into at least a more pleasant zoo with enhanced landscaping and much-improved conditions for the animals. Good stop for kids. It's set down in a hollow basin with thick vegetation and plenty of songbirds around. Just north of the park is **Spyrogyra** *(☎ 506/222-2937)*, a butterfly farm that features live butterflies and hummingbirds in a natural garden environment. It's accessible from El Pueblo shopping plaza on the hill above. The zoo and butterfly farm are open Tuesday through Friday, 8 am-4 pm, and on weekends, 9 am-5 pm. Admission, $2. There is a larger butterfly farm in Alajuela and others around the countryside, but this is convenient and inexpensive if you're in town.

Head back south on 15 to the **Parque Nacional** •23•, San José's largest urban park, which was remodeled in 2000-2001. It had a worn-down look from over-use until replanting and improved paths brightened it up. It's a popular place for students and strolling lovers under its tall tropical trees. Important statues include the 1856 **National Warrior Monument**, cast in Rodin's Paris studio, which commemorates the battle against William Walker. Also in the southwest corner is the statue of

Juan Santamaría, the boy-hero who helped rout Walker's army. Farther east from the park is the **National Train Museum** •28• set in the old Atlantico train station of the line that once went to Limón.

Due south from the park, across Av Central, is the **Plaza de la Democracia** and the **Museo Nacional** •26• *(☎ 506/257-1433, www.museocostarica.com)*, housed in the historic old **Fuerte Bellevista**. Only in peace-loving Costa Rica would they name a military installation "Fort With a Beautiful View." The Plaza has a good flea market daily, but it is especially big on weekends. The old yellow Castilian fort was constructed in 1887 and used as the nation's military headquarters until the abolition of the armed forces in 1948. Notice the side walls and look up at the balustrades to see all the gun shot holes around the gun slits, evidence of the serious nature of the 1948 Civil War. The museum has four sections in *salas*, or large rooms. The first features archeological artifacts from Costa Rica's pre-Columbian peoples and their history (most explanations are in Spanish). Precious gold figures and jewelry are housed in a separate *Sala de Oro*. Another section of colonial life features artifacts and displays of the first Spanish conquistadors and early *mestizos*. Historic photos and implements from the recent past are featured in the last area. A stroll around the fort's interior, towers and old jail shows the conditions faced by turn-of-the-century soldiers. Located in the central courtyard you'll find a good example of the traditional brightly painted oxcarts (*carretas*), as well as several varied-size, mysterious stone spheres – made by a long-forgotten people in the southern zone. Entrance is on Calle 17 between Av Central and Av 2. Open Tuesday through Sunday, 8:30 am-4:30 pm; $5 for adults, children under age 10, free. Hungry? Nearby is the restaurant, **Ay Sofya**, on Av Central & Calle 21.

Go 2½ blocks south on Calle 17, which brings you to the Court Administration buildings, specifically the Organismo de Investigación Judicial, between Av 6 & 8. What better place for a **Criminology Museum** *(☎ 506/295-3850)* – a fascinating graphic history of crime in Costa Rica – and perfect for people like us who look for more uncommon things to do and see. We hang out there (hang there – get it?). Free.

Other Downtown Sights

The Children's Museum •5• (☎ *506/258-4929; www.cccc.co.cr.)*, Museo de Niños, is located in a huge converted old fort that served as the city's prison for many years. This large yellow fortress with crenellated walls and towers is at the north end of Calle 4, on a hill above the Río Torres. It has free, hands-on interactive exhibits and activities. Art exhibitions and concerts for adults in the new National Auditorium are also held here. Open Monday-Friday, 8 am-3 pm, weekends, 10 am-4 pm. Adults, $2, children under 18, $1.50.

The **post office**, *Correo*, is another baroque building (on Calle 2, between Av 1 & 3). It is home to an interesting **stamp museum** with free admission. It's on the second floor and a bit of a challenge to find.

Parque la Merced •3•, nicknamed Nico Park, in front of the Hospital San Juan de Dios, is where Nicaraguan nationals congregate. Overlooking it is the **Iglesia de la Merced**, Mercy Church. Damaged in the 1991 earthquake, the church boasts an Italian marble altar, magnificent stained-glass windows and a vaulted wooden ceiling.

At the west end of downtown, at the end of Boulevard Paseo Colon, lies the former airport that is now a large wooded park called **La Sabana**. On weekends it's full of families, sports players, runners, exercisers, kite fliers, picnickers, ice cream vendors and more. La Sabana serves the recreational needs of the city's west end and surrounding *barrios*. It also contains the **Costa Rican Art Museum** *(☎ 506/222-7155, www.crc.co.cr/musarco)*, housed in the former control tower. Open Tuesday-Sunday, 10 am-4 pm, weekends. Adults, $2, children free. Free admission to all on Sunday. The museum offers a comprehensive overview of Costa Rican art, from its early days to contemporary times, through a host of fine Tico painters and sculptors. Ironically, the most significant work is by a French artist, Luis Ferrón, whose metal relief mural depicts the history of Costa Rica on all four walls of the Salon Dorado. The coffee shop in back, **Café Ruiseñor**, serves typically great joe and Tico food.

In the southwest corner of the park is the **Natural Sciences Museum** *(☎ 506/232-1306, www.lasalle.ed.cr)*, in the old La Salle college building. It contains varied displays of zoology, archeology, geology and mineralogy, plus the only paleontology exhibit in the country. Open 8 am-4 pm, Monday-Saturday, 9 am-5 pm on Sunday. $2.

Last but not least are the two old cemeteries in Barrio San Bosco. **Cemeterio General**, on the south side of Av 10, Calles 22-28, contains the Italianate mausoleums and graves with sculptures of many Costa Rican artists, writers, politicians and coffee barons. Ask to see "La Novia," the tomb of a woman who died on her wedding day. On All Souls and All Saints days, families come and spend time at the grave sites, leaving fresh flowers. Between Calles 18 & 20 is the **Foreigners' Cemetery**, which dates to the 1840s. Railway workers and immigrant entrepreneurs from Europe, North America and Arabia are interred here.

Above: Folklore dancers in San José.

Below: Girl refreshing herself under the waterfalls at Tabacón Hot Springs, Arenal.

Above: Whitewater rafting on the Rio Pacuare.
Opposite: A verdant avenue at a palm oil plantation.
Below: Irazú Volcano crater lake.

Above: The Costa Rican flag.
Opposite: Angels on top of the Teatro Nacional in San José.
Below: White-faced monkey, Manuel Antonio National Park.

bove: This rainforest resident is one of many butterfly species found in Costa Rica.
Opposite: The pulling power of oxen is used to haul carts.
Below: Smiling surfer, Playa Tamarindo.

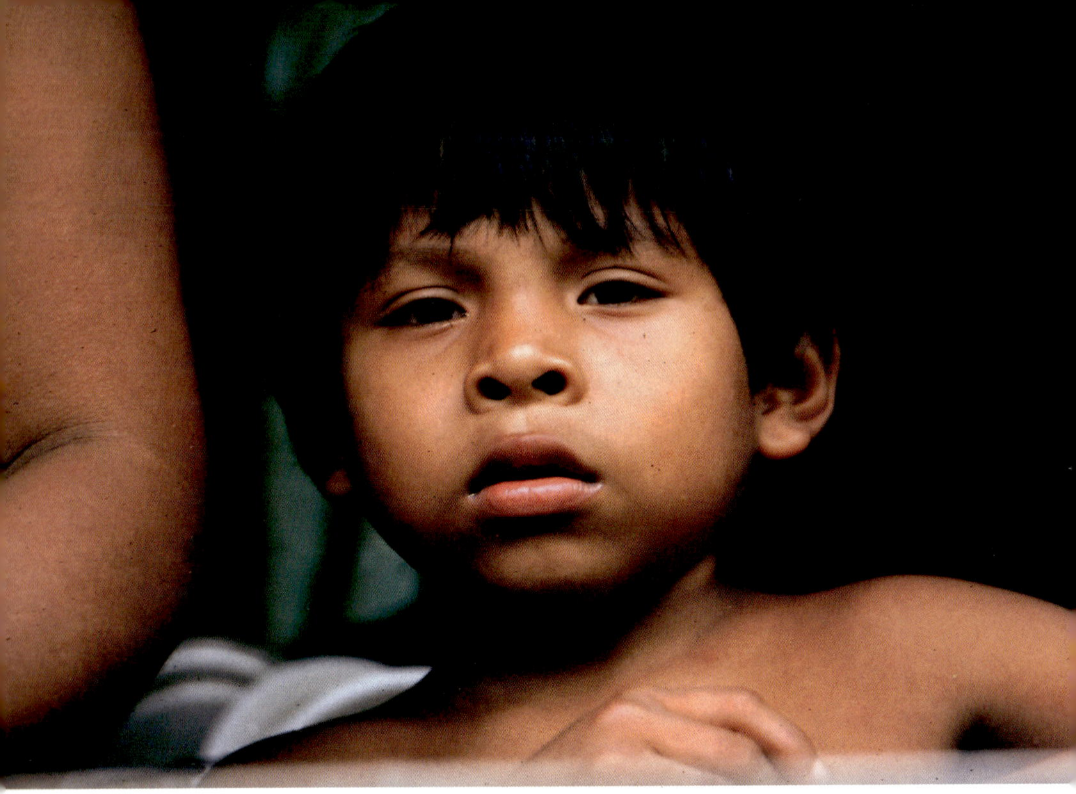

Above: *Bribri boy, Puerto Vieja.*

Below: Brightly painted oxcarts brighten Costa Rican streets and marketplaces.

Adventures on a Shoestring

If you can count your money, you don't have a billion dollars.
~ J. Paul Getty, American billionaire

A number of things can be enjoyed in San José without paying an arm and a leg – some are even free! We've already mentioned the two town cemeteries (cemeteries are our personal fascination) and La Sabana Park (bring your sneakers just in case you join a game). Plus, we thrive on wandering around fruit and vegetable and flower markets, or window shopping in town *mercados*.

Ripe tomatoes fill a vendor's cart.

Saturday is the big day for the **Pavas Farmer's Market**, which take up five blocks from Av 2 north to Pavas Boulevard. Pavas is a suburb west of San José.

Some museums offer free admission on Sundays – you can't beat that for a deal. There are also numerous band concerts in squares and parks; just walk up and listen. Professional concerts, such as the Symphonic, are a fraction of the cost of European or American performances.

One of our favorite budget adventures is just getting on a bus and going somewhere in a different neighborhood, or to nearby towns such as Heredia or Cartago. It's cheap and fun.

■ Hot Springs

Get cool in the hot spring swimming pools of **Orosí** by taking the bus to Cartago, then catching the Orosí bus. There are locker rooms and a restaurant. Orosí is a small town on the floor of a deep valley – we like it so much that we nearly bought a vacation home here. Its dramatic topography appears as you turn a corner – on the floor is the meandering Orosí River, while the steep valley walls are lined with deep green coffee bushes. The view from the overlook just before the valley entrance is spectacular. This trip can be combined with visits to Cartago, Lankester Gardens (on the way), and the Orosí Valley overlook park. We can't think of a better way to spend a day.

The same type of attraction at a different location is **Ojo de Agua** in San Antonio de Belén, not far from San José. It features five swimming pools and a park for sports and picnics, plus a restaurant. Hourly buses from the Coca-Cola terminal, Calle 10, Av 4 & 6. Admission is about US $2.

■ Road to Nowhere

Walk and ride. We took a *Tico Times* reporter's advice for a cheap date and caught the public bus (US 80¢) to **Bebedero**, also marked "Vista de Oro," from Av 6 at Calle 14, just behind the Hospital San Juan de Dios. The local bus runs through wealthy Escazú and up a steep mountain through rural suburbs of adobe and wooden houses, many with the old oval brick ovens scattered among terraced farms and gardens. It's a good way to look down on San José spreading through the Central Valley below you. The bus route ends in the middle of nowhere on the side of the mountain at a quiet church. Rather than just hang out until the bus was ready to return – if you do, be sure to get a front seat for the view – we walked down the dirt road to the side of the chapel and followed it downhill. We were rewarded with some stunning vistas of the countryside.

Living fence on the way to Guanacaste.

AUTHOR NOTE: If you follow in our footsteps, be aware this route takes a lot of them. Wear comfortable walking shoes.

On either side of the road are verdant farms enclosed by "living fence posts" – tree branches or trunks that have been cut to post size and pounded into the ground, where they sprout roots in the rich soil and grow into new trees. Nearly halfway down the mountain is a small store, where you can stop for snacks and cold soda. Go just a bit farther and you reach a great little restaurant (**Tiquicia**, reviewed on page 162) that offers stunning views. You can also do this trip on a Saturday, the only day the eatery is open for lunch. (Alternately, a little soda open daily just past the restaurant offers snacks and drinks.) This excursion could also be done in the early evening, which would allow you to have dinner at Tiquicia, overlooking the valley's twinkling lights. You'll need determination, a good flashlight and sure feet! At night, have the restaurant call a taxi for the ride to San José or Escazú. The walk from Tiquicia back to where we could finally catch a bus back to San José took us one hour and 15 minutes. Huff and puff.

Local Adventures, Tours & Day-Trips

It is good to be out on the road and going one knows not where.
Going through meadow and village, one knows not whither or why.
~ *Tewksbury Road*, John Mansfield, 1878-1967

Whether you stay in San José or the surrounding suburbs or cities, the Central Valley affords plenty of adventures and side-trips – enough to fill a vacation. And that's before you head off to the beaches, volcanoes or rainforests. Much of Costa Rica's adventure and cultural destinations are within a day's drive, so we'll list only easy day-trips here. Overnight destinations are listed in their individual region's chapter.

Some attractions are best enjoyed by booking a tour, although several sites can be combined if you take a public bus or drive a rental car.

■ Tour Companies

Full-Service Tour Companies

Among the major local full-service tour companies offering day and overnight trips operating out of San José are **Valle Dorado** *(☎ 506/228-9933, vdorado@sol.racsa.co.cr)* in Escazú; **Horizontes** *(☎ 506/222-2022,* *www.horizontes.com)*; **Costa Rica Sun Tours** *(☎ 506/233-6890, www.crsuntours.com);* **Ecole Travel** *(☎ 506-223-2240, ecolecr@sol.racsa.co.cr)*, next door to 7th Street Books; and **Green Tropical Tours** *(☎ 506/229-4192, www.greentropical.com)*, who offer a Los Juncos Cloud Forest hike daily; **Centralamerica.com** *(☎ 800/401-7337, CR ☎ 506/221-3912, www.centralamerica.com)*; and **Marbella Travel** *(☎ 866/251-4461, CR ☎ 506/227-0101, www.marbellatours.com).*

One-on-One Tours

And if you're looking for a more personalized experience, call **Carlos Solano** *(☎ 506/226-1349)*. Carlos is a Costa Rican-American who lives in San José and has proven to be a very dependable, bilingual, personal guide; one who has worked for several VIPs. He will drive you in his van or your rental car anywhere in the country and knows all the ins and outs. Carlos is related to Juan Santamaría, national hero, and his grandmother posed for a portrait that was printed on Costa Rican bills in the 1930s and 1940s. A reader recommends **Oscar Chavarria Mora** *(☎ 506/373-2736)*, who speaks excellent English and will drive you in his car.

■ Adventures in the Air

If you ever thought of being a bouncer, here's your chance: **Tropical Bungee** *(☎ 506/290-5629, www.bungee.co.cr)* provides an inexpensive opportunity to bungee jump off the 90-meter/296-foot Río Colorado bridge, near Naranjo, into a deep gorge. The setting is close enough to San José that you'll have time to do several jumps in a day. Cost runs about US $50 for the first jump, and $25 for subsequent jumps. The steel bridge was built in 1948 and now sees mostly tractors and farm equipment going to and from surrounding plantations. You go first; we'll watch. After the jump you get to plunge into the Colorado for a swim.

A slightly less down-to-earth excursion is **hot air ballooning** with **Serendipity Adventures** *(in US & Canada, ☎ 877/507-1358, in CR ☎ 506/558-1000, www.serendipityadventures.com)*. Lift-off is very early in the morning. They pick you up from your hotel and drive you out to the lift-off location. The pilot will show you how to guide the balloon for ascents and descents over the bucolic countryside. It's a feeling of euphoria that is hard to beat. Around US $900 for up to five people, or a special of about US $500 for two. It's a lot of money, but the trip of a lifetime. Serendipity also arrange tours for disabled travelers.

The **Aerial Tram** *(☎ 506/257-5971, www.rainforesttram.com)* is a very popular day tour from San José. Located in a rainforest on the eastern edge of the Braulio Carrillo National Park (see page 326), the converted ski lift tram was developed and built in 1994 by Don Perry, a famous North American biologist who pioneered forest canopy research. It features a slow-moving, 90-minute trip in a cable car at various altitudes above the forest floor. A "naturalist" guide accompanies the group in each car. This tour was one of our most highly recommended and was extremely popular with tour groups, but Dr. Perry is no longer involved in its daily function and has expressed serious reservations to us about the staff's emergency response training. Let's hope that issue is resolved. The tram has opened an agreeable lodge for overnighters.

A common next stop with day tour groups is their affiliated **Banana Tour** of one of two Dole banana plantations in the Caribbean lowlands. Make arrangements for this intriguing educational trip with a local travel agent in your hotel or contact them directly *(☎ 506/768-8683. www.rainforesttram.com).*

■ Adventures on Horseback

In nearby Santa Ana, a country village with roadside veggie stands, you can horseback ride at **La Caraña Riding Academy** *(☎ 506/282-6106)* or **Centro Equestre Valle Yos-Yo** *(☎ 506/282-6934).*

■ Adventures on Foot

We found the best and most convenient cloud forest hiking opportunities in the **Los Angeles Cloud Forest Reserve** in San Ramón. It receives far fewer visitors than Monteverde and is much closer to San José. Los Angeles offers unspoiled quiet and mystical misty paths up a mountain slope. Arrange this as a day-trip or overnight excursion with the **Villablanca Hotel** *(☎ 506/228-4603)*, located at the edge of the forest. The hotel was recently purchased and remodeled by the same people who own Si Como No in Manuel Antonio. They'll pick you up in San José and bring you back for about US $80, including breakfast, lunch, transportation and camaraderie. The knowledgeable hotel staff serve as guides.

> **AUTHOR NOTE:** For hiking in the forest, wear clothes that you don't mind getting dirty and bring some spares. Leather shoes or boots are best.

Consider hiking a historical path on a day-trip to **Guayabo National Monument**, 19 km/12 miles northeast of Turrialba. This is Costa Rica's most auspicious pre-Columbian archeological dig. Pleasant, natural and quiet, the area offers cobbled streets, ruined aqueducts, bridges and rocky building foundations that have been uncovered from an early indigenous settlement. It's a serene and completely un-touristy attraction. See our write-up under *Turrialba*, page 176.

Café Britt Coffee *(☎ 506/260-2748, www.cafebritt.com)* is located at the heart of the coffee *finca* (farm) area near Heredia. It features a multimedia show about coffee growing. Entertaining live actors walk with you around some growing coffee bushes. A coffee-tasting lesson is then held. The large gift shop features lots of coffee souvenirs, including a delicious coffee liqueur, and it has a pleasant little cafeteria serving typical Tico food. Britt Coffee is arguably Costa Rica's best (and most expensive) brand – but worth every penny. The beans are not grown by them, but rather purchased elsewhere (top quality beans only) and roasted here. Book with your hotel or Tico travel agency. This tour is often combined with other area attractions such as a coffee plantation, the Aerial Tram, K&S Microbrewery, Butterfly Farm or Poás Volcano.

My idea of great coffee is when it tastes as good as it smells.
~ Peggy Veeder, Costa Rican traveler

A different coffee tour on a working farm is the **Doka Estate** *(☎ 506/449-5152, www.dokaestate.com)* off the road from Alajuela to Poás. Their excellent roasted coffee, called Tres Generacions, is available on-line or by phone (toll-free) in the US *(☎ 877/789-3652)*. Near the Doka Estate is a cute B&B called **Siempre Verde** *(☎ 506/449-5134)*. This small luxury hotel borders the estate's fragrant coffee fields.

La Paz Waterfall *(☎ 506/482-2720, www.waterfallgardens.com)* is a pleasant natural wonder on the slopes of the Poás Volcano (admission about US $16) in Montaña Azul de Heredia. Not a single waterfall but a series of five powerful falls and cataracts that crash dramatically down a steep, heavily wooded gorge. You can climb many steps and walk several paths along the river. The orchid gardens attract scores of hummingbirds. There's also an indoor/outdoor buffet-style restaurant. Most visitors arrive with a group tour, usually one that combines other attractions such as the Butterfly Farm or a visit to Poás Volcano. La Paz, which means "Peace," is located six km/four miles north of Vara Blanca.

Two fascinating and distinctly different semi-active volcanoes flank San José and both are not-to-be-missed day-trips. **Poas Volcano** (see page 184) rises into the clouds up from Heredia and Alejuela where the air is pure and cool and the scenery is memorable. To the west of San José, up from Cartago, lies **Irazú Volcano** (see page 169), a grand and majestic mountain that offers a spectacular drive up its slopes. Combine these with other attractions or destinations such as La Paz waterfalls for unforgettable day trips. Mucho gusto.

Take a new luxury trip on the **Tico Train** *(☎ 506/233-3311, www.americatraveler.com)*, which departs on weekends from Santa Ana at 7 am for Orotina. Once there, you can take on any number of local tours. Total time, 13 hours. Rail travel is in restored 1940s vintage cars, including a flatbed converted as an observation car. Round-trip cost is between US $60 and $85, depending on which tours you select in Orotina. A less fancy "Popular" trip leaves San José at 6:30 am to Caldera, where passengers can swim. US $25. The ticket office is located in the cavernous Ferrocarril Pacifico Terminal, Av 18 between Calle 2 and 4. A travel agent should also be able to book this for you.

InBio Parque *(☎ 506/244-4790, www.inbioparque.org)* is an ambitious project that functions to categorize the bio-diversity of the country's ecosystems, as well as educate visitors about ecology. It's fascinating, but fairly pricey at about US $15. Ask to see the numerous trays of insects, which include a gold scarab beetle, made famous by Edgar Allan Poe. It's located between Heredia and Santo Domingo; take a taxi from San José.

Catch the Dulce Nombre de Coronado bus (ask the driver for the stop) from Av. 7, Calle Central to the **Clodomiro Picado Institute** *(☎ 506/229-0344)*, where they research snake venom. You can watch them extract venom on Fridays at 2 pm. Open daily, 9-4.

In Costa Rica, **birding** has a low impact on the environment and offers a high degree of satisfaction. A huge number of native and migratory birds (870 recorded species) call Costa Rica home, at least for a while each year. At **La Selva Biological Station** *(☎ 506/766-6565)* near Puerto Viejo de Sarapiquí (not to be confused with the better-known Puerto Viejo near Cahuita), over 400 bird species have been seen. Two other outstanding areas for migratory species are **Caño Negro Wildlife Refuge** in the north, and **Palo Verde National Park** in the northwest. Brilliantly colored toucans and scarlet macaws can be seen along the western shore from near Jacó at **Carara National Park** down to the Dominical area and over to **Corcovado National Park**, on Osa.

> Dedicated birdwatchers should pick up ***Birds of Costa Rica*** by Gary Stiles and Alexander Skutch, Cornell University Press. You can listen at home to the sounds of Costa Rican nature with a stereo CD called the *Costa Rican Bird Song Sampler*, an audio guide to recognizing forest bird songs. Both are produced by David Ross at Cornell's Laboratory of Ornithology (www.birds.cornell.edu/lab_cds.html).

Costa Rica's Birding Club offers organized day and overnight birding trips *(☎ 506/230-2258, http://crbirdingclub.tripod.com)*. If you're spending Christmas in Costa Rica (instead of staying home and eating that other big bird), you can participate in the annual **Christmas Bird**

Count by getting in touch with naturalist Richard Garrigues *(☎ 506/ 293-2710, gonebirding@mailcity.com).*

■ Adventures on Water

Nothing means adventure more than **whitewater rafting** on one of Costa Rica's famous rainforest-fed rivers. It's the one trip we take every time we're in-country. Challenging curls and whirls of whitewater – rolling, roiling rapids – are enjoyed in six- to eight-person self-bailing rafts. These raft trips – offered by a select few safety-conscious tour operators – are good for everyone, from late pre-teens to dexterous senior citizens (depending on the river levels).

RAUCOUS RAPIDS

Our mini-bus strains uphill through the cloud-shrouded Braulio Carrillo National Forest, a sprawling nature preserve that bridges the Pacific and Caribbean halves of the country – across the mountainous Continental Divide. An hour later, on the other side, we pile out of our bus and into a rugged, mud-splattered, four-wheel-drive vehicle. Heading down a precarious road hacked out of the jungle, we drive as near as we can get to the river. Then we hike the last 200 meters/658 feet downhill, past enough colorful heliconia flowers to stock an expansive florist. The rainforest here is thick, pungent and alive with sounds. It speaks its own language – lyrical but somewhat dangerous – much different than the woods near our home.

At the bottom of a ravine, we don yellow helmets and orange life jackets and push off into the Río Pacuare. The raft in front of us is sucked into the thick misty shadows ahead and its occupants' screams echo up the steep verdant walls of jungle on either side. Michi, our guide, looks at us and smiles with anticipation. Suddenly, our bright orange raft rushes after the first one – tossing and twisting in the water as we paddle furiously in the rain-swollen rapids. The raft slides down into the swirling water below and, just as suddenly, we emerge from our first rapids triumphant. Everyone is whooping in excitement.

Several big river-rafting operators own their own lodges and also work with other adventure expeditions. These are: **Costa Rica Expeditions** *(☎ 506/257-0766, www.costaricaexpeditions.com)*, the largest of its kind; **Ríos Tropicales** *(☎ 506/233-6455, www.riostropicales.com)*; **Adventuras Naturales** *(US ☎ 800/514-0411, CR ☎ 506/225-3939, www.toenjoynature.com);* and **Horizontes** (*☎ 506/222-2022, www.*

horizontes.com). **Coast to Coast Adventures** *(☎ 506/280-8054, info@ctocadventures.com)* runs the most physically challenging adventures. They also sponsor a grueling, two-week race across Costa Rica each year. Trips combine mountain and road biking, hiking, climbing, rafting, swimming and much more.

The most popular rivers to raft during a day-trip from San José are the **Reventazón** and the **Pacuare** (pa-QUAR-ree). The former used to include Class IV & V rapids until a dam in Turrialba softened the river's currents. Today it features Class II, III and a few IV rapids. The Pacuare, however, offers much more spectacular scenery as it flows down through primary as well as secondary deep green rainforest with Class III and IV rapids. When you're not shooting rapids you can watch brilliant blue morpho butterflies drift by and jungle birds along the riverbank. The Pacuare is by far the classiest trip and boasts two rustic luxury lodges, one owned by Adventuras Naturales and the other by Ríos Tropicales. You get to stay overnight at one of the lodges on the two-day trips, which are filled with hiking and nature activities. Companionship comes from fellow adventurers. The Pacuare offers the river ride that most people see only in brochures.

Whitewater rafting on Río Pacuare.

The **Sarapiquí** is a more gentle raft float river strictly for beginners or birders. One-day tours offer breakfast and lunch and often combine the trip with a visit to the La Paz Waterfalls en route. Book with **Costa Rica Fun Adventures** *(☎ 506/290-6015)* or with any of the other rafting companies.

Most day-trips start around US $75 per person, including transportation, breakfast and lunch. Other rivers rafted around the country include the Corbicí, General, Río Naranjo, Peñas Blancas, Río Bravo and Río Toro – proof that Costa Rica boasts more whitewater rapids per square mile than any other country in the hemisphere.

A pleasant day-long cruise on a catamaran to **Isla Tortuga** in the Gulf of Nicoya can be arranged with **Calypso Tours** *(☎ 506/256-2727, www.calypsotours.com)*. A bus ride to Puntarenas and back, light breakfast, onboard drinks and lunch on the small island are included in the $100 fee.

Speaking of overseas, **Cruise West** *(☎ 888/851-8133, www.cruisewest.com)* has a fun, nine-day cruise route up and down the Pacific coast of Costa Rica and Coiba Island off Panama. An 11-night sail on one of their intimate luxury ships includes passing through the Panama Canal. Vessels carry (100 passengers in 50 ocean-view cabins. List prices in 2005 were US $2,000-3,000. If you're already in Costa Rica and tempted to sail, local Tico travel agents can book this for you.

Seven-night sailing yacht cruises are offered by **Windstar Cruises** *(☎ 800/258-SAIL, www.windstarcruises.com)*. Ships depart from Puerto Caldera, near Puntarenas, and sail to San Juan del Sur, Nicaragua, then down the Costa Rica coast as far south as Quepos.

If you are a serious diver or naturalist, a voyage into the Pacific to **Cocos Island**, 535 km/327 miles offshore in the Pacific, could be the highlight of your trip. See the Puntarenas entry, page 282.

■ Adventures for Children

If your parents didn't have any children,
there's a very good chance you won't have any.
~ Clarence Day, 1874-1935

KID FRIENDLY

There are many things kids can enjoy in Costa Rica. Anything to do with water comes to mind – look for swimming pool parks at Ojo de Agua and Orosi, in addition to many small towns. But for other distractions, look to **Pueblo Antiguo** ("Old Town," open Fridays and weekends), two km/1.25 miles west of Hospital Mexico in La Uruca, on the way to the airport. In its 4.86 hectares/12 acres, it attempts to re-create Costa Rican life as it was between 1880 and 1930 in three sections: City, Coast and Country. Live actors display and explain crafts. Rides and fast food. Proceeds benefit the hospital, built in 1964 in response to the polio epidemic of the 1950s. ☎ *506/231-2001.*

Of course, San José offers the **Children's Museum**, **Bolivar Zoo** and **Spyrogyra** (see *San José Walking Tour*, above), and the **Butterfly Farm** and **Ave Zoo** near Alajuela (see *Alajuela* section for details).

Catch the Dulce Nombre de Coronado bus from Av 7, Calle Central, to the **Clodomiro Picado Institute** *(☎ 506/229-0344)*. Here, kids can research snake venom and watch the staff extract it on Fridays at 2 pm. Ask the driver for the stop. Free.

The **Tico Tren Trolley Tour** (see page 120) is a fun half-hour ride around San José on a historic sightseeing trolley. The **Tico Train Trip** (see page 133) is a fascinating day-trip over the mountains to Puntarenas on a real train.

Costa Rica Expeditions *(☎ 506/257-0766; www.costaricaexpeditions.com)* offers active Family Tours, which they've done for years. This might be a good choice for your family group.

Shopping

If there were dreams to sell, what would you buy?
~ *Dream-Pedlary*, Thomas Lovell Beddoes, 1803-1849

Although never known for its indigenous crafts like those in Guatemala or Mexico, Costa Rica has generally high quality and reasonably priced goods. Coffee and coffee-related products – coffee liqueur (yum) – clothing, souvenirs and art abound. Artisans in Guanacaste have recreated the delicate and delightful pre-Columbian Chorotega pottery style and make some wonderful pieces. Don't pass them by. Costa Rica's great woodcrafts are centered around the town of Sarchí. Leather craft has its creative home in the San José suburb of Moravia. If you don't make it to either place, local gift shops are full of leather items, belts, distinctive Costa Rican hats, woodcarvings, boxes, miniature furniture and even full-size furniture.

■ Woodworks

Two notable North American artists have made an impact in wood design. **Barry Biesanz** works out of a new, open-to-the-public showroom and tourist attraction in Escazú *(☎ 506/228-1811, www.biesanz.com)* creating exquisite hardwood bowls, intricate boxes and furniture. **Jay Morrison**, another ex-pat, specializes in creative custom hardwood furniture. Upscale San José shops feature these and other fine artists and craftsman. Try **Atmósfera** (Calle 5, Av 1), **La Galería** (Calle 1 Av, Central), **Magia** (Calle 5, Av 7) and **Suraska** (Calle 5, Av 3).

■ Silver & Gold

Silver jewelry is ubiquitous in Central America, but Costa Rica's unique contributions include reproductions of its pre-Columbian jewelry, often incorporating semi-precious stones and jade. Images of ancient deities, which resemble animals, are made into earrings, necklaces, pins, bracelets and pendants. They are called *huacas*, and the most famous is a frog-like image with flat feet. Our favorite, however, is a human-like figure that is anatomically correct. Great for people who rub their earrings. These figures are also available in gold, replicating those on show in the Gold Museum.

■ Shops You'll Love

San José

Choices for shopping around San José range from downtown department stores and traditional marketplaces, *mercados*, to North American suburban indoor malls and member superstores such as Price Club. San Pedro has an upscale outlet mall now, but retail shops like those that once anchored the disappearing downtown areas in the United States still offer their wares on the pedestrian walkway of San José.

Downtown shops generally open around 9 am and close at 9 pm or later. Here are some of the best places to get a bargain and find some worthwhile things to lug home.

Farmer's Market. By 6 am on weekends, local farmers have set up tables at outdoor locations in the *barrios* of San José and small towns and villages throughout Costa Rica. An array of beat-up pick-up trucks and trailers haul fresh fruits and vegetables into town for sale. These markets, called *ferias*, are wonderful. You can taste samples of exotic fruits and practice your Spanish. Besides fruit and veggies, you'll find a cornucopia of cheeses, honey, flowers and plants, aromatic herbs and spices, plus homemade breads and tamales. In San José, the best Saturday *ferias* are in Pavas, Tibas and Escazú. On Sunday, head to Hatillo.

El Pueblo *(in Barrio Tournón, just across the Río Torres)*. This imitation colonial village – complete with a warren of narrow walkways and small shops – has an engaging charm all its own. El Pueblo (which means "The Town") is a complex that offers intimate dining in numerous restaurants and funky little bars, as well as shopping in art studios, craft shops, eclectic and gift stores. It even has a popular disco. We like it best in the evening when El Pueblo's narrow cobblestone plazas fill with people and it really feels as if you are in an old colonial village. Plus, we love the intimate bars (see our review in *Nightlife*, the next section). Don't miss it.

Mercado Central *(Calle 6 and Av Central)*. At the western end of the pedestrian walkway is the indoor complex of flea market-like stalls of the central market, a fixture here since 1880. Not large by Central American standards, it's still a fun place to wander. Flower stalls, clothing and shoe stores, gift shops, fabric vendors, butchers, restaurants, and much, much more. Be wary of pickpockets in the narrow aisles. Outside on Calle 6 there is a coffee roasting and grinding shop. It's one of our favorite places to buy – or just smell – the dark beans. Always buy the best, not the cheapest.

Hotel Don Carlos *(Calle 9 between Av 7 & 9, ☎ 506/221-6767)*. Many of the better hotels have a modest gift shop, but people come to the Don Carlos Hotel just to buy things in the Annemarie Shop. It has a large selection (two floors) of jewelry, gifts, clothing, arts and crafts, and souvenirs. It's always well stocked, clean and modestly priced. A best shopping bet and well worth a visit.

La Gloria *(Av Central walkway, between Calles 4 & 6)* is a department store right out of the 1950s or 1960s. Fabrics, clothes, household goods in an old-fashioned multi-level store.

Guzman Guitars *(Cinco Esquinas, Tibas)*. The Enrique Guzman family have been making acoustic guitars here since 1833, so if you are into music, a pilgrimage to the Guzman factory is a must. Spanish and classical guitars are their specialty and you can buy one – or order one custom made – in their small showroom. You can also ask permission to visit the factory in back and see guitars being made. Tell a taxi driver *cingo esquinas*, it's not far.

Guzman Guitars has been in business since 1833.

Mercado de Artesanías Morazán *(Calle 7, half-block north of the Balmoral).* A souvenir shop with lots of examples of one of Costa Rica's most notable crafts – woodworking. Some very finely crafted boxes made of iron and rosewood, polished so highly that the grain is eye-popping.

Mercado Nacional de Artesanías *(Calle 11 behind the Soledad Church, ☎ 506/221-5012, www.mercadoartesania.com).* This large store is the "official" national arts and crafts gift shop in San José. It's just one block off Av 2. They have many of the same handcrafted gifts and souvenirs that other downtown stores feature – which include beautiful woodworking and original paintings – but claim that because they are "more or less" related to the government, they pass on tax savings to you, making their prices lower. We're not too sure about the price claim, but the selection is very good (masks, ceramics, T-shirts, etc.) and the prices reasonable. Other locations are in the National Museum and Puerto Caldera in Puntarenas.

La Casona *(Av Central and Calle Central).* Open daily from 9 to 7, La Casona is San José's largest souvenir shop – warehouse-sized on two big floors, in the middle of downtown. It has a big selection. A good place to do comparison shopping, so start your shopping day here.

Galería Namu *(Av 7 & Calle 5, ☎ 506/256-3412, www.galerianamu.com).* The word "Namu" signifies jaguar, a sacred animal, in the language of the Bribri, an indigenous Indian people. This gallery presents a unique collection of art and crafts from the six surviving communities of Costa Rica's indigenous people. Owner Aisling French (who is Irish) searches out some of the best works from all over Costa Rica, including art produced by a San José outreach program that benefits local street children and orphans. Though her prices are a little higher than you may find in San José souvenir shops, you won't find these things elsewhere. Her masks are from the Huetar, Chorotega and Boruca cultures. The gallery also offers tours to indigenous communities, strongly recommended for a very unusual experience. Next door is **Arte Latino**, a gallery with paintings by local artists.

Books, maps and nature guides can be found at **7th Street Books** *(Calle 7, Av Central & Av 1, ☎ 506/256-8251)*, the largest English-language bookstore in town. It has a fair selection of travel guides, Latin American studies, literature, science and nature, reference books and posters. If they're not carrying our guide, please ask them to. The same goes for the other big bookstore in town that sells a wide assortment of English-language books, as well as used titles. **Mora Books** is in the Omni Building *(Av 1 and Calles 3 & 5).* Please ask them if they stock our guide.

Flea markets spring up in two main locations: on the small street in front of the National Museum *(Av 2, Calle 15)*, daily rain or shine, and on weekends along the eastern end of the pedestrian walk, downtown.

Out of Town

In the rising foothills of the valley, **Escazú** can be said to look down on San José in more ways than one. Upscale shops and restaurants, a car wash, English-language movie theater, private schools, health food stores, art galleries, jewelry shops, B&Bs, a country club, bowling alley, and the home of the American Ambassador are all in this San José suburb, a few kilometers southwest of La Sabana Park off the *autopista*.

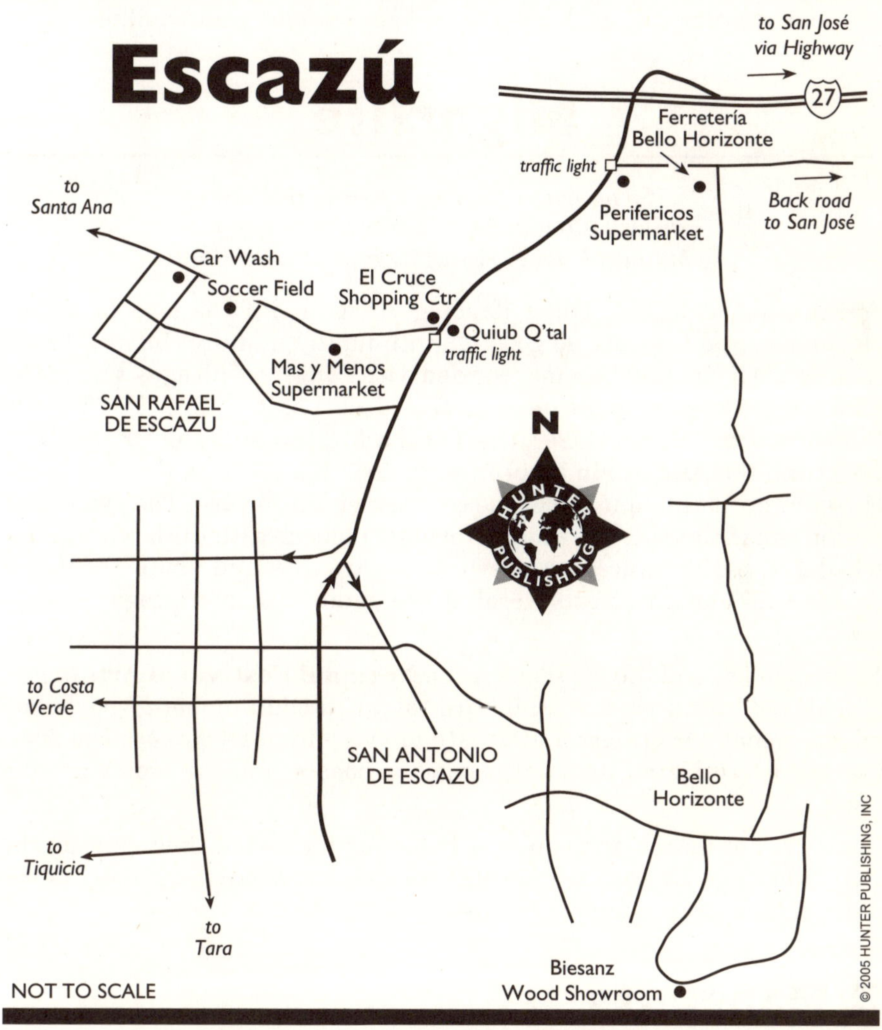

Escazú is actually two towns in one, San Rafael de Escazú and San Antonio de Escazú. The main intersection in town has a right turn for the San Rafael section and the town of Santa Ana (six km/four miles more), or you can go straight uphill for the San Antonio district. Follow the old road toward Santa Ana to Alto Palomas to visit **Cerámica Las Palomas**, where you can watch potters make ceramics available for sale in their showroom.

Sarchí is the hometown of Costa Rica's **craft** industry. A pleasant day-trip from San José, this Central Valley village features woodworking, furniture and traditional hand-painted oxcarts (ask for Joaquín Chaverri's oxcart factory) as well as a wide selection of pottery, fine art, arts and crafts, gifts and souvenirs. A drive through the town reveals a number of different storefronts – pick one that takes your fancy and pull over.

Nightlife

The hangover became a part of the day
as well allowed for as the Spanish siesta.
~ *My Lost City*, F. Scott Fitzgerald, 1896-1940

Thinking nightlife? Think dancing. Ticos and Ticas love to *bailar*, dance, and they are so good it's intimidating. Many clubs and bars have dance floors that become crowded with dancers of all ages, but it is by no means the only nightlife that San José has to offer. The **Teatro Nacional** *(☎ 506/221-1329)*, the **Teatro Melico Salazar** *(☎ 506/221-5341)* and the **Auditorio Nacional** *(☎ 506/223-7003)* in the Children's Museum lay claim to much of the country's high culture. The symphony season runs from March through November (check with the box offices for scheduled performances). Modern dance troupes and small theatrical groups find audiences at several of the smaller theater stages around town.

The crown jewel of the art scene is the **National Festival of Arts** *(www.festivalcostarica.org)*, featuring Tico talent in odd-numbered years and international performers and artists in even-numbered *años*. The festival runs every March, all month long, and boasts multiple venues around town.

But if it's not March and you're still looking for a good time, pick up the *Tico Times* and *La Nacion* (one of several Spanish-language daily newspapers) for listings of movies, music and mischief.

■ Hangouts

El Pueblo *(Barrio Tournón, just north of the downtown, go by taxi)* was built as an imitation Spanish colonial-era village. The buildings contain gift shops, art studios, funky little bars and intimate restaurants. As the big draw, El Pueblo features **El Infinito**, a popular dance/disco spot that offers varied musical styles on its three dance floors. No cover. We always merengue here because we can't salsa. Another good but small dance club in the complex is **Cocolocos**. But El Pueblo offers much more than disco for an evening's entertainment.

Our sentimental favorite, **Tango Bar**, is hidden among the nooks and crannies of El Pueblo's warren of shops and bars. Like a step back in time, you enter the evening-only, hole-in-the-wall tango bar through an old orange curtain. Inside, the walls are lined with black-and-white photos of ancient singers and musicians from Argentina's golden age of tango. Tables crowd the tiny space. The bar maid is geriatric and so is the accordion player/singer, who performs tango standards and teary South American ballads. They take requests. If we only knew the words as well as the inebriated expatriate Argentine regulars at the bar, we'd sing along too. It's a wonderful place to sit and listen, nurse a beer or mixed drink, and play a part in a life drama you won't find anywhere else – even Buenos Aires!

> "The Tango sums up the feelings of a people – the pain, the happiness, the irony and the roughness of life," Maria Julia Berdé, a Tango dance producer, told A. M. Costa Rica News. "There is a tango for each situation in life and its followers exalt it above any other existing rhythm."

If you'd like to dance, or see the tango danced, (the bar in El Pueblo is a bit too small) find **Sabor y Sueños** Restaurant in Barrio Escalante. Located 25 meters/82 feet west of the Rotanda del Farolito (a traffic circle with a street lamp) and two blocks northeast of St. Teresita's Church. They also feature professional dance shows.

When you've had enough Argentine nostalgia in El Pueblo, slip out and wander among the little bars and artsy shops, have a late-night snack at one of the many little eateries that pop up around every corner. Or dine at **La Cocina de Leña** *($$-$$$)*, which specializes in typical Costa Rican food. Try the three-meat soup, *Olla de Carne Típica*.

Across the street from El Pueblo is another happening dance/disco hall. **La Plaza** has a very large dance floor and lots of young people. Very popular. Be sure to have your feet check it out.

Sadly, **La Esmeralda** (on Av 2 between Calles 5 & 7, near the Teatro) has closed. It once was a bar/restaurant where you could get a beer, *boca*, and be serenaded by the numerous mariachi groups, some of which still hang out in front.

El Cuartel de la Boca del Monte *(Av 1, Calle 21 & 23)* is one of the east end's hottest singles bar and the place to be seen. Live music on Monday, Wednesday and Friday night means wall-to-wall people. It still attracts some of the artsy crowd that made its name, as well as hard bodies who work as tour and rafting guides. Very interesting mix. Bar-hop across the street to **Cielo**, or to the corner for **Bash**, for a more hard rock atmosphere.

Risas *(Calle 1 between Av Central and Av 1)* is a downtown storefront dance bar and restaurant occupying three floors of a beautiful old brick building. The restaurant has a pleasing ambiance and reasonably priced food. The dance floor throbs with heavy techno, Top 40 rock, and Latin beats. It attracts mostly a young crowd and has a second-floor peanut gallery for those who want to watch. A slide runs from the second story down to the exit. We went out for a night here with some British Embassy staff members and had a real blast (after all, they have three bars). US $2 cover buys a beer.

Salsa 54 *(Calle 3, Av 1 & 3)*. Since we can't salsa very well and Ticos can, we like to watch them whirl on the raised stage at Salsa 54. If you're in town for a while and want to improve your steps, ask about lessons in Latin dance here. This dance floor is crowded with locals, couples and singles. One adjoining hall offers slow, close-dancing romantic *bolero* music, while another hall blasts techno rock. Some of the best dancers head here or to El Tobogan.

El Tobogan *(200 meters/650 feet north, 100/300 east of the La República offices; take a cab)* offers a huge dance floor that fills with gyrating bodies moving to the beat of live Latin music on the weekends. Good dancers. In case you stay all night, they offer a family swimming pool from 9-5 on Saturday and Sunday. Near El Pueblo in Barrio Tournón.

The **Shakespeare Bar** won't make you smarter. But it's smart to have a drink or eat at this quiet but popular place under the Laurence Olivier Theater. *(It's next to the Sala Garbo complex, Av 2 at Calle 28.)* Perfect for a before- or after-play hangout. Jazz is the music of choice.

El Tunel del Tiempo *(Av Central, Calle 11 & 13)*. We hear the "Tunnel of Time" is the spot to go after hours – 2 am until 10 am the next morning. Well past our bedtime.

■ All-Night Eateries

You may choose to people-watch, drink and eat at some famous all-night eateries. Two places vie for top position on the list of all-nighters. **La Parisian** is a very good outdoor/veranda restaurant in front of the Gran Hotel. Everyone passes by here sooner or later. **Chelles** *(Av Central walkway at Calle 9)* was a favored spot of local Ticos. Sadly, it was burnt to the ground. Let us know if it has been rebuilt by the time you visit.

Other 24-hour eating-places include **Manolo's** *(Av Central)*, **Pollo Frito Pío Pío** *(Av 2, Calle 2)*, **Restaurant don Amado** *(Av 2, Calle 6)* and the open-all-weekend **Soda Tapia**, on the southeast side of Sabana Park. Lastly, if you are desperate for late-night bland American food, there's a new **Denny's** at the Best Western Irazú.

■ Gay & Lesbian Nights

I'm a practicing homosexual...but bisexuality immediately doubles your chances for a date on Saturday night.
~ Woody Allen, comedian.

La Avispa *(Av 1, between 7 & 8)* is lesbian-owned bar that features three dance floors, big screen TV and pool tables. It's a "rolled-up-shirt-sleeves" kind of place with good music. La Avispa also draws gay men, especially Tuesday nights, and local lesbian women to dance. Closed Mondays. Monthly "Fem" night.

A lively place for people of all sexual persuasions to go clubbing is **Deja Vu** *(Calle 2, Av 14 & 16)*. With loud techno, salsa and dance music, Deja Vu is a magnet for gay men and straight party people – especially on the weekends when a well-dressed crowd fills the two dance floors. Café and souvenir shop. Drag shows.

One neighborhood gay-friendly, restaurant/bar is **Kasbah** *(Calle Central, Av 7 & 9)*, which features a faux Moorish décor and an Internet café.

Places to Stay

Oh bed! Oh bed! Delicious bed!
That heaven upon earth to the weary head.
~ *Her Dream*, Thomas Hood, 1799-1845

These are the best and most interesting of the downtown San José and neighborhood accommodations. They vary from traditional establishments and welcoming B&Bs to luxury small hotels or cheap digs. The common denominator is that they are near the center of town, have charm and inviting attributes, and they are secure, clean and respectable. If you need more amenities, check out the luxury accommodations toward Escazú and the airport, such as the Marriott, Vista del Valle, Finca Rosa Blanca or Alta. Accommodations in these areas are listed later in the suburbs section.

■ Prices

We made every effort to be as thorough and as accurate as possible. However, things change in Costa Rica – sometimes *muy rapido* – and prices change faster than anything. To help you stick to your budget and find a place that meets your needs, we indicate relative price levels of various accommodations using dollar symbols. Use it as a guideline, and always check rates.

Moderate Accommodations

Hotel Prices

Prices are per night for two people and do not include 16% tax.

[No $]	less than US $20
$	US $20-$40
$$	US $41-$80
$$$	US $81-$125
$$$$	US $126-$200

Hotel Santo Tomás *(Av 7 between Calles 3 & 5, ☎ 506/255-0448, fax 222-3950, www.hotelsantotomas.com, 20 non-smoking rooms, safety deposit boxes, pool & jacuzzi, restaurant, telephone, cable TV, tropical breakfast included, $$-$$$).* It's easy to miss the Santo Tomás because of its relatively nondescript outward appearance, right on the sidewalk of Av 7 in Barrio Amón. But if you do walk on by, you'll miss one of San José's most pleasing little B&B inns. Owner Thomas Douglas, a North American, saved this beautiful home from demolition 15 years ago. Built in 1910 by a former coffee baron, it features antique Louis XV reproduction furniture, hand-painted tile, highly polished wood floors, Persian rugs and 14-foot-high ceilings. Most rooms rent on the first floor, well back from any traffic noise. A very helpful staff provide assistance with travel plans, including car rentals. There's a beautiful solar-heated pool with a waterslide, a hard-to-find amenity in the downtown area. Its restaurant, **El Oasis**, is set in the garden courtyard patio and is one of the city's best dining choices. The hotel has an ideal location close to everything downtown. Free Internet access is included in the room rate, plus local telephone calls are free.

TROPICAL BREAKFAST

You'll often see a "Tropical Breakfast" offered on the hotel menus. Offering a healthy start to the day, tropical breakfasts feature fresh fruit, cereal and yoghurt.

Hotel Grano de Oro *(Calle 30 between Av 2 & 4, ☎ 506/255-3322, fax 221-2782, www.hotelgranodeoro.com, 35 rooms, jacuzzis, restaurant, telephone, cable TV, gift shop, mini-bar, room safe, non-smoking, $$$).* The Grano de Oro, grain of gold, is an excellent choice if you desire luxury without ostentation. Set in a quiet residential section west of the Hospital de los Niños, the Grano was once one of the grand old mansions that dot San José's better neighborhoods. The public areas, lighted by skylights, are decorated with plants and foun- tains, original artwork and carpeting. The large rooms are universally well appointed and comfortable, decorated in blues and whites and warmed with oriental rugs and natural woods. Spotlessly clean. Ask to see several rooms (such as the ones with French windows facing the patio) and you'll be unable to choose. They offer a simple two-bedroom family suite as well as the Garden Suite and the Vista de Oro Suite, both of which are stunning. All rooms are non-smoking. Rooftop jacuzzis and sun deck. Even local residents know the Grano because its lush garden patio restaurant is famous for gourmet meals in an intimate setting. It's worth eating here on a special occasion (lunch or dinner), even if you don't stay here. Reservations suggested. Say hello to Marco for us.

Bird's-eye view of San José city.

Hotel Fleur de Lys *(Calle 13, between Av 2 & 6, ☎ 506/223-1206, fax 257-3637, www.hotelfleurdelys.com, 30 rooms plus master suite, telephone, cable TV, parking, mini-bar, restaurant, bar, breakfast included, $$-$$$).* The French-flavored Fleur de Lys is a high-quality conversion of a beautiful Victorian mansion into a pleasing hotel. All bedrooms are named after native Costa Rican flowers and the 12 rooms in a new addition are jr. suites with jacuzzis. The original 19 rooms in the old house are artfully restored with rich wooden ceilings, original paintings and sculpture and comfortable beds with floral duvets. Marble bathrooms have hair dryers. Public areas in this pink and purple hotel sport the original architectural details, including magnificent tile floors. Forget your umbrella? An umbrella stand at the front door is full of colorful ones for guests' use. The intimate restaurant overlooks a green patio and features French food, and the cozy bar includes a terrace. Fleur de Lys is off the busy streets, but is still close to everything. Just around the corner on Av 6 is a public parking area with a super clean, open-air, diner-style counter eatery, **La Melee**.

Hotel Britannia *(Calle 11 & Av 3, ☎ 506/223-6667, fax 223-6411, in US, ☎ 800/263-2618, www.hotelbritanniacostarica.com, 24 rooms and jr. suites, restaurant for breakfast and dinner, cable TV, telephone, $$$).* The Britannia is a step above most mansion-into-hotel conversions and it caters especially well to North American tastes. More formal and regal, this downtown gem is a grand pink house with a wrap-around porch and large central lobby lighted by stained glass above the atrium. All the rooms in the old house section (built 1910) are large, with high ceilings, Mosaic tile floors, rugs, fine wallpaper and generally pleasing décor. If you're looking for dependable luxury in the historic northern section of town, Hail Britannia.

Hotel Don Carlos *(Calle 9 & Av 9, ☎ 506/221-6707, fax 255-0828, www.doncarlos.co.cr, 36 rooms, cable TV, restaurant, gift shop, room safe, parking nearby, continental breakfast included, $$-$$$).* The Don Carlos is celebrated as an attractive, clean and comfortable colonial hotel that takes up an entire corner block in a residential area of downtown. You'll know you're there when you pass beautiful hand-painted tile episodes from Cervantes' *Don Quixote* set into their brick wall facing Calle 9. Typical Costa Rican scenes grace the wall on Av 9 and bronze statues of *campesinos* (country folk) rest near the entry. A wall at the reception is completely tiled with colorful hand-painted tiles of village scenes (tiles are for sale individually in their well-stocked, well-known, and very large gift shop). The hotel's bedrooms are in two buildings divided by a patio breakfast area with two bubbling fountains. The "colonial" section has the older rooms with typical high ceilings. The newer rooms on two upper stories feature parquet wood floors and floral bedspreads with complimentary curtains. There is original artwork throughout, most for sale. In a previous life, the Don Carlos hotel was the mansion of a Costa Rican president. The attic suite, with two queen beds, is a good value at under US $90 because it has a large amount of space and lots of privacy. No charge for children 12 or younger sharing a room. Lunch and dinner are served in the Pre-Columbian Lounge, where stone reproductions of Maya statues fill a waterfall fountain. Unfortunately, we found the restaurant's food somewhat disappointing.

Hotel Aloki *(Calle 13 between Av 9 & 11, ☎ 506/222-6702, fax 221-2533, in US, ☎ 770/660-1503, www.traveltocostarica.com, 6 rooms plus 1 suite, restaurant, pub, telephone, cable TV, breakfast included, $$$).* The name Alóki may sound Hawaiian but it is Bribri, an indigenous Costa Rican language, and it means "ray of sun." Perhaps the North American owner was inspired by the morning sunlight that filters through the huge skylight covering the courtyard dining area. As you enter, this formal central dining area almost steals the show from the magnificent hand-painted tile floors, antique furnishings, fresh tropical flowers and colonial ambience found in the guest rooms. It is a turn-of-the-century Spanish home in the true colonial style with rooms built around a center courtyard and a hall walkway surrounding. In the case of the Alóki, the courtyard has been tiled and made into a restaurant with starched white linen and fine crystal. At back is a balcony overlooking the woods near the zoo. Here you'll find a more informal but equally pleasing pub restaurant. Well worth a special meal out, even if you're not in the hotel. Formerly called L'Ambiance.

Hotel Presidente *(Av Central walkway & 7, ☎ 506/222-3022, fax 221-1205, www.hotel-presidente.com, 110 rooms and suites, cable, telephone, restaurant/café, parking, $$$).* The Presidente is a pleasant downtown hotel right on the pedestrian walkway at the heart of San José. We stayed here twice and really enjoyed its convenience and comfort. The public areas were remodeled late in 2001, when the restaurant and coffee shop became the **News Café**,a super popular eatery. It's owned by the same people who run the Tabacón Resort at Arenal, and each property will gladly make reservations at the other. Helpful staff.

Hotel Balmoral *(Av Central, between Calle 7 & 9, ☎ 506/222-5022, fax 221-1919, www.balmoral.co.cr, 110 rooms, air, cable TV, telephone, restaurant, parking, business center, $$$).* The Balmoral sits at the end of the pedestrian Av Central. The formal entry foyer is impressive, with sand marble floors, leather furniture and natural cherry wood trim. This hotel, which caters to business travelers, remodeled its public areas and rooms in late 2000. The standard room – which most tourists opt for because of the price – is a little smaller with a less fancy bathroom than the superior room. (The fancy bathroom is attractive, but worth nearly twice the price.) Two queen-sized beds in each. Excellent little restaurant out front.

Hotel Le Bergerac *(Calle 35 & Av 8, Los Yoses, ☎ 506/234-7850, www.bergerac.co.cr, 26 rooms, cable, restaurant, breakfast included, $$-$$$).* On a curve, on a quiet street in the Los Yoses section of San José, sits this intimate, French-flavored hotel that attracts business, tourist and even diplomatic clientele. Separate old homes have been beautifully remodeled and connected together with verdant gardens to form the core of one of the more distinctive Costa Rican hotels we've had the pleasure to stay in. Superior rooms, with period furniture and hardwood floors, offer private garden terraces or pocket patios, a welcome addition to already comfortable accommodations. We eschewed the cable TV for hours spent reading, writing and drinking wine on our private patio. The continental restaurant, with an outstanding gourmet menu and French and Chilean wines, is closed Mondays. *C'este magnifique!*

Hotel Milvia *(100 meters/300 feet north and 100 meters/300 feet east from Centro Comercail Muñoz y Nanne, San Pedro, ☎ 506/225-4543, fax 225-7801, www.novanet.co.cr/milvia, 9 rooms, cable TV, e-mail, breakfast included, $$-$$$).* The Milvia is set on the corner of a quiet side road a short walk from the two universities in San Pedro. This intimate, sophisticated inn occupies a turn-of-the-century restored Caribbean villa. Each of the large rooms is non-smoking, tastefully furnished and infectiously bright and cheerful. Lunch and dinner are available in their polished dining room, and cheap eateries abound near the college. Coffee and tea is available all day long – *gracias*. Children stay free when sharing a room with their parents. Best bet in San Pedro, or perhaps San José for that matter.

Raya Vida B&B *(Calle 15, Av 11 & 13, ☎ 506/223-4168, www.rayavida.com, $$$)* is a luxury B&B run by North Americans in a quiet area of large residential homes north of the Casa Amarillo, in Barrio Otoya, near the Bolivar Zoo Park. It's an old, small colonial-style villa that has been restored and decorated with artwork from around the world. One of the four guest rooms features masks from African and Latin American cultures hanging on the walls. Raya Vida is very special, but very hard to find in what feels like a maze of side streets. Best to get a cab.

Hotel 1492 *(Av 1, Calle 31 & 33, ☎ 506/256-5913, fax 280-6206, www.hotel1492.com, 10 rooms, breakfast, telephone, cable, safe, $$-$$$)*. The 1492 is an inviting, lavishly decorated Spanish-style mansion converted into a charming little hotel on a one-way street in Barrio Escalante. It has been in the same Costa Rican hotelier family since 1941 and they treat each guest like visiting royalty. High ceilings original Portuguese tile floors, polished hardwoods, tropical gardens and lovely large rooms make this a fine place to stay in the eastern section of San José. Guests get to mingle at happy hour when the convivial hosts bring out wine and cheese in the garden. Several appealing eateries grace this trendy neighborhood and the local movie theater is down the block. Formerly known as the Jade y Oro, you'll find it almost in back of the Toruma Youth Hostel. Ask Sabrina about their all-inclusive packages that include at least six tours, meals, entrance fees, lodging and bilingual guides. The packages are a good way to get an informed overview of Costa Rica in a week.

■ Budget Accommodations

Hotel Petite Victoria *(Av 2 & Calle 28, ☎/fax 506/233-5193, 16 rooms, restaurant/bar, parking, breakfast included, $-$$)*. If you'd love to stay in one of San José's old Caribbean-style wooden Victorian mansions but think they are beyond your budget, check out the economy/moderate prices at the Petite Victoria, near the Grano de Oro, one block off the Paseo Colon from the Pizza Hut. This corner B&B has a fascinating sitting area near the reception, with original hand-painted floral tile floors, Victorian latticework room dividers and a circular center settee. The rooms are very basic, without much appeal other than that they're in an old, old house with tongue-and-groove painted wood walls and high ceilings. A young clientele keeps the bar/restaurant busy.

Hotel Doña Inés *(Calle 11, between Av 2 & 6, behind the Iglesia La Soledad, ☎ 506/222-7443, fax 223-5426, hoteldonaines@rasca.co.cr, telephone, local TV, parking, restaurant, breakfast included, $$)*. The Italian-owned Doña Inés stands out as a quaint and quiet B&B in immaculate condition. Located behind the Soledad Church and across from the Mercado Nacional de Artesania, this little colonial-style villa is painted white with lime green trim. It's well worth checking out. The smallish rooms (some without exterior windows) are centered around tropical patios with green plants everywhere. We describe the rooms as "old-fashioned sweet" – perhaps like the bedroom you had at your grandmother's house – with a 1950s clock radio, shiny wallpaper, floral bedspreads and a chintz armchair. The dining area and darling little bar are semi-formal, with heavy Spanish-style furniture. The hotel serves full breakfasts (included in the rate) and cook-to-order dinners by reservation. Secure behind a graceful wrought-iron gate, and with little or no traffic noise in the rooms, we found the Doña Inés homey and appealing.

Gran Hotel Costa Rica *(Av 2 at the National Theater, ☎ 506/221-4000, fax 221-3500, www.granhotelcr.com, 110 rooms, with 3 suites & 4 jr. suites, 2 restaurants, casino, cable TV, room safe, coffeemaker, breakfast included, $$-$$$).* In the tradition of Central American town planning, the center of a city houses a "grand hotel," usually the oldest and most prestigious. San José's Gran Hotel, which misses being the oldest by a hair, was built in the first decade of the 1900s. The standard rooms are a good size (some have a king-size bed) and painted in light colors or off-white with striped bed covers. The Presidential Suite takes up a huge corner of the hotel overlooking the square. It has a large bedroom, sitting and dining rooms with a wet bar, and two bathrooms. This hotel is in the center of everything. The inside restaurant and bar are quite good, and the outdoor café **La Parisien** – open 24 hours – is San José's best place to sit and eat or have a drink while the world passes by your table. Check out our review under *All-Night Eateries*, page 144.

Taylor's Inn *(Av 13 near Calle 3, ☎ 506/257-4333, fax 221-1475, taylor@catours.co.cr, 12 rooms, cable TV, non-smoking, tropical breakfast included, $-$$).* The appealing Taylor's inn is a quiet B&B in the Barrio Amón neighborhood that's sure to please. A block down from the Britannia and several other good hotels, Taylor's has an artsy charm both inside and out – and it's all non-smoking! Just across the Río Torres from the El Pueblo complex. The front of the one-story building is exposed brick with painted flowers gracing the white wooden trim and wrought iron window grills. Adriano Arie, an Italian immigrant and skilled worker on the National Theater, built the original house in 1908. Unlike the grandiose theater building, the home he built on a side street is simple Spanish colonial. Where reception and the breakfast area are now was once open as a courtyard garden. The bedrooms that rent out are the original that faced the courtyard along the hallway. Up front are several rooms that also feature sleeping lofts. Full kitchen privileges. Good family hotel. Tasteful works of art are displayed throughout and no hooch or hookers allowed.

Hemingway Inn *(Av 9 & Calle 9, ☎ 506/257-8630, www.hemingwayinn.com, 17 rooms, cable TV, telephone, room safe, jacuzzi, tropical breakfast included, $$).* In an old Spanish colonial mansion high on a corner across from the Don Carlos sits the Hemingway. It looks more formal than it is, with moderately priced accommodations in the heart of the historic district. The only pretension that Hemingway Inn expresses is that each room is named for famous authors, including the friendly corner room graced with Ernie's moniker (#4); number 16 (Pamplona) may be the larger. Pleasant public areas. The rooms are appealing, but basic. Eric Robinson, an environmentalist from Canada who offers tour services, owns the popular inn. A refreshing jacuzzi is set in the garden.

Hotel La Amistad *(Av 11 & Calle 15, ☎ 506/221-1597, fax 221-1409, www.centralamerica.com/cr/hotel/amistad.html, 40 rooms including 17 deluxe rooms, cable TV, in room safe, telephone, breakfast included, $$).* The Amistad is simple and clean and has recently added a new wing. It features natural wood walls in the bedrooms and is a popular corner hotel in the quiet Barrio Otoya section of San José. Triply attractive for cheery accommodations, low price and friendly staff – plus, it draws interesting clientele from around the world. Amistad means "friendship" in Spanish. They'll store your luggage. Dependable, pleasing place to stay.

Cinco Hormigas Rojas *(Calle 15 between Av 9 & 11, ☎/fax 506/257-8581, 6 rooms, 2 shared baths, full breakfast included, $$).* Cinco Hormigas Rojas translates from the Spanish as "Five Red Ants," reflecting owner Mayra Güell's love of nature – even nature shunned by city dwellers. Her small, funky B&B is tucked in a thriving, overgrown jungle garden front yard, all behind a secure iron gate. Mayra is an artist and her accommodations definitely show it – wait till you see the bathrooms! Her sense of style combines a Haight Ashbury "Summer of Love" artistic attitude with a Tica naturalist's love of all creatures great and small. "People tend to look only at the macro world," she explains. "I want them to appreciate the beauty of the micro world." Each basic room has a different color scheme, with bright curtains, plants and a very eclectic décor. Her paintings and art are everywhere – she has an obvious feminine theme – and are all for sale. Cinco Hormigas offers Costa Rica's natural beauty in the middle of a city, and Mayra's home attracts birds, butterflies and human wanderers from all over, including five red ants.

Hotel Rincon de San José *(Av 9 & Calle 15, ☎ 506/221-9702, fax 222-1241, www.hotelrincondesanjose.com, 29 rooms, bar, cable TV in most rooms, telephone, continental breakfast included, $$).* The quaint Rincon hotel resembles an Alpine chalet. It's bright white outside and cheerful inside, with custom-designed furniture. It sits on a somewhat busy corner in the Otoya Barrio. The hotel features a big open patio bar with lots of greenery under the skylight and comfortable accommodations woven together in winding hallways. Light sleepers should avoid rooms that face the street, even though the corner room, #14, is particularly interesting. We fancied the garden-like setting of #8. Formerly the Eidelweiss.

Hotel Europa *(Calle Central & Av 5, ☎ 506/222-1222, fax 221-3976, europa@rasca.co.cr, 72 rooms plus 3 suites/apartments, pool, air, cable TV, telephone, room safe, restaurant, room service, $$).* The Europa is San José's oldest hotel (1903), but it doesn't show. It hosts annual North American Elder Hostel programs when they're in town. (Elder Hostels are organized learning vacations, primarily but not exclusively for retirees, and they are great values.) The Europa is a grand old dame of a hotel, with an old-fashioned feel, but with modern rooms and good prices. The color scheme is rather dark, in bold blue/green and wine, but the rooms are spacious and comfortable. There is an outdoor pool and an impressive formal restaurant. Make sure your room doesn't face the busy street corner. Several suites/apartments are for rent. One has three bedrooms and costs around US $450 (negotiable), plus tax, for up to four people staying six nights, seven days.

Dunn Inn *(Av 11 at Calle 5, ☎ 506/222-3232, fax 221-4596, www.dunninn.com, 23 rooms plus one suite, mini-bar, cable TV, restaurant/bar, telephone, barber shop, $$).* The Dunn Inn, owned by Jennifer Hurley, has kind of a relaxed California atmosphere, especially in the cheery skylight-covered bar and restaurant. Lots of hanging plants, terra cotta tile, exposed brick, and an American clientele add to that impression. Good-size rooms in the original 100-year-old mansion have warm natural wood walls, which can make it a tad dark in the evening. There are also newer rooms, some slightly smaller, in an addition under a bright hallway. Modern bathrooms throughout. All bedrooms are named using an indigenous or Spanish word and the meanings are posted on wall plaques. The hotel's large appealing suite features a jacuzzi. Skylights are effectively used throughout the hotel to open up the interior. Look for the giant-stained glass window of birds and flowers outside; it is part of a private apartment. There is an English-speaking barbershop attached, where owner Roy has over 50 years experience.

Kap's Place *(Calle 19 between Av 11 & 13, ☎ 506/221-1169, fax 256-4850, www.cabinas-playa-cacao.com, cable, phone, breakfast included, $$).* A mother-daughter team of Tica hoteliers offer this wonderful little bed and breakfast guesthouse on a quiet street in Aranjuez, one of San José's oldest *barrios*. Walk downtown in 10 minutes or take a local bus. The lovely home boasts three nicely decorated rooms with private baths, plus an apartment that sleeps four. Guests have full kitchen privileges and there's a garden terrace with hammocks. A pleasant place to stay. Long-term rates offered.

Hotel Gran Via *(Av Central, near Calle 3, ☎ 506/222-7737, fax 222-7205, www.granvia.co.cr, 32 rooms, cable TV, room safe, desks, telephone, café-restaurant, $$).* The long Gran Via sign runs vertically up the side of this interesting multi-story hotel. It has a central location on the pedestrian walkway, within sight of the Plaza de Cultura. Take the skinny elevators to the café on the third floor, where you can sit on the balcony and people watch or choose a more comfortable little tables inside. Some floors have a water cooler in the hallway. The rooms are fairly large with either a queen or two twin beds. Small bathrooms. Clean and bright.

■ Other Choices

A low-cost hotel just before Los Yoses and San Pedro is the tiny **Ara Macao Inn** *(Calle 27 & Av Central, 50 meters / 164 feet south of Pizza Hut La California, ☎ 506/233-2742, www.hotels.co.cr/aramacao, tropical breakfast, $)*. Well worth considering if you're looking for cheap digs. On a quiet side street in a restored, early century home with colonial styling. The congenial hosts here will hold your luggage while you wander the countryside – a big plus.

Toruma Youth Hostel *(Av Central East, across from KFC, ☎ 506/224-4085, www.hi-cr.org)* has a congenial European quality. It's located in San José, not far from Los Yoses, San Pedro, and the U of C campus. The sprawling old building houses the largest hostel in the country, with 95 beds, segregated by sex. Each dormitory room has four to six beds and a large, high-ceiling central lounge that makes a nice place to relax and meet with fellow hostellers. If you need a membership card, get one here for US $12 (a passport photo is needed).

Casa Ridgeway *(Calle 15 & Av 6, ☎/fax 506/233-6168)* is a small but popular, people-friendly hostel. This no-smoking establishment is adjacent to the Quaker Peace Center, which it helps support. Reservations recommended.

Clean and convenient, the **Green House Hostel** (*Calle, % 506/258-9084, www.greenhousehotel.altavista.org, breakfast, $*) boasts comfortable airy rooms that surround its tiled public areas in a central courtyard configuration. Newly opened and run by the owners.

A popular low-budget hotel in downtown San José is the **Pension de la Cuesta** *(Av 1, Calle 11 & 15, ☎ 506/255-2896, breakfast included, $)*. You can't miss it – it's a pink and purple Caribbean-Victorian mansion, with a sunny communal area that fills with fellow cultural travelers.

The hostel whose name doesn't say it all is **Costa Rica Backpackers Hostel** (*Ave 6 between Calle 21 & 23, ☎ 506/221-6191, pool, Internet*), a surprisingly good budget oasis in town. It has three garden areas, a swimming pool, laundromat and, best of all, free coffee and tea all day. With dorms and rooms, CRBP is a step above the image of a backpacker's hostel.

Hotel Cacts (*Av. 3 bis between Calle 28-30, ☎ 506/221-2928, www.hotelcacts.com, 25 non-smoking rooms, cable, pool, secure parking, garden, sun deck, $$*) is one of San José's more venerable small hotels with an outstanding reputation for service and quite pleasant digs. It's a long-time personal favorite, too, and has a large and reputable travel agency for tours. Cacts is an excellent value that would be even better if the area around it in Barrio Pitahaya hadn't become somewhat seedy.

An excellent choice near the north side of La Sabana Park is the **Hotel Sabana B&B** (*☎ 506/232-2876, www.costaricabb.com, $*), "run by a Costa Rican family who will shower you with attention." In a modern suburban home in a quiet neighborhood, this B&B is super clean, pleasant and stylish. From **Restaurant El Chicote** (a famous steakhouse that's well worth a visit) go one block north, a quarter-block west and two blocks north. It's the next to last house on left.

■ Escazú

Escazú has replaced Rohrmoser as the neighborhood of choice for expats living near the capital. If you'd like to stay in Rohrmoser, look for the very appealing B&B **Hotel Casa Roland** (*☎ 506/231-6571, www.casaroland.com, jacuzzis, cable, breakfast, bar, $$*), a short distance from Tobias Bolanos Airport. Casa Roland offers 20 rooms tastefully decorated with a collection of works by local artists. Tropical plants fill the lobby and patio areas, inviting you to pull up a stool at the handcrafted bar. An excellent choice if you're taking the morning Nature Air flight to explore other parts of the country.

The best values in gentrified Escazú are the small hotels and luxury bed-and-breakfast rooms that dot the hillside villages. Escazú is actually two towns in one, San Rafael de Escazú and San Antonio de Escazú. As you come into town, the road forks (at an area known as El Cruce). Go right for San Rafael or straight up the hill for San Antonio.

Costa Verde Inn (*San Antonio de Escazú, ☎ 506/228-4080, fax 289-8591, www.hotelcostaverde.com, tennis, pool, jacuzzi, restaurant, fireplace, breakfast included, $$*). The Costa Verde in Escazú is a private mansion that belongs to the owner of the beautiful Costa Verde Hotel in Manuel Antonio. Now, after a remodeling, it lives up to its claim as a "country inn with all the amenities of a small resort." An apartment is available and one of the big guest rooms has a unique stone waterfall-like shower. Filling breakfasts are served. Costa Verde is located in a quiet residential area above the city, near the graveyard. Great value.

Hotel Alta (*Las Palomas on the road through San Rafael, US ☎ 888/388-2582, CR ☎ 506/282-4160, fax 282-4162, www.altatravelplanners.com, 23 rooms, gym, pool, restaurant, $$$-$$$$*). *Alta* means "high" in

Spanish, and at this hotel the quality matches the altitude. Hotel Alta overlooks the Central Valley from high on a mountainside where the service and accommodations are nothing less than first class. Built in cascading levels down the mountain, its old-world allure blends with modern-day extravagance – tile and polished hardwood floors; ocher-tinted, maize-color stucco walls; cedro, mahogany, and caoba wood accents; and a 500-bottle wine cellar. Dine in style downstairs at **La Luz**, a gourmet restaurant.

The hotelier's hotel in Escazú is **Hotel San Gildar** *(next to CR Country Club, ☎ 506-289-8843, fax 228-6454, www.hotelsangildar.com, pool, restaurant, breakfast included, $$$)*, a beautiful inn that opens from its entryway to a hacienda-style central courtyard with a big blue pool and three levels of rooms on a garden hillside. It impressed and pleased us.

Apartotel Villas del Río *(next to the Country Club in San Rafael de Escazú, ☎ 506/289-8833, fax 289-8835, www.villasdelrio.com, 40 apartments, pool, kitchen, daily maid service, private parking, $$$$).* For long-term stays of a week or more, there's no better way to save time and money than to stay in an "Apartotel," a hotel consisting of various size apartments. We stayed here for a week and used the kitchen quite a bit, plus the separate sitting room to read or watch cable TV. And staying in one spot in a luxury apartment allowed us to feel as if we were coming "home" each evening after our day-trips. It's especially good for a family or two or more couples. A comfortable base.

B&Bs

Not to be overlooked in Escazú are the plethora of B&Bs, many owned by North Americans. In the San Rafael section there's **Casa le las Tias** *(☎ 506/289-5517, fax 289-7353, www.hotels.co.cr/casatias.html, $$-$$$).* Xavier Vela and Pilar Saavedra-Vela (great names, aren't they?) welcome guests to their charming country town house, one of Escazú's finer guest quarters. Elegant, airy rooms and big breakfast. The name means "The Aunts' House."

Villa Escazú *(☎ 506/289-7971, fax 228-9566, www.hotels.co.cr/vescazu.htm, $$)*, owned by Inez Chapman, is a comfortable Swiss chalet-style B&B with an additional studio apartment available on the pretty garden level. Gourmet breakfast included. Central Escazú, 900 meters/2,961 feet west of Banco Nacional. Very pleasant.

On the crest of a hill in a suburb known as Bello Horizonte is **Posada El Quijote** *(☎ 506/289-8401, fax 289-8729, www.quijote.co.cr, $$-$$$)* a sprawling, colonial ranch house that serves as a B&B inn. It has wonderful large open areas with a patio and terrace that overlook the valley. Very homey and warm, yet with plenty of class. Big, bright rooms and walled gardens. Turn left before the light at Trejos Montealegre shopping center, stay left, and look right for the sign to Bello Horizonte Barrio.

ZARATE, THE WITCH OF ESCAZÚ

Escazú has its share of new age followers among the North Americans living there – perhaps because locals know Escazú as the home of Zarate, an enchanted *bruja* (witch). Zarate lives in Piedra Blanca, the rock formation on Pico Mountain above the town. Known for her acts of kindness, she once healed a poor *campesino* and gave him a sack of grapefruit. Tired of carrying it, he kept one and dumped the rest on the road. He regretted his laziness later. When he took the fruit out of his pocket, it had turned to gold. One year, Zarate fell in love with the provincial governor who spurned her. In anger, she turned him into a peacock. The lovelorn, cigar-smoking *bruja* is said to roam the countryside in the misty hills above Escazú. Listen for her laugh in the mist.

■ Airport & Beyond

If you don't want to stay in San José proper, or you are priced out of Escazú, there are many very good hotels in suburban or rural communities within easy driving distance of downtown and the airport. Be sure to read about the worthwhile accommodations, including boutique hotels, in Heredia, Alejuela, Atenas and Grecia.

Places to Eat

Con pan y vino, se anda el camino.
With bread and wine you can walk your road.
~ Spanish proverb

If nothing else, San José offers a great variety of eating establishments for all tastes. One of the most enjoyable things we do on our vacation is to find new "favorite" places to eat. Our personal criteria are quite flexible – sometimes we look for ambiance, service, the character of the place, or even the characters *in* the place. On other days we search for the best view or value. In every case we look for quality food. Don't forget, several of the hotels above have very fine dining rooms for all three meals; we also mention some good places to nosh in the *Nightlife* section. The following eateries should provide you with a pleasing combination of good food, good vibes and good times. We organized the restaurants by location to the downtown, not by cuisine or price, so read through them all before you decide. Most of the various *barrios* are within easy walking or cab distance. *Buen provecho!*

Except where noted, reservations are not usually necessary unless you have a large party or specific needs.

■ Prices

Although we made every effort to be as thorough, complete and as accurate as possible, things change in Costa Rica – sometimes *muy rapido*. Prices change faster than anything, so we indicate relative price levels here, using dollar signs.

Restaurant Prices	
Prices are for an average entrée	
$	less than US $5
$$	US $5-$10
$$$	US $11-$20

■ Center City

El Balcon de Europa *(Calle 9 near Av Central walkway, ☎ 506/221-4841, $$, no credit cards).* This is the oldest restaurant in town – and certainly the most ambient. It opened as an Italian eatery in 1909, and has an Old World atmosphere with wood paneling. Glass plate photos of San José hang on the walls, along with framed inspirational quotations such as "Truth is the child of time," portraits of Costa Rican presidents and postcards from *todo el mundo*. Green-and-white linen tablecloths and a wooden plank table stacked with Italian cheeses add flavor to the inviting dining room. We always eat here and we have never been disappointed (although the chef's rendition of Italian "gravy" is not ours!). The house wine is very pleasing. Closed Saturday because the Costa Rican-Italian owner/chef, Franco Piatti, is Jewish.

Tin Jo *(Calle 11, Av 6 & 8, ☎ 506/221-7605, $$-$$$).* In 1972, the Tin Jo, (the name means "The Best") restaurant opened to little fanfare – after all, there is a plethora of Chinese restaurants in San José. But by 1989 it had lived up to its name when a reader's poll in *La Nacion* called it the "Best Chinese Restaurant in Costa Rica." Originally set up in an old house that belonged to a colorful Tico character named Cucu Arieta – a money lender and the first person in Costa Rica to be kidnapped for ransom – in 1988 it expanded into the building next door, now decorated in a variegated Asian style. The delicious and diverse menu features dishes from China, Japan (including sushi), Thailand, Indonesia, India, Cambodia, the Philippines and Malaysia. The owner, Maria, who has an MBA from UCLA, and her German-born husband, Robert, met while working in refugee camps in Thailand and spent time all over Southeast Asia developing an appreciation for the food. Attentive wait staff, double damask linens and fresh flowers on the tables. This is one of San José's best restaurants, but with entrée prices averaging under US $10, it is not one of the most expensive. Treat yourself.

Manolo's Restaurante *(Av Central between Calle Central & 2, ☎ 506/221-2041, $).* Under the awning and through the portal to Manolo's place we go. This well-known hangout has a fast-food eatery on the ground floor that spills out onto the pedestrian part of Av Central with tables under a short awning. It's open 24 hours. The restaurant seating continues upstairs on two floors. The décor is heavy, with natural wood insets and tile work. The open kitchen with a charcoal grill serves up good Costa Rican dishes. Try their *churro*, a deep-fried pastry filled with a sweet condensed filling that is a like a milky butterscotch. It's high in fat and calories, but delicious!

Chelles Bar & Soda *(Av Central walkway at Calle 9, open 24 hours, $).* This favorite spot of local Ticos, at the end of the pedestrian walkway, is equally appealing to tourists. Despite the heavy traffic on the corner, the dining area is clean and neat, with red chairs, a white tile floor and hardwood walls. The food is cheap – try their generous Cuban sandwich. If you're looking for more, or a bit of greenery, try the **Restaurant Mariscar** next door. It has a large, open garden dining level past the bar. The menu is varied but, as the name implies, they specialize in seafood. Note that Chelles (and the place next door) was badly damaged in a fire; at press time, it remained closed.

Spoons *(Av Central walkway, as half-block east of the Plaza, plus various locations, $).* With a name like Spoons, this place sounds like it should be an ice cream parlor, but it is actually a popular eat-in restaurant chain (also offering takeout). Sit in the dining area and fill out your order on a pre-printed form. If you're not sure what things are, the helpful waiters will happily explain. They serve a varied menu (three meals a day), as well as bakery sweets and excellent coffee.

Triego Miel *(Calle 3 & Av Central, next to Cine Omni and opposite Lehmann's bookstore, $).* This pleasing large bakery has a tiny eating area, where you can breakfast, lunch or dinner on delicious pastries and sandwiches, both sweet and savory. Good coffee.

Restaurante Pollo Campesino *(Av 2 & Calle 7, $-$$).* Forget the Colonel, the best roasted chicken in the world is slow-roasted over a smoky, coffee-root wood fire. It's tough to walk past this small storefront and not succumb to your innermost need: food. The air is redolent of roasting meat, a golden aura coming from the wood fire. Great for take-out.

Rosti Pollo *(various locations, $-$$).* For sit-down dining, especially if you're ordering their wood-roasted chicken, Rosti Pollo is a good choice. They offer much more than their mouth-watering chicken specialty on the menu, with a good selection of traditional Costa Rican desserts, such as *tres leches*. A friendly ambiance, pretty good food and very good service. The eatery in Escazú is almost upscale.

Vishnu Restaurant *(various locations: Av 1, Calle 1 & 3; Av 6, Calle 7 & 9; Calle 1, Av 4: and Calle 14, Av Central & 2, $-$$)* is an appealing vegetarian restaurant chain that sometimes lines table dividers with exotic fruit and fresh vegetables. Opens early and closes late. Draws a crowd at lunch and dinner. Another veggie hotspot is **Shakti** (☎ *506/222-4475*) on Av. 8 and Calle 13. It offers good vegetarian lunches and great juice drinks in attractive surroundings.

Restaurant Del Mar (*Parque Morazán, $$*). Look for the Del Mar behind incredible stained glass bow windows in the former Gurdián Mansion, behind the iron gates facing the park. Handsome hardwood floors and an impressive genteel décor make up for its lackluster cuisine. Worth a meal just for the setting, the ambitious Del Mar is owned by the Del Rey Hotel, *sans les dames de la nuit*.

■ Barrios Amon & Otova

El Oasis (*Av 7 between Calles 3 & 5,* ☎ *506/255-0448, fax 222-3950, www.hotelsantotomas.com, $$-$$$*). This restaurant is, as its name implies, a welcome oasis of fine dining in a desert of continental cuisine opportunities downtown. With big windows with iron grills and walls that push right to the sidewalk on Av. 7, El Oasis is the pride of the Hotel Santo Tomas. The back of the restaurant faces the hotel's pocket pool and garden. The food here is particularly tasty and the presentation attractive. The dining room features quality china and linens and soft music. There is a bar for cocktails, open for dinner only. Decadent desserts.

JR's House of Ribs (*next to the zoo entrance,* ☎ *506/223-0523,$$*). Open seven days a week. Owner Ron Nickelson Rose, a WWII vet, barbeques up country-style ribs with a Cajun influence. The rest of the menu is definitely Tex-Mex and definitely delicious.

Café Mundo (*Av 9, Calle 15, Barrio Otoya,* ☎ *506/222-6190, $$$*) could be called "Tico yuppie" for its international nouveau cuisine, but it resists becoming too upscale and manages to maintain a bohemian ambiance and clientele. It's at once a hangout, gourmet restaurant and lively bar. Housed in a colonial mansion, Café Mundo meets all our criteria for a special night out, or just a quick bite and a drink. Put on your mock turtleneck and mingle with San José's eclectic elite.

La Cocina de Leña (*El Pueblo Shopping Center,* ☎ *506/255-1360, $$*). The "Firewood Kitchen" offers traditional Costa Rican country food in a rustic dining room with soft lighting. It's located near the center gazebo in the middle of El Pueblo's complex of hip bars and artsy shops (see page 138). Tico rico.

Amon Plaza Café (*Av. 11 & Calle 3, $$*). Under a large overhang up from the street, on a corner, is the new café restaurant of the Hotel Amon Plaza. It offers terrace dining and an interesting and generous all-you-can-eat buffet dining for breakfast, lunch and dinner. Pleasant, relaxed eating.

■ West End

Grano de Oro (*Calle 30, between Av. 2 & 4, ☎ 506/255-3322, $$$*). Rich food in a lush garden patio setting is a good start in describing the renowned restaurant in the Hotel Grano de Oro. And it's renowned for good reasons – with a gourmet international menu, excellent wines and a soothing romantic ambiance, the "Grain of Gold" is the perfect place for nocturnal gastronomic delights. We eat here every time we're in town and recommend it very highly; reservations suggested.

Lubnan Lebanese Restaurant *(Paseo Colon, between Calle 22 & 24, across from the Mercedes dealership, ☎ 506/257-6071, $$).* For vegetarians and lovers of Middle Eastern food, the Lubnan is the place to go for Lebanese cuisine. Their appealing décor in a storefront restaurant makes clever use of low-cost burlap and cork – à la 1960s. They offer very generous servings of delicious Middle Eastern food. Look for the John Cleese look-alike waiter wearing a traditional fez. Try the platter of hummus (chick peas), baba ghanouj (eggplant with garlic), labne (strained yogurt), and a huge falafel with lettuce and tomato wrapped in a thick pita. Say hello from us to Mario, the convivial host and owner.

Fellini Restaurant *(Av 4a at Calle 36 in the west end, ☎ 506/222-3520, $$).* Long before there was a Latino "La Vida Loca," there was an Italian "La Dolce Vita," or sweet life, epitomized by a sultry Anita Ekberg and mocked by Italian film director Frederico Fellini. This restaurant borrows his name and decorates its walls with posters and pictures from the heyday of the Italian film industry in the 1960s and 70s. With linen tablecloths and formal crystal, Fellini's offers a romantic meal (120 different dishes) inside a fascinating converted home from the 1920s. On the corner of the small Mata Redonda Park.

San José

Mil Sabores (*Paseo Colon, near the overhead walkway, $-$$*). The pagoda-styled "Thousand Flavors" may not offer quite that many – but it is a large and very popular Chinese restaurant with a brisk take-out business and daily specials. We are particularly fond of the chicken-vegetable soup with a side order of white or fried rice. Ahh-so good.

■ East End

Aya Sofya *(Av Central & Calle 21, ☎ 506/221-7185, ayasofiacr@yahoo.com, $).* Though the Turkish community numbers in San José can be counted on two hands, Mehmet Onuralp opened a traditional Turkish café. It's popular with people who enjoy ethnic foods. Aya Sofya is set in a one-story colonial building on a corner. Its interior is bright and modern, with large windows on two sides and handwoven Turkish rugs on the walls. The café has a large menu in Spanish and Turkish. Not fluent in either? Ask for the Turkish/English menu. Lunch and dinner specials. Doner kebap is a gyro, pide is thick homemade pita bread. There is homemade yogurt, and a glass of strong tea after your meal is on the house. Or have a sweet cup of Turkish coffee made in their samovar. Don't stir it, let the coffee grinds settle. Non-smoking. Open from 11 am to 10:30 pm.

Olio Pub (*Av. 5 & Calle 35, Barrio Escalante, ☎ 506/281-0541, $$*). Olio's is a hip tapas bar/restaurant, just like they have in Spain, with an intimate atmosphere and cubby-hole seating. A trendy crowd of students and young professionals flock here for the food, conversation and *vino*. Across the railroad tracks from Olio's is a lower-priced alternative, the **Bolero Bar**, serving typical Tico grub with beer and drinks, either inside or on a street-level terrace.

Parrillada El Churrasco Steak House (*Calle 31, Av 9 & 11, ☎ 506/225-0778, $$$*). We decided to eat here the first time on a whim, hoping for a good restaurant to celebrate the publication of a new edition of our *Adventure Guide to the Yucatán* on our last night in Costa Rica. What a great choice El Churrasco turned out to be! Like every Argentine *parrillada* it's a meat-lovers delight; but with excellent Chilean wines and an inviting hacienda décor, it offers a thoroughly enchanting evening meal. Open for dinner at 8 pm.

■ Escazú

Open seven days a week, and weekends too!
~ sign in a restaurant

You'll never go hungry in Escazú, where there seems to be a good restaurant on every corner. We'll skip the more predictable chain offerings such as Tony Roma's so you can explore the many *sabors* offered by smaller, local establishments. If it's a real dinner you're craving, look up the mountain above the *barrio* of Paco to **Le Monastere** (*☎ 506/289-4404, $$$*), which offers romantic French-inspired dinners in a multi-level former monastery featuring a panoramic vista of the towns below. They're open from 6 pm for dinner (closed Sundays) and offer a bar downstairs called **La Cava**. They've received mixed culinary reviews, but nothing beats the fabulous view of the valley.

Another eye-popping view – at a lower cost – is from **Tiquicia** (*☎ 506/289-5839, $$*), a rustic restaurant perched on the edge of the mountain overlooking the valley. We enjoyed a long evening here, eating typical Tico fare and watching a thousand points of light twinkle below. Take a taxi or follow the many signs on the road up through San Antonio. Call before you head out to make sure they are open.

Back down at **El Cruce**, where the road forks, almost hidden in the little strip mall behind a building, is a delicious Italian restaurant on two floors called **Sale e Pepe** *(☎ 506/289-5750)*. The pizza good and the place gets packed with hungry noshers who come for the reasonable prices, fresh pasta dishes plus good selection of wines. Popular for lunch and dinner.

Across the street is another quintessential Escazú eatery, **Quiubo** *(☎ 506/228-4091, $$)*. They serve good Tico food and generous drinks in a faux-rustic dining area. Late-night diners can wander over to the attached nightclub, **Requete**, or mellow out in the piano bar, **Q'tal**. In fact, there's seemingly no end to nightlife, finger-food kinds of bar/restaurants that cater to local ex-pat crowd and hip visitors. And for exotic ice-cream aficionados (or just lovers, like us), look for **Mondo Gelato**, located in Multicentro Paco, San Rafael de Escazú. They serve mouth-watering unconventional and traditional flavors.

Brit culture (Manchester style) mixes with the revolution at **El Ché Pub** in San Rafael. Have a drink and lament that Man United gets so much more coverage than City, even though Paulo Wanchope played there! Don't look at us, we root for the Argyle.

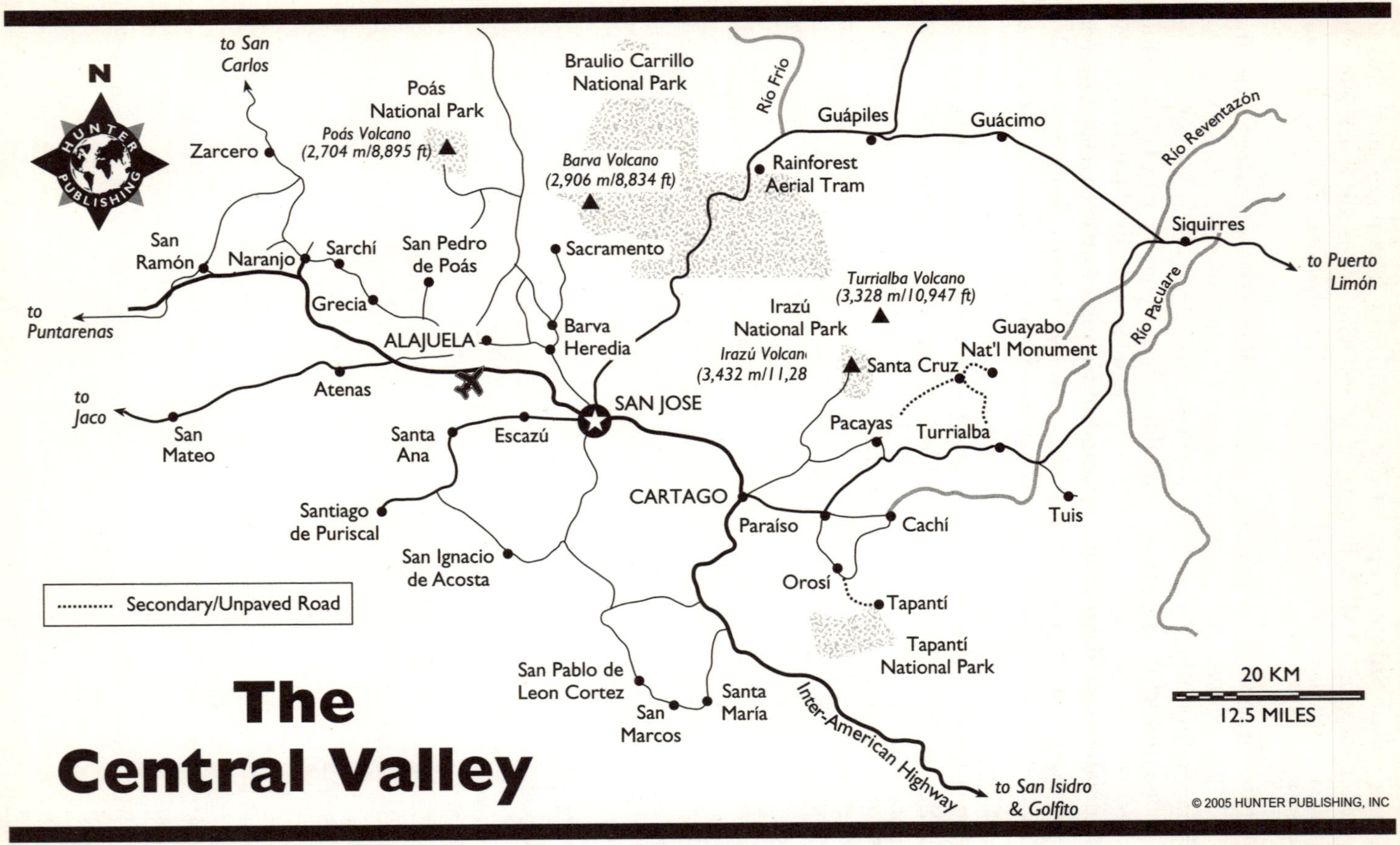
N
HUNTER PUBLISHING
to San Carlos
Zarcero
San Ramón
Naranjo
Sarchí
Grecia
to Puntarenas
Poás National Park
Poás Volcano (2,704 m/8,895 ft)
San Pedro de Poás
ALAJUELA
Atenas
to Jaco
San Mateo
Braulio Carrillo National Park
Barva Volcano (2,906 m/8,834 ft)
Sacramento
Barva
Heredia
SAN JOSE
Santa Ana
Escazú
Santiago de Puriscal
San Ignacio de Acosta
Río Frío
Guápiles
Guácimo
Rainforest Aerial Tram
Río Reventazón
Siquirres
to Puerto Limón
Río Pacuare
Turrialba Volcano (3,328 m/10,947 ft)
Irazú National Park
Irazú Volcan
(3,432 m/11,28
Santa Cruz
Guayabo Nat'l Monument
Pacayas
Turrialba
CARTAGO
Paraíso
Cachí
Tuis
Orosí
Tapantí
Tapantí National Park
San Pablo de Leon Cortez
San Marcos
Santa María
Inter-American Highway
to San Isidro & Golfito
Secondary/Unpaved Road
20 KM
12.5 MILES
The Central Valley
© 2005 HUNTER PUBLISHING, INC

The Central Valley

O grant me, Heaven, a middle state
Neither too humble nor too great
More than enough, for nature's ends
With something left to treat my friends.
~ *Imitation of Horace*, David Mallet, 1705-1765

What makes the Central Valley, or Meseta Central, surrounding San José so fascinating are the many smaller communities where you can absorb a real slice of Costa Rican life. We often base ourselves in the capital and take day-trips to explore and enjoy the attractions in nearby cities. If you have the time and self-discipline to relax and wander on a vacation (rather than jam everything in at once), towns like Cartago, Heredia, Zarcero and Orosí are like treasure chests to open. You'll find gems of shops, bars, restaurants, art galleries, flower and vegetable markets, little churches and community affairs. The next time you come to Costa Rica you may find you want to stay in one of these instead of San José. We look at locales southeast of San José first, then northwest.

IN THIS CHAPTER

East of Town

In the East my pleasure lies.
~ *Anthony & Cleopatra*, Shakespeare

■ Cartago

Cartago lies about a half-hour by bus (less by car) and 23 km/14 miles east of San José via the Inter-American Highway. It was the original capital of Costa Rica for 300 years until it lost the power struggle to San José. An additional obstacle to its capital ambitions is that the town rests at the base of the Irazú Volcano, which makes it prone to sporadic *temblores*, earthquakes, and volcanic eruptions. Its elevation is 1,440 meters/4,737 feet. When President Kennedy visited Costa Rica eight months before his assassination in 1963, Irazú welcomed his Alliance for Peace efforts with an eruption that blanketed Cartago and San José with thick volcanic ash. Because of that, Kennedy left earlier than planned... or perhaps it was the lack of a good hotel in town.

In the center of town, the ruins of the roofless **St. Bartholomew Temple** (most commonly known as **Las Ruinas**) dominate the central square, with its peaceful garden of bougainvillea and sweet pine trees. First dedicated in 1575, the church was destroyed in that same year by an earthquake. It was rebuilt, and a second severe earthquake destroyed it in 1910; legend claims it was divine punishment for the actions of an amorous priest. People from Cartago took that as an omen and left the church ruins standing.

Ruined church, Cartago.

Cartago's other famous church is **La Basílica de Nuestra Señora de Los Angeles**, where, on August 2 each year, thousands of Costa Ricans from all over the country make a pilgrimage to honor the black-skinned Virgin Mary, **La Negrita**. The town all but shuts down as pilgrims head to the *basilica*. Many walk on their knees as they near the church, a sign of humility.

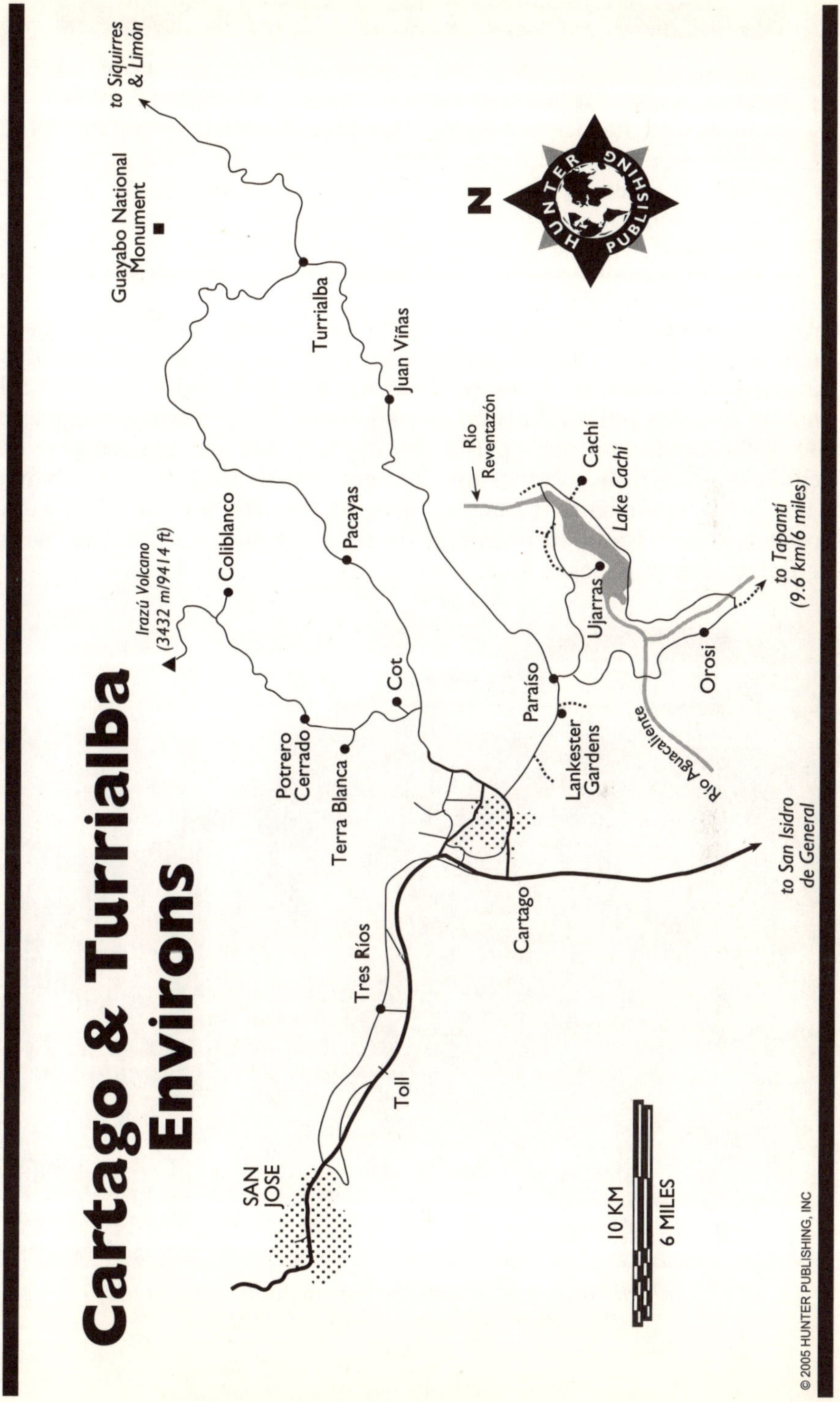
Cartago & Turrialba Environs
to Siquirres & Limón
Guayabo National Monument
Turrialba
Juan Viñas
Río Reventazón
Cachí
Lake Cachí
Coliblanco
Irazú Volcano (3432 m/9414 ft)
Pacayas
Ujarras
to Tapantí (9.6 km/6 miles)
Orosi
Cot
Paraíso
Potrero Cerrado
Terra Blanca
Lankester Gardens
Río Aguacaliente
to San Isidro de General
Cartago
Tres Ríos
Toll
SAN JOSE
10 KM
6 MILES
N
HUNTER PUBLISHING
© 2005 HUNTER PUBLISHING, INC

THE STORY OF LA NEGRITA

A peasant girl, Juana Pereira, discovered the statue of La Negrita, a black-skinned image of the Virgin, on August 2, 1635, on what was then the outskirts of Cartago. After a couple of miracles were attributed to the statue, Church authorities authorized a *basilica* to be built on the site of its discovery. The unusual Byzantine style of the Basílica dates from 1926 after the original church was badly damaged in the 1920 earthquake.

Follow the steps behind the altar down to the **Cripta de Piedra** to see the rock where Juana first found the statue. The adjoining room contains hundreds of miniature silver trinkets (mostly legs and arms) and all kinds of charms left by faithful parishioners. They signify alleged or wished-for healing miracles performed by the Virgin. The offerings even include trophies from soccer teams that supposedly won their match due to her help. Water bottled from a spring at the site, purported to work miracles, is for sale in surrounding shops. La Señora de Los Angeles is the Patron Saint of Costa Rica.

This historic photo shows Cartago in ruins after the 1963 earthquake and subsequent flood.

■ LEAVING CARTAGO

If you've taken a public bus to Cartago, you can ask for the bus stop that will take you to Orosí and Lankester Gardens, or opt to take a taxi in a different direction up to Irazú Volcano. If you're planning your visit a day in advance, be sure to get to Irazú early as clouds roll in by late morning (park opens at 8 am).

■ Irazú Volcano

The voyage of discovery lies not in finding new landscapes – but in having new eyes.
~ *Remembrance of Things Past*, Marcel Proust, 1871-1922

The drive up from Cartago to Costa Rica's highest volcano is magnificent. Volcanic eruptions over the millennium have blessed the slopes with fertile soil and farmers try to use every inch of land. The route is quite long, 32 km/20 miles of winding road above the valley, to a high altitude, 4,332 meters/11,260 feet above sea level. Bring warm clothes to layer as you get higher, it can get very cold at the summit. Each switchback offers a photo opportunity and there are numerous pull-overs. As you get closer to the summit, the temperature drops and the crops change from semi-tropical to temperate to cool/cold-weather produce.

Near the park entrance (US $8, open 8 am-3:30 pm) you'll find restrooms in a small coffee and gift shop next to the parking lot. The landscape up here is lunar; nothing but lichens and the odd plant grow in the gray cinder soil. Follow the path to the long wooden fence that separates you from the deep active crater. In its center is a round lake of bright pea-green water with a couple of *fumarolas* that emit hot sulfur gas and steam. Although it's not spewing lava now, eruptions of Irazú have been recorded since 1723.

We've been assured there are days when you can see both oceans from Irazú's summit, but unless you're very early and very lucky, chances are the Atlantic side will be too cloudy. Bring your camera just in case. Buses to the volcano direct from San José leave at 8 am weekends and holidays from Av 2 across from the Gran Hotel. They return at 12:30. Call **Buses Metrópli** *(☎ 506/272-0651)*. The US $5 price beats the cost of going with a travel agent tour.

A few restaurants are noteworthy on the mountain's slopes. The first, **Restaurant 1910** (open daily from noon to 10 pm), offers buffet dining with a generous selection of Tico and international dishes. The walls have old photos from the 1910 earthquake that destroyed so many of Cartago's old colonial buildings. Good eats.

Hotel Montaña is a family restaurant and old hotel with a great deal of character on the bend half way up the mountain. The grounds that surround its metal-clad exterior are manicured and the atmosphere inside the restaurant is like stepping back in time. The Victorian-era rooms upstairs provide adequate accommodations – basic but clean – with lots and lots of wood. This is a grand place that has retained its grandeur, but one that progress and people have passed by.

Closer to the summit is the funky and rustic overhanging restaurant/bar, **Linda Vista**. It's larger than it looks from the outside and its entry foyer, bar, walls and trim are covered with business cards, expired licenses, photo IDs and postcards from the thousands of visitors. Stop in for a typical Tico lunch or drink and enjoy the view. Try to find one of our old Press Cards hanging around.

■ Lankester Gardens

Over the long haul of life on this planet
it is the ecologists, and not the bookkeepers of business,
who are the ultimate accountants.
~ Stewart L. Udall, former US Secretary of the Interior

In 1917, 40-year-old British botanist Charles Lankester laid out 10 hectares/24.7 acres of gardens six km/four miles east of Cartago on the road to Paraíso that is now known as Lankester Gardens *(☎ 506/552-3247)*. Surrounded by tall pines, Lankester worked to catalog and develop new species of orchids, a delicate bloom treasured by flower aficionados.

Costa Rica is world renowned for its beautiful orchids. Exotic blooming orchids are the largest family of flowering plants in the world and they thrive in warm, humid climates. The many varieties have different styles of flowers, from tiny delicate petals running along the stem to bold blossoms and big thick green leaves. Costa Rica claims up to 1,500 different varieties, 75% of which are epiphytes.

EPI-WHATS?

Epiphytes (from the Greek for "upon plants") attach to host trees and gain their nourishment from airborne dust and rain. These are not parasitic relationships because they do not feed on their hosts. In the amazing world of nature, many tree orchids make use of variety-specific pollinators – bees, ants, hummingbirds, wasps and moths – for their fertilization.

By the time Lankester died in 1969, at the age of 90, he had cataloged 80 native species previously unknown in Costa Rica, and developed another 110 hybrid species. Upon his death the gardens were taken over by the

University of Costa Rica. Marked paths point out the over 800 different species on display as well as other flowering native plants. The gardens have orchids in bloom all year, but blooms are most prolific between February and May. Open 8:30 am-3:30 pm; US $5.

If you're on a public bus from Cartago toward Orosí (marked with Lankester as a stop) ask the bus driver to let you off at the gardens. Follow a short dirt road past a go-kart racetrack to the gardens Across the main road at the bus stop is a good restaurant with hotel rooms, **Casa Vieja** *(☎ 506/591-1165)*, in a Caribbean-style main house. At press time, it was closed for renovations.

A lavender orchid.

LOST LOVE AT CASA VIEJA

This former hacienda was the home of Doña Ana Cleto Arnesto, who had a passionate affair with General Francisco Morazán, one of Costa Rica's dictators. The Honduran General had seized power from Braulio Carrillo and pushed for Costa Rica to rejoin a Central American Federation. When he was in turn overthrown, 14 September 1842, he fled the capital and spent the night there with Ana. The next day he was arrested and executed by firing squad in San José's Central Park.

■ Orosí

I have been overcome by the beauty and richness of our life together,
those early mornings setting out,
those evenings gleaming with rivers and lakes below us,
still holding the last light.
~ *War Within and Without*, Anne Morrow Lindbergh, 1906-2001

Once you zigzag through the town of Paraíso, you'll be only a short distance from the green, dramatically beautiful Orosí Valley. The best way to appreciate it is to stop at the **Mirador de Orosí** public park, just outside of town. Climb the steps or wander the path but get to the hilltop

Coffee fields line Orosí Valley.

knob that offers a breathtaking vista of Orosí. Its steep valley walls are dark green with coffee bushes and shade trees and the Río Orosí, which feeds the larger Río Reventazón, meanders along its flat floor. Once you turn the corner after the Mirador, it's all downhill – steeply downhill – from there. Opposite the park entrance is **Restaurant La Cajeta**, closed Monday through Wednesday, with shows on Friday nights. Is that Celtic music you hear?

The peaceful little village of Orosí, founded in 1561 by Franciscan monks, boasts several minor attractions, a couple of spring-fed swim clubs, *balnearios*, as well as an important colonial church.

The **Iglesia de San José de Orosí** sits on the far side of the soccer field in the middle of town, at the bottom of a steep coffee-covered hillside. It is Costa Rica's oldest church in continuous use, built in 1735 to replace one that had fallen into ruin when villagers abandoned the town for 36 years after a plague decimated the population. It is styled like an adobe church you might find in Taos, New Mexico. The original large wooden roof, terra cotta floor and heavy wood pews have survived. The interior period paintings are by Mexican artist Miguel Cabrera.

Next door in the former Franciscan cloister is the **Religious Art Museum**, which was founded in 1980. It contains religious artifacts, paintings and a restored monk's cell. It's open 9-5 pm, closed Mondays, US $1.

Just south of town is **Balneario Los Patios**, one of two major swim clubs in town, fed by slightly warmed (10°C/50°F) springs heated by the Turrialba Volcano. Our favorite is **Balneario Termal**, up a side street next to the B&B Orosí Lodge (look for signs on the right). Each club offers a restaurant and several swimming pools, changing rooms, showers, basketball courts and picnic areas. We often made it here as a day-trip or an impromptu overnight from San José during weekdays. On weekends and school holidays the swim clubs get very crowded.

On your way in, just before the Mirador, you pass the pine tree-wooded driveway to **Sanchiri Lodge** *(a short distance south of the Orosí Mirador Park, ☎ 506/533-3210, www.sanchiri.com)*, a very appealing cabina hotel with an inexpensive restaurant overlooking the town and the valley. In fact, if you look back from town you'll see the name painted on their roof atop the hillside. The lodge has been in the same family for five generations. Their dining room serves good typical Costa Rican food at very reasonable prices. Plus, there's a wonderful view. Individual cabins come with hot water and balconies – very good for a stay. Take local tours with **Aventuras Turísticas Orosí** *(☎ 506/573-3030)*.

Good accommodations in town start with the **Orosí Lodge** *(next to the* balneario*, ☎ 506/533-3578, owww.orosilodge.com)*, a charming B&B that features new rooms with verandas and a pleasing café and snack bar. It's sandwiched between a plant nursery and the pool, and backs up to the mountainside. This wonderful little gem, originally built as a vacation home by a Canadian couple from Québec, is super clean and friendly. On the same street is **1/2 Libra** *(☎ 506/533-3838, $$)*, a new hotel, bar and restaurant with motel-like rooms. Breakfast is included in the room rate.

If you're on a slim budget, another choice could be **Montaña Linda** *(☎ 506/533-3640, www.montanalinda.com)*, a rudimentary backpackers' youth hostel that also offers conversational language courses (the cheapest we've found). They'll arrange a homestay with a Tico family for increased immersion, the key to learning more quickly. Another low-cost place to stay, located on a bend on the road into town, is **Las Torrejas** *(☎ 506/533-3534)*, owned by friendly Freddie Mora.

> **DID YOU KNOW?** Tropical plants are the gene source for many familiar foods, including coffee, chocolate, tomatoes, sugar, spices and bananas.

Two km/1.25 miles east of town (there's just one road that goes in and out) is a turn for a partially paved road (not so good in the rainy season) to Tapantí, a 6,000-acre reserve that has recently been joined with the Macizo de la Muerte Park to create the **Tapantí-Macizo de la Muerte National Park**. There's a cloud forest entrance on the Inter-American Highway, south of Cartago, but you may like the adventuresome "back way," along the river past the Río Macha Electrical Plant. The park has hiking trails and gets a significant amount of precipitation each year, even for Costa Rica, resulting in an impressive biodiversity. A good place to stay while investigating the many trails is **Kirí Lodge** *(☎ 506/551-3746)* near the park entrance. Five km/three miles short of Tapantí you'll pass **Purisil Park** *(☎ 506/228-6630, www.purisilpark.org)*, a trout farm. Here, you can fish in either of its three lakes or wander along the three hiking trails. Call in advance for an early morning or teatime guided birding tour. Restaurant on site. *Purisil* comes from an indigenous word meaning "clear water." The nearby picturesque town of the same name was once the coffee plantation of former President, Rafael Angel Calderón.

As you leave Orosí village and follow the main road over the single lane, steel suspension bridge, you'll come to the **Restaurante Río** in Palomo, a large, well-known restaurant with rustic décor and good fresh fish. If you continue on this road, it is the long way to Lake Cachí. On the way (be patient) is the rustic art studio known as **Casa del Sonador**, Dreamer's House *(☎ 506/533-3297)*. There, woodcarving brothers Miguel and Hermes Queseda carry on the art tradition of their father, Macedenio. Born in 1926 near Cartago, Macedenio was one of six children of a poor farmer. The family never had enough money to buy toys for their children so the boy carved his own. Eventually he became Professor of Art at the U of Costa Rica. Macedenio began teaching his sons his craft when they were very young and their current work is reminiscent of his well-known *primitivista* style. Using only scrap coffee roots and woods gleaned from the countryside and rivers, they create wonderful images of people in their beloved country – all without the use of stains, varnishes or sealants. "We feel very lucky to be here," Hermes told us. "It's a very special place." Find their rustic art studio on the right, just before a small bridge and stop sign. Only 100 meters/330 feet or so away is the popular **La Casona del Cafetal** *(☎ 506/577-1414)*, where they offer elegant lakeview dining, with an international and Tico menu, in an relaxing garden setting. Ask about horse rides and tours of their family-run coffee plantation and be sure to check out their cool jungle-painted restrooms.

This trip can be made into a giant loop by going around Lake Cachí and heading back to Cartago through Paraíso and Ujarrás – or clockwise from Ujarrás to Cachí to Orosí and back.

■ Lake Cachí & Ujarrás

Father, each of your sermons is like water to a drowning man.
~ anonymous churchgoer, quoted by the Episcopalian bishop of Chicago.

The more common way to Lake Cachí is not through Orosí, but straight through Paraíso for seven km/four miles through **Ujarrás**, a simple village with a great little ruined colonial-era church (1.5 km/.9 miles down a side road). Before the village stop at a dramatic lookout, **Mirador de Ujarrás**, at the top of the valley walls. It offers an impressive view of the valley and lake, with a quiet picnic area.

The colonial church in Ujarrás, built of limestone between 1575 and 1580, resides where an indigenous Huetar Indian fisherman claimed to have found a box with an image of the Virgin Mary. Unable to move the box, local people built a classic **Spanish Colonial church**, (Nuestra Señora de la Limpia Concepción) on the site. Over time, the spirit of Mary is said to have performed several miracles for the villagers, including a spontaneous bell ringing that warned them of an impending flood. In 1666 she is credited with helping a hastily organized militia repel an invasion of the Caribbean coast by the English pirate Henry Morgan. But she failed to save Ujarrás from a devastating flood in 1833 that damaged the village so badly the people abandoned it and rebuilt on higher ground. Today, the church is beautifully landscaped and makes a worthwhile visit for people like us who love gardens and old church ruins. Across the road is one of two spring-fed *balnearios* (swimming pools) open to the public.

The big hydroelectric dam of **Lake Cachí** is three km outside of Ujarrás on the northeastern side of the lake. From here the Río Reventazón ("Bursting River") spills down the mountainside to the Caribbean. Weekenders fish, hike lakeside trails and windsurf here.

Cool off in the highlands above the lake in the delightful **Cabañas de Montaña Piedras Albas** *(in US ☎ 305/279-2468, in CR ☎ 506/577-1462, http://cabinas.co.cr, $-$$)*. This airy hotel offers pleasing spacious one-bedroom cabañas perched on the mountainside overlooking the Orosí Valley. At 4,250 feet, the temperature averages 70°F.

■ Turrialba

Costa Rica currently has five active volcanoes and seven "sleeping" ones. Over the ions, more than 100 other volcanoes have gone extinct on land and another 100 can be found under the surrounding seas.

The agricultural town of Turrialba, at the foot of the dormant volcano of the same name, used to be a crossroads town on the only route from San José to Limón until the Guápiles highway replaced it in importance in 1978. Before the powerful 1991 earthquake destroyed the Atlantic Railroad, it also served as an important rail junction for shipping produce by train. Down now, but not out, Turrialba (pop. 30,000, 55 km/34 miles from San José) has reinvented itself as the whitewater capital of Costa Rica, thanks in part to the interest of eco-adventure tourism. To get here by car, go to Cartago, then Paraíso and make a left at the park in the center of town. If you're going to and from the Caribbean side, consider taking this scenic route at least once.

Adventures

Proximity of the Río Reventazón, Río Pacuare and Lake Cachí make Turrialba a good, inexpensive base to kayak, raft and windsurf. See the *Adventures on Water* section in San José (page 134) for more details. If you're staying in town, rafting prices are cheaper, even from the companies in San José. In addition, there are some local operators: **Loco's Tropical Tours** *(☎ 506/556-6035, www.whiteh2o.com)*; **Costa Rica Rios** *(☎ 506/556-9617, www.costaricarios.com);* and **Tico River Adventures** *(☎ 506/556-1231, www.ticoriver.com).*

Equally thrilling is canyoning, an active adventure that involves rappelling down a waterfall. **Explornatura**, a part of Costa Rica Canyoning (☎ *506/556 4032, www. costaricacanyoning.com*) offers raft trips and rappelling down four waterfalls, zip lining down four canopy cables through the forest, descending a tall tree and walking across a 132-foot hanging bridge – an unforgettable adrenaline rush.

Since you're in the area, perhaps you'd like to hike the dormant Turrialba Volcano, explore the Guayabo National Monument, or just wander free.

BASEBALL TRIVIA

Turrialba claims to be the town where they made the baseball that Barry Bonds hit out of the park to gain the major league record of 73 home runs in one year. Babe who? For anyone who is a real fan, note that Josh Gibson hit 75 in 1931 in the Negro Leagues.

On the road east of town heading toward **Siquirres** there's a swimming pool, **Balneario Las Americas** (US $1), and also **CATIE** (**Tropical Agronomic Research and Education Center**), a major hemispheric educational research facility on 2,000 acres. The grounds have several pleasant trails (great for birdwatchers); a hassle-free entrance is made easiest by arranging a low-cost guided tour in advance *(☎ 506/556-6431).*

Farther east of town in the small town of **Pavones** is **Parque Viborana** *(☎ 506/538-1510)*, an educational serpentarium and wildlife rehabilitation center. Owner Minor Camacho's specialty is snakes, from which he extracts venom for study and medicinal purposes – quite a sight! Don Minor worked for years at the university of Costa Rica until setting up his own center. Large terrariums house Costa Rica's most deadly reptiles and there's a walk-in cage to get close to boa constrictors. Open daily, and well worth the US $5 admission.

Guayabo National Monument, 19 km/11.5 miles northeast of Turrialba, is Costa Rica's most auspicious pre-Columbian archeological dig. Pleasant, natural and quiet, the cobbled streets, ruined aqueducts, bridges, and rocky building foundations that have been uncovered from an early indigenous settlement make for a serene and completely untouristy attraction. A path resembling a Maya *sacbé* (a raised road believed used for holy processions) can be distinguished pointing toward the top of the Turrialba Volcano in the distance.

Archeologists believe the site was inhabited by as many as 10,000 people from around 1000 B.C. until 100 years or so before the Spanish landed.

No one knows why it was abandoned. Admission is US $6. Guides are available and camping is permitted. The only convenient connection to the ruins from Turrialba is on Sunday, when a bus leaves the main terminal at 9 am and returns at 4 pm. By car, watch for the park sign on the left (it's a rough road), east of downtown. In the general area (but not open to the public because it's being excavated by the National Museum) is **Angostura Archeology Area**, an 11,000 year-old camp used by pre-Columbian peoples. Please e-mail us if you find this is now open.

Turrialba Volcano is now long dormant. The name Turrialba means "White Tower" in old Spanish, and the volcano was presumably named after the columns of steam that once rose from its core. With much effort you can hike to the top and overlook its three craters. To get here, make your way to the little town of **Santa Cruz** from Cartago or Turrialba, and turn at the Bar Cañada. The last 12 km/seven miles of this road must be done on foot. Perhaps arrange your trek or horseback ride with one of the local tour companies.

Places to Stay & Eat

Our favorite place to stay in Turrialba is **Turrialtico** *(8 km/5 miles east of town, ☎ 506/538-1111, www.turriatico.com, 14 rooms, restaurant, $$)*, a venerable hotel and restaurant on a hill overlooking the valley on one side and Río Reventazón and its new dam on the other. Since 1968 the García family has run this rustic, two-story wooden lodge that has been a favorite of rafters and tourists who return year after year. There's nothing fancy about the small rooms, but the wonderfully delicious restaurant (filled with carved artwork) downstairs draws many locals at lunch or dinner. The best room is #1.

Hotel Prices	
Prices are per night for two people	
[No $]	less than US $20
$	US $20-$40
$$	US $41-$80
$$$	US $81-$125
$$$$	US $126-$200
$$$$$	over US $200

Since 1984, the brothers in the García family have run the **Pochotel** *(☎ 506/384-7292, restaurant $-$$)*, much higher on a nearby mountain, which offers truly impressive panoramic views from its private *cabinas*. All accommodations have good views (best is #10), and the new cabins are especially appealing inside, with warm cedar woods for cool evenings.

Restaurant Prices	
Prices are for an average entrée	
$	less than US $5
$$	US $5-$10
$$$	US $11-$20
$$$$$	over US $20

Eighty percent of the buildings here were made from wood left as "seconds" by loggers. The entire family combines art, ecology and accommodations very well.

More luxurious digs can be had in **Casa Turire** *(☎ 506/531-1111, www.hotelcasaturire.com, $$$-$$$$)*, an elegant new plantation mansion/hotel down in the farmlands next to the Reventazón dam and Angostura Lake. The new Swiss owners of this magnificent property have made appealing improvements to the outdoor patio dining area and small figure-eight pool, so an extended stay here is both comfortable and desirable. The bedrooms are all first class, with fine linens and drapes, marble baths and lovely wooden furniture. The international dining offerings are of gourmet quality. Even if you're "roughing" it on the Pacuare River or hiking the volcano, you can relax in comfort at this boutique hotel. A short stretch of very pot-holed road gets you there from the main route out of town.

A little out of town is **Guayabo Lodge** *(Santa Cruz, ☎ 506/538-8492,www.guayabolodge.com, $$$, breakfast included)*. All 20 rooms here have private baths and simple but attractive wrought iron furniture. Second- and third-floor quarters have panoramic views. This place is spotlessly clean.

West of Town

Out where the handclasp's a little stronger,
Out where the smile dwells a little longer,
That's where the West begins.
~ *Out Where the West Begins*, Arthur Chapman, 1873-1935

■ Heredia

Just 11 km/6.8 miles northwest of San José is the pretty city of Heredia (air-REY-dia), nicknamed the City of Flowers. Founded in 1706, Costa Rica's fourth-largest city was first known by its Huetar Indian name, Cubujuquí, but that was soon changed to Heredia in honor of a Guatemalan president, Fernadez de Heredia. It is home to four Costa Rican universities. There are good road signs into town from the highway next to the airport. **Buses Heredianas** *(☎ 506/261-7171)* get you there from San José on Av 2 between Calles 10 and 12.

The town's central park is filled with trees and is a great place to stroll or sit on a bench and watch people go by, or to attend the band concert every Sunday at 10 am. To the east of it is the impressive **Church of the Immaculate Conception**, which took 30 years to build (1767-1797). The

Water crashes down through the forest at La Paz Waterfalls.

wait was worth it – its massive towers and walls proved stronger than the earthquakes that have rocked the country since it was built. It has stained glass from Europe and church bells from Cuzco, Peru. Nearby is another imposing fixture, **El Fortín**, a brick fort with peepholes built backwards. Good thing it never saw action.

The **Casa de Cultura**, just north of the park, is the 1843 home of ex-president Alfredo Gonzalez. It re-creates the life and home of a coffee baron, the life of peasants during that time, and it also presents local art shows. Another architecturally interesting building here is the 1915 neo-classical, **post office** (*correo*).

Heredia doesn't have a thriving flower market as its nickname suggests, but does have a wonderful fruit and vegetable ***mercado***. It also boasts a number of Spanish-language schools and the **University Nacional**, a rival to UCR.

Outside of Heredia, the coffee town of **San Rafael de Heredia** boasts a contemporary gothic-style church (built in 1962), inspired by Notre Dame cathedral, with spires visible from far away. A bit farther on is **Monte de la Cruz Park**, a privately-owned preserve with hiking trails and sweeping vistas.

On the way from Heredia to the historic town of **Barva**, you'll find the **Café Britt** showroom, office and store (*☎ 506/260-2748*). Tours of the roasting plant are offered. There's a good cafeteria and gift shop that are open to the public without taking the tour. What delicious coffee! Open 9-4 daily. Most visitors arrive with organized tours originating in San José; check with a local travel agent.

San José de la Montaña is an idyllic little village full of 18th-century wooden houses. Originally, it was established as a health resort and people came here to breathe the fresh mountain air. Nine km/5.6 miles beyond that you can improve your health by tackling some of the little-known hikes from **Sacramento**, a village gateway to the **Barva Vol-**

cano, now a part of **Braulio Carrillo National Park**. Take the "de la Montaña/Sacramento" bus from behind the *mercado* in Heredia. Hike two hours to the park entrance then to the peak and down into the lake of the forested volcano. Let the ranger station know you're hiking.

La Paz Waterfalls near Vara Blanca are easily accessible from Heredia, as is the **Poás Volcano** (see page 184).

Places to Stay & Eat

In Heredia you can choose **Hotel Valladolid** *(Av 7, Calle 7, ☎ 506/260-2905, fax 260-2912, valladol@rasca.co.cr, breakfast included, $$-$$$)*, an elegant Spanish-influenced luxury hotel. **Apartotel Vargas** *(750 meters/2,467 feet north of Santa Cecilia College, ☎ 506/237-8526, fax 260-4698)* is not as fancy, but offers apartments and is recommended for extended stays. Budget-conscious travelers can try **Hotel Las Flores** *(west of the central market, ☎ 506/237-1616)*, which is clean and neat.

Hotel Prices	
Prices are per night for two people	
[No $]	less than US $20
$	US $20-$40
$$	US $41-$80
$$$	US $81-$125
$$$$	US $126-$200
$$$$$	over US $200

Three hotels in the heart of downtown offer excellent value: the largest is **Hotel America** *(☎ 506/260-9292)*, then **Hotel Ceos** *(☎ 506/262-2628)*, 100 meters/330 feet from Central Park, and finally **Hotel Heredia** *(☎ 506/260-9292)*, two blocks north of downtown. Prices run from US $45 (America) down to $15 (Heredia). All have restaurants and cable. See www.hamerica.net.

Hotel Chalet El Tirol *(☎ 506/267-6222, fax 267-6229, www.chalet-tirol.com, $$$)* offers a Swiss experience in a misty cloud forest above the city, near Monte de la Cruz Recreation Area. The ride uphill through a cool, heavily scented evergreen forest is exhilarating – and it seems perfectly natural to encounter the chalet's alpine cottages and restaurant at 1,800 meters/5,900 feet above sea level. The vine-covered chalets are cute; some have a slight musty smell, but are charming nonetheless. One of the best features is the French restaurant. It was awarded the Châine des Rotisseurs from the French Gourmet Society and is open daily at noon. Hiking trails lead into the cloud forest and guided tours are available.

The best accommodations in the area is **Finca Rosa Blanca** *(Santa Barbara de Heredia, north of town, ☎ 506/269-9392, fax 269-9555, www.finca-rblanca.co.cr, hot tub, free-form pool, $$$$)*. The "White Rose Farm" is a very special, luxury country inn with gorgeous, individually themed rooms. The main house is a contemporary pueblo design, with curved walls covered by original artwork. For the ultimate in sensuality, check out the fantastic, waterfall-themed bathroom in the master suite. Dinner is by reservation in their intimate dining room, and breakfast is included. Seven units in the main house and two private villas, all with superb views of the countryside.

Many visitors to Costa Rica base themselves out of suburban Santo Domingo at the **Hotel Bouganvillea** *(☎ 506/244-1414, fax 244-1313, www.bougainvillea.co.cr, restaurant, pool, tennis, cable, $$$-$$$$)*. The Bouganvillea is a boutique hotel with high standards of service, pleasingly large rooms and gorgeous gardens. The pool and tennis courts require a hike through the peaceful flower gardens. This is also a popular hotel for first and last nights in Costa Rica and it's near INBio Park (see page 133).

Le Petit Paris *(Calle 5, Av 2, ☎ 506/262-2564, $$)* features French food, especially crêpes, in two fascinating dining rooms – one indoors decorated with art posters and paintings, the other a relaxing covered garden setting. Art shows by aspiring Central American painters are sometimes featured. This recommended restaurant lies in the heart of Heredia at the edge of the college campus.

Restaurant Prices	
Prices are for an average entrée	
$	less than US $5
$$	US $5-$10
$$$	US $11-$20
$$$$$	over US $20

Bar food can be found at any number of small eateries around the university. One block north is **Fresas** *(☎ 506/262-5555)*, featuring good, inexpensive Tico/Italian food and a wood-burning brick oven for pizza. More Tico food is served at **Las Tinajas**, downtown. A great compliment to a day spent in Heredia is the outdoor, Euopean-style coffee shop, **Café Heladería Azzurra** at the Casa de Cultura, a little north of the peaceful Central Park. *Mucho* types of coffees, sandwiches and homemade ice cream.

La Luna de Valencia *(☎ 506/269-6665)*, in San Pedro de Barva on the road from Barva to San Rafael, is a trendy Spanish/Mediterranean restaurant that is attracting a diverse crowd of academics, intellectuals, food aficionados, foreign workers, expats, bohemians, couples and families. It serves up political rhetoric and literary discussions, along with outstanding *paellas*. The tradition of open discussions is called *tertulia* or *charla* (chat). Reminiscent of Greenwich Village in the old days. Closed Mondays.

■ Alajuela

Four-wheeled oxcarts and four-wheel-drive Land Rovers punctuate Costa Rica's second-largest city. Just three km/1.8 miles from the international airport on the slopes of the Poás Volcano, Alajuela's elevated position makes it slightly warmer and sunnier than San José. And, although a big city in its own right, Alajuela is quieter and calmer as well. Buses leave every few minutes for the airport and Alajuela from Av 2, opposite La Merced Park. The red ones are TUASA and the yellow ones are SWA.

Huge mango trees in the central park reflect the pride Alajuela takes in the delicious tropical fruit. It even hosts a nine-day **Mango Festival** in July *(contact the Tourist Board for exact dates, ☎ 506/223-1733)*. But the city's most famous citizen is cause for a celebration on April 11 of each year – **Juan Santamaría**, the little drummer boy who helped defeat the invasion of William Walker, was born in Alajuela. His birthday brings parades with brass bands, parties, dancing in the streets and drinking in the bars. Juan's statute is in a park that bears his name, two blocks south of Parque Central. **Central Park** is a good place to start nosing around. It has a big squirrel population in the trees if you don't get enough of them in your home birdfeeders. The church **cathedral** (restored after the 1991 quake) and garden grounds on the east side are open to the public and, on the west side, is the **Casa de Cultura**, in a municipal palace that dates from 1914. It has lots of local information. On the north side you'll find the **Juan Santamaría Museum**, which contains art galleries, exhibits from the 1856 War, orchid gardens and clean restrooms. The **Museo Historíco Cultural**, in the old city jail, is on Calle 2, Av 2. The **Central Market**, two blocks west of the park, dates from the 1930s and is a good place to shop, especially on Saturdays when the market becomes like a fiesta. Try some of the many exotic fruits for sale.

One of the best coffee plantations in Costa Rica is 15 km/50 miles outside town on the road to Poas, near La Fraijanes. Watch Starbucks' coffee grow before you drink it at the **Doka Coffee Tour** *(☎ 506/449-5152, www.dokaestate.com, reservations required, see page 132)*.

On weekends, *Josefinos* head to **Balneario Ojo de Agua** (swimming park) in San Antonio de Belén (south of San José's airport, not in the town of Ojo de Agua, in case you find it on the map).

A treat for the eyes is an evening performance of Spanish Lippazaner stallions in an "equestrian fantasy" of choreographed riding with dressage, quadrilles, fancy costumes and thrilling horsemanship by both men and women. Ask a local travel agent about performances at **Rancho San Miguel**, in Alajuela's La Guácima suburb *(www.ranchosanmiguel.co.cr)*.

Also in La Guácima is **The Butterfly Farm** *(☎ 506/438-0400, www.butterflyfarm.co.cr)*, a popular place to see and learn about some spectacular butterflies native to Costa Rica (as well as stingless bees). They offer a two-hour guided tour through enclosed tropical gardens. The **Zoo Ave** *(☎ 506/433/8989, www.zooave.org)*, which means "bird zoo," is in La Garita, on the road from Alajuela to Atenas. Its aviaries house a wide variety of birds and also serve as rehabilitation stations for a diversity of native animal life intended for release into the wild. Open daily, 9-5; US $10. Also nearby is **Orchid Alley** *(☎ 506/487-7086, orchid@sol.racsa.co.cr)*, a nursery of more than 100,000 delightful, colorful epiphytes. They offer gift-wrapped orchids mailed worldwide from US $50.

Places to Stay & Eat

Lots of lodgings are available around Alajuela's Central Park and museum. **Hotel 1915** *(275 meters/904 feet north of Central Park, ☎ 506/441-0495, breakfast included, $$)*, built in 1915 and furnished in period, is the best of the bunch. Quiet, with a tasteful ambiance, this newly opened hotel is our absolute in-town favorite. *La dueña es muy amable* (the female owner is very friendly).

A good value is **La Guaria Inn** *(100 meters/329 feet south and 125 meters/411 feet east of the cathedral on Av 2, ☎ 506/440-2948, breakfast included, $)*, a B&B with pleasing rooms. The **Hotel Miraflores** *(a half-block west of the Santamaría Museum on Av 3, ☎ 506/441-7116, breakfast included, $)* has seven rooms. **Hostel Mango Verde** *(diagonal from the Santamaría Museum on Av 3, ☎ 506/441-6330, www.mangoverdehostel.com, $)* is budget but is also a good value. It has a small lounge.

East of the park is **B&B Vida Tropical** (*100 meters/330 feet east, 300 meters/1000 feet north of the hospital, across Academia de Natacíon, % 506/433-9576, www.vidatropical.com, $)*. The "Tropical Life" is pleasant and popular; proof is pinned on its walls in the form of postcards and pictures of guests. It has a lounge and kitchen for guest use. **Hotel Caña Dulce** (*Av 5 & Calle 8, ☎ 506/440-8619, ask for Mitch*) is a B&B that offers five rooms in a house full of musicians. It's homey and acoustic, with a Jacuzzi and small restaurant. Perhaps the best place to stay in Alajuela, just outside the downtown, is the **Pura Vida Hotel** *(☎ 506/441-1157, www.puravidahotel.com, restaurant, $$-$$$)*, a very special and delightful colonial-style hotel set on a former coffee farm (*finca*). It features a large main house known as the Casa and four private casitas set in what guests refer to as a secret garden. There's a studio with a view and rooms in the main house are enhanced with a wonderful garden restaurant. It's a great place to begin or end your country wanderings, with free airport pick-up.

On the way up the mountain to Poás Volcano, several hotel properties are also well worth a look. American-owned **Hotel Buena Vista** *(Pilas, 8 km/5 miles north of Alajuela, in US ☎ 800/506-2304, in CR ☎ 506/442-8595, fax 442-8701, www.hotelbuenavistacr.com, $$$)* has comfortable rooms with *buenas vistas* – great views – of the valley from their balconies. Across the street is the appealing Mediterranean restaurant, **Vina Romantica**, open from 5-10 pm. Five km from Alejuela is **Hotel Orquídeas Inn** (*on the old route to Grecia and Poas, % 506/433-9346, www.orquideasinn.com, $$-$$$*), a congenial hotel that features beautiful grounds, attractive rooms and personal attention. Geodesic dome rooms or apartment accommodations. On a bus line.

Alajuela's cuisine is mostly *Tico típico*, so choose a simple place that appeals to you. The **Ambrosia Café** (*Av. 3 & Calle 6*) is a pleasant outdoor Italian café, open from 9 to 6, closed Sundays. Veggie types can nosh at **Mixto Vegetariano**, on the second floor of a building on Av Central, 50 meters/150 feet west of the park. Or the popular **Soda Alberto**, one block north of Llobet's Department Store (check out the snack bar on the fourth floor). Alberto's offers generous veggie dishes to Tico taste at very reasonable prices. At **Gabriel's Soda** (*Av. 3, between Calle 4 & 6*), they speak English and serve good food, including *casados* at very good prices. Always dependable is **Spoon's**, open all day until evening, and the **Trigo Miel Bakery Café**, across from Santamaría Park, which features sweet treats. Check out the **Cafeteria Lemirage** (*125 meters/400 feet west of the Juan Santamaria Museum*), where they serve a traditional dish, *tamal asado*, as well as other tasty Tico cuisine. If in-town is too much for you, head up to **Hotel Pura Vida** for their quiet garden *rancho* restaurant.

■ Poás Volcano

The laws of nature are written deep in the folds and faults of the earth.
~ John Joseph Lynch, President, NY Academy of Sciences, 1963

As volcanoes go, Poás is quite a good one – and a definite must-see on your trip. This broad, tall, semi-conical mountain is a part of the Continental Divide – its eastern slope drains to the Atlantic, and the western slope goes to the Pacific. Poás is only 60 km/37 miles from San José, but a world apart in atmosphere. Like Irazú, it has a fine paved road to the top.

As you climb its broad slopes, plantations with rows of coffee give way to strawberry and onion farms and finally to green dairy pasture before the 5,319-hectare/13,138-acre park covered by moss, bromeliads, epiphytes and wild gorse. It gets much colder as you approach the 2,708-meter (8,883-foot) summit. A new welcome center greets visitors at the path to the 1,320 meter-wide (that's 4,342 feet) active crater (reportedly the widest in the world) that surrounds a deep, turquoise-blue lake. A wooden viewing platform lets you look down at the surreal landscape in the crater – great for photos – and the windswept lake.

Poás is semi-active, with several vents that send steam and gas rising from the edge of its crater lake. We once watched a volcanologist in his "spacesuit" taking samples deep in the crater where a vent steams almost continuously. Its last serious eruptions were in 1989, when a sulfur blow caused acid rain that destroyed 75% of Grecia's coffee crop. In 1994 fear of an eruption caused the evacuation of tourists.

The crater that holds the lake is the fifth vent created during Poás' active life. Hiking trials fork off from the paved road, leading to **Lake Botos**, a deep lake formed in the third crater and surrounded by thick vegetation. Clouds sweep across your path suddenly, enveloping you in thick fog and blowing away just as quickly with the breeze. Dress for the cold, but use layers, as it can also get warm in the bright sun.

Poás is a sleeping giant that offers visitors an educational and fascinating day-trip with a truly beautiful ride to its summit and lots to do in the area. The fast way to the top from San José is straight through Alajuela from the highway near the airport. Follow the road through Pilas to Poasito and up you go. There are signs, so you won't get lost (if you're going downhill, you're going the wrong way).

Since you probably rushed up to arrive before clouds socked in the mountaintop, consider going back down a more scenic route by turning left at Poasito through Vara Blanca, then down to Birrí, Barva to Heredia. This route passes close to **La Paz Waterfalls** (see page 132).

Places to Stay

The last hamlet on the way to up the volcano is **Poasito**, 11 km/seven miles from the summit, where a right turn leads to La Paz Waterfalls and, eventually, Heredia. A kilometer or less before the turn, on the left, is a cute little restaurant with lace curtains, **La Casona de Doña Julia** *($)*, our favorite place for *batidos*, especially when strawberries are in season.

Hotel Prices	
Prices are per night for two people	
[No $]	less than US $20
$	US $20-$40
$$	US $41-$80
$$$	US $81-$125
$$$$	US $126-$200
$$$$$	over US $200

On our return trip back down we usually turn left here for a scenic route through Heredia back to San José. At the junction is the meaty **La Churrasco Restaurant and Lodge** *(☎ 506/482-2135, $$)*, a pleasingly rustic and popular stop. Follow our car wheels on this route to stay at the stunning **Poás Volcano Lodge** *(on the road from Poasito to Vara Blancas, ☎ 506/482-2194, fax 482-2513, www.poasvolcanolodge.com, $$)*, built by an English family as the principal house for their dairy farm on the cloud-shrouded slopes. Stone walls, reading areas and a sunken fireplace in this large but intimate home make it feel as if it's your own comfortable house. The simple, immaculately clean and appealing rooms are decorated with souvenirs from travels around the world. The sociable owner converted the former stables to wonderful intimate rooms and expanded the accommodations in 2005 to include individual chalets. This is one of favorite lodges (and he makes his own wine, too!). Say hello to Michael for us.

As you continue you'll come to the turn for **La Paz Waterfalls** (see page 132), a short distance down a road that would take you across the surprisingly small country to Caribbean side and Puerto Viejo de Sarapiquí. It's a fascinating drive. However, if you're hungry for fine food, keep going 400 meters/1200 feet east of the junction to **Restaurant Colbert** *(☎ 506/482-2776, $$)*, nestled on a rise above the road, overlooking dairy farms on the volcano's slopes from wide picture windows. It has great views, linen napery, excellent presentation and superb French food from chef Frances. Another spot worth considering for a bite to eat or an overnight in an individual hillside rustic chalet room (with romantic wood burning stoves) is **Mi Casita Restaurant & Cabanas** (*Vara Blanca, ☎ 506/482-2629, $*). The owner, Emily, speaks English because she is English.

You can do this route in reverse order if you come up to Poas from Heredia. Hold the book upside down.

Coffee Country

Good coffee is like friendship: rich and warm and strong.
~ Inter-American Coffee Bureau advertisement

The agricultural area west of San José is rich farmland and, because of its climate and altitude, a great coffee-growing region. The towns scattered along this route are friendly, quaint and un-touristy – yet with much for tourists to see. We often make this trip – Alajuela, Grecia, Sarchí, Naranjo, then either Zarcero or San Ramón or both – into a day or overnight journey to wander the country lanes. You can continue to the scenic route to La Fortuna and Arenal. Reach these towns by side roads or go quickly via the Inter-American Highway. Although not on a direct driving loop through coffee country, we include the town of Atenas here, which is located southeast of Grecia on one of the routes to the Pacific Coast, because it is very attractive and because it doesn't fit anywhere else.

COFFEE PRIMER

Coffee should be hot as hell, black as the devil,
pure as an angel, and sweet as love.
~ Charles Maurice de Talleyrand, 1754-1838.

La Carpintera Hill coffee offers an exquisite balance between body, aroma and acidity. It's grown on a hillside at 1,200-1,400 meters/3,947-4,605 feet in the Tres Rios zone of the Central Valley. The combination of fertile soil and year-round pleasant climate here produces a high quality coffee. Harvest: November through February.

From the Tarrazú zone of the Talamanca mountains, come **Café Atarazu** coffee beans, grown at an altitude of 1,200-1,750 meters/3,947-5,757 feet in an area known as Los Santos. When ground, they produce a coffee with a very rich body, pleasant aroma and excellent acidity. Harvest: November to February.

Café el Gran Vito is grown in Coto Brus, a remote southern area colonized by Italian immigrants in 1941. Surrounded by mountains and forests at 800-1,200 meters/2,632-3,947 feet, their coffee bean is light and delicate, yet has a strong taste. Harvest: October thru December.

The beautiful, peaceful province of **Heredia** is home to the oldest coffee plantations in Costa Rica. **Cafe Zurquí** is grown close to the Volcano Range on the slopes of Zurquí Hill at an elevation of 1,200-1,500 meters/3,947-4,605 feet. The beans are high in acidity and very aromatic. Harvest: November thru February.

The lush, fertile green **Reventazón River Basin**, near Cachí, is where the rich **Café Ujarraci** beans grow. Nurtured by the pleasant Atlantic climate and rich soil, the beans have unique qualities much prized by Europeans, Americans and Japanese alike. Harvest: October through December.

Valle de General is the fastest-growing agriculture area in Costa Rica. It has an average daily temperature of 24°C, which means the coffee bushes flourish. **Café Buenavista** delivers a full-bodied, pleasantly acidic brew with a delightful aroma. Harvest: October through December.

■ Atenas

Atenas, the Spanish word for Athens, is a semi-rural farming community with what *National Geographic* named as the "Best Climate in the World." You'll stop here if you take the Tico Train Tour from San José to Puntarenas. Besides its great climate, Atenas' claims to fame include the

Monument to the *Boyero* (ox cart driver), an earthquake fault that runs through town, a stuffed grapefruit dessert specialty called *toronja rellena*, and spectacular, twisty mountain roads in and out of town. Rent a manly motorbike at **Motorcycles Costa Rica** *(☎ 506/446-7237, www.motoscostarica.com)* or go on one of their popular guided tours. Eat facing the main park at **K-puchinos**, an Internet café with a decent menu and good coffee, *por supuesto*.

We thoroughly enjoyed staying north of town, in Santa Eulalia on the road to Orotina, at **El Cafetal Inn** *(☎ 506/446-5785, www.cafetal.com, pool, breakfast included, $$)*. The congenial owners and hosts, Romy and Lee Rodriguez, built a modern hacienda mansion surrounded by pretty gardens and coffee plants (their "La Negrita" roast, named after their daughter, is excellent). Photos of the house's construction show that the land around it was once cleared for cattle, but hard work and a few years have made it verdant again. It's a pleasure to relax here. El Cafetal is close enough to the airport to make it a good place to get your feet wet in the country or to stay for a last goodnight. Good food, too (see below).

The road through Atenas to Orotina and the Pacific coast twists and turns along steep mountain sides. Don't miss the cliffside restaurant, **Mirador del Cafetal**, owned by the innkeepers of El Cafetal Inn. It offers a well-stocked gift shop, spectacular views of the valley below, excellent food and soups to die for. We mean that – absolutely super soups – especially *sopa Azteca*, a chicken vegetable soup with strips of tortilla.

■ Grecia

There are pleasant back roads from Alajuela, through Tacares, to Grecia, but if you're heading directly from San José, take the Inter-American Highway for 34 km/21 miles and turn at the Greek columned sign. The residents call themselves Greeks (*Greigos*) because of the town's unusual name, and many of them have added columns to their porches in an imitation of Greek architectural styles. However, what distinguishes the village of Grecia most is its award for being the "Cleanest City in Latin America." It is. The center of town is dominated by the metal-clad **Cathedral de la Mercedes**, a red brick church made of metal panels that were imported from Belgium in 1897. The interior is equally ornate.

Five minutes east of town is the well-run **Mundo de las Serpientes**, World of Snakes *(☎ 506/494-3700)*, in the village of Poró. Admission to this interesting stop is about US $12. It has dozens of snake varieties behind glass in outdoor cages. Guides will allow you to handle some of the less deadly species.

Northwest of town is **Los Trapiches** *(☎ 506/444-6656)*, a factory that makes *tapa dulce*, sweet hard brown sugar. A waterwheel drives a 19th-century Scottish cane press that squeezes out the sugar juice. The juice is collected and boiled, then poured into molds. Open weekends. There's swimming too, and you can bring a picnic or eat Tico food in their restaurant.

Places to Stay

In town, accommodations start and end with the worthwhile **Hotel Aeromundo** (*one block behind the church, ☎ 506/494-0094, www.aerotess@rasca.co.cr, cable, breakfast included, $)*, with rooms as well as three furnished apartments. They also offer a very good travel agency. Tell them we sent you.

Hotel Prices	
Prices are per night for two people	
[No $]	less than US $20
$	US $20-$40
$$	US $41-$80
$$$	US $81-$125
$$$$	US $126-$200
$$$$$	over US $200

In the surrounding hills are several good hotels. Follow the many signs to **B&B Posada Mimosa** (*off the main road, ☎ 506/494-5868, www.mimosa.co.cr, cable, pool, breakfast, $$)*. The location has mountaintop views and peaceful farmland. They offer rooms, a four-person suite and cabins, plus a house that's available for groups or long-term rental.

A true gem of a hotel is the **Vista del Valle Plantation Inn** *(Rosaria, ☎/fax 506/451-1165, www.vistadelvalle.com, tennis, pool, breakfast included, $$$$)*. This luxury inn is surrounded by coffee *fincas* and orange groves bordering the Río Grande Canyon Nature Reserve. It features sweeping views of the valley, as its name implies. Japanese-style cottages are simple but elegant, with private balconies. The comfort, peace and tranquility they exude is enhanced by lush tropical gardens.

■ Sarchí

If anywhere in Costa Rica can claim to be an arts and crafts center, it is the town of Sarchí. This small town, divided into two by the Río Trojas, is located in a deep valley 30 km/18.6 miles northwest of Alajuela. Coffee and farm fields line the valley walls. Its name may be from the Aztec word for "volcano," *xalachi*, but its fame is definitely founded on the manufacturing of oxcarts (*carretas*). Back in the 18th century, oxen were used to haul multi-purpose, heavy duty, two-wheeled carts up and down Costa Rica's rough roads and farm trails. In Sarchí, artisans perfected the design and began painting them – solid wood wheels and all – in bright, multicolored intricate designs. The carts have become a national symbol and are still in use in rural areas.

A local craftsman works on a cane chair, Sachí.

Today, Sarchí woodworkers make furniture and endless useful and decorative items from tropical hardwoods. Leather craftsmen have joined them and the town now fills with tourists and Tico shoppers who move from workroom to gallery to store, searching out bargains. If you see something you like, a rocking chair or a full-size oxcart (they also make small-scale versions and miniatures), they'll pack it and ship it for you.

LA CARRETA SIN BUEYES

If you wake up early in the pre-dawn and hear the creak of oxcart wheels pass in the dark, it may be the Carreta sin Bueyes, the Oxcart without Oxen. When San José was first being built, a young man stole the lumber meant for the city's first church. He built himself a beautiful oxcart, but didn't count on the divine intervention of St. Joseph, the carpenter. When the thief died he was condemned forever to ride his oxcart alone through the early morning countryside.

Fábrica de Carretas Joaquín Chaverri has been making hand-painted oxcarts since 1903. They now have a souvenir shop at the southern end of town, as well as **Restaurant Las Carretas**, serving traditional Tico food. The idea is to wander and check out the many shops scattered around the city center. Try to see the town's **church**, which looks like an eye-popping cake, painted pink and peacock-blue.

Restaurante El Rio, in the Centro Turistico gift shop, serves typical Tico food. Mexican dishes can be had at **La Troja del Abuelo**, but the

Restaurant La Finca, at the north end of town, may do it best. Pleasant, serene surroundings. Stay in town at **Villa Sarchí Lodge** *(☎ 506/ 454-4596, 800 m north of the Shell gas station, pool, cable, $+).*

A side trip from Sarchí to the **Catarata del Toro** (Bull's Waterfall) can be refreshing. Take the road across from the Chaverri Oxcart Factory and follow the signs. The 200-meter/658-foot falls, allegedly the highest in Costa Rica (according to the owner), cascades down on your left, seven km/4.3 miles past the town of Bajos de Toro. You can't see it from the road, but there are three overlooks on the half-mile hike downhill to its base (entry fee is around US $1). The first half of the route is an easy walk, but the second half is fairly strenuous. There's a very appealing little restaurant here with inexpensive, clean rooms above. They always have rooms available.

■ Naranjo

Naranjo is the next town you'll reach. Sleepy, small, agricultural, rural, a crossroads, a gateway – Naranjo is all of these. But its also a jumping-off point for **Tropical Bungee** *(☎ 506/248-2212, www.bungee.co.cr)*, where jumping off the nearby Río Colorado bridge is exhilarating by any stretch of the imagination. Tropical offers transportation from San José, as do Costa Rica Bungee and **Rappel Adventures** *(☎ 506/494-5102).* Try a night jump – wow! Or rappel deep into the Colorado gorge with either company for about US $20 (equipment and instructor included). This is great adventure – technically thrilling yet not too difficult for beginners. Afterward, you can ascend the cliff face or hike back up a trail. The **Serendipity Tours'** hot-air balloon leaves from this area as well *(US & Canada ☎ 877/507/1359, CR 506/558-1000, www.serendipity-adventures.com).*

Thirteen km/eight miles west of Naranjo on the road to Zarcero is **Restaurante El Mirador**, which offers good Tico food and fabulous views of Tico land. From here your choice is visiting the large towns of San Ramon and Palmares (famous for its fiesta in January) or the topiary town of Zarcero to the north. You can cut your circuit short by getting back on the Inter-American Highway in San Ramon, or take the twisty road from there up through cheese country north to La Fortuna/Arenal. From Zarcero you can push on to Cuidad Quesada, then to Arenal, or loop around Platamar and Poás volcanoes on a little-used paved road, back past La Paz waterfalls to Heredia/Alejuela. It's a long ride, but very different.

■ San Ramón

The heavily agricultural city of San Ramón is famous for its Saturday **Farmers Market** in the center of town. Farmers and buyers from all over the valley come here to barter, and the atmosphere is festival-like when the weather is good.

The **Catedral de San Ramón**, on the square, is modern, despite its pseudo-Baroque style. On the north side of Parque Central is the 1893 **Palacio Municipal** (admission is free, open 9-4), built in the style of a Spanish mansion and complete with high ceilings and an inner courtyard. Admission is free to next door's student-run **San Ramón Museum**, and interesting and quaint distraction. If you're around in January, take a detour into the nearby town of **Palmares** for the rodeos, celebrations and parades of the **Fiesta de Palmares**.

The area's biggest attraction is the private, 809-hectare/2,000-acre **Los Angeles Cloud Forest**, 20 km/12.4 miles north of town. Owned by ex-President Rodrigo Carazo, and his wife, Estrella, this little-known ecological paradise is a good alternative destination to the long hard ride to Monteverde. It's close enough to be done as a day-trip from San José. LA is a delight for hikers, eco-minded tourists and birders interested in seeing a quetzal, the colorful national bird. Cloud forests, a type of rainforest, get moisture more from the fog-like clouds that sweep up the mountainside, depositing water in the form of dew. A naturalist offers tours and explains the diverse eco-system, in English and Spanish, twice daily (9 am and 1 pm, US $25).

Places to Stay

Hotel Villablanca *(☎ 506/ 228-4603, www.villablanca-costarica.com, $$$)* borders the preserve and serves as a magnet for seniors as well as young students (who sometimes stay dorm-style for US $35, including meals and forest entry – a real bargain). The *casitas*, bungalows, are built in adobe style and beautifully decorated inside, each with its own fireplace – warmly reassuring when the temperature drops. Floors are tiled, as is the vaulted ceiling of the striking **La Mariana Chapel** here. Meals are offered in the central lodge, where you can mingle with fellow travelers. The hotel can arrange day-trips from San José or, better yet, stay overnight and go back the next afternoon. Horseback riding and other adventures too. In 2004,

Hotel Prices	
Prices are per night for two people	
[No $]	less than US $20
$	US $20-$40
$$	US $41-$80
$$$	US $81-$125
$$$$	US $126-$200
$$$$$	over US $200

this hotel was bought by the same people who own So Como No in Manuel Antonio. We expect it to go upscale as a result.

In a hidden valley, 30 minutes north of San Ramón on the way to Arenal Volcano, is **Hotel Valle Escondido** *(San Lorenzo, ☎ 506/577-1000, www.hotelvalleescondido.net, 31 rooms, pool, restaurant, jacuzzi, canopy slide, $$-$$$)*. This 101-hectare/250-acre private reserve sits on the floor of a steep valley whose walls are lush with vegetation. Great for a get-away-from-it-all vacation. Indulge in nature activities or relax and let nature come to you. Tours. Ask about the package rate.

■ Zarcero

North from Naranjo is Zarcero, a little valley agricultural town famous for its cool weather, pine trees, organic veggies, peach jam, dairy farms, homemade cheese and the charming topiary garden in front of the town's white and red church.

In 1960, gardener **Don Evangelisto Blanco** began to trim the bushes in Parque Francisco Alvardo into fanciful shapes – elephants, birds, rabbits, oxen and cart, a monkey on a motorcycle and even a tunneled walkway with sides like giant Hershey's Kisses. Don claimed God told him which shapes to create. He deserves a medal for his outstanding work. Take a walk through this verdant fantasy land.

Trees have been trimmed into whimsical shapes in Zarcero's topiary garden.

Tiny **Hotel Don Beto** *(☎ 506/463-3137, www.donbeto.com)*, next to the church, is good for an overnight in town. It's very clean and comfy, with two rooms that share a bath and two with private baths. On a hilltop south of town roadside stands offer homemade cheeses and fruit preserves. Look there for **La Cocina de Turno de Doña**, a rustic kitchen that features homemade soups and stews, slow cooked over a wood-burning stove.

SPRING ALL YEAR LONG!

When you're in Zarcero, remember you are in the highlands (1,700 meters/5,592 feet), where the climate is spring-like all year round.

One half of the world cannot understand the pleasures of the other.
~ *Emma*, Jane Austen, 1775-1817

Arenal and the central north are reached by two common routes: the first runs north through Zarcero and Quesada (San Carlos) and then on to La Fortuna. Another way is from San Ramón north through San Lorenzo and La Tigra to La Fortuna.

■ Ciudad Quesada (San Carlos)

You may delay, but time will not.
~ Benjamin Franklin, 1706-1790

The northern region around Quesada, formerly known as San Carlos, lies southeast of Arenal, where mountain slopes transit from the drier north to the hot and humid tropics of the Atlantic side. A loop from San José – through the coffee towns to Quesada, over to Aguas Zarcas and the eastern side of Poás Volcano, and then finally back through Heredia past La Paz Waterfalls – can be a great drive done in a single long day, or two or more if you want to stop and explore.

Quesada is down in a lush grassland valley at only 650 meters/2,138 feet above sea level. It is the agricultural capital of the north, famous for its delicious string-like cheese. However, it was not renamed for the Spanish word for cheese, *queso*, but to honor the poet, Napoleon Quesada. Several very enticing properties offer bases from which to explore the area, including Arenal and the north, or Rara Avis and La Selva (see the *Caribbean Coast* chapter, pages 328-331) to the east.

AUTHOR NOTE: All of the delightful hotels below are set outside of town in the native territory of the beautiful green macaw, now in danger of extinction.

Places to Stay

Ten minutes northeast of the city, **Termales del Bosque** *(☎ 506/460-1356, termales@racsa.co.cr, restaurant, breakfast included, $$)* offer what their name implies: relaxing hot springs in a forest. The natural spa consists of several warm mineral pools in the shade next to a bubbling stream. There's a canopy tour too. Non-smoking cabins. Good value with simple rooms. **Hotel La Garza** *(☎ 506/*

Hotel Prices	
Prices are per night for two people	
[No $]	less than US $20
$	US $20-$40
$$	US $41-$80
$$$	US $81-$125
$$$$	US $126-$200
$$$$$	over US $200

475-5222, www.hotellagarza.com, pool, restaurant, $$-$$$), north of Ciudad Quesada, is a 304-hectare/750-acre working hacienda that offers horseback riding, tours of the working ranch and hearts of palm plantation, plus nearby La Garza Biological Reserve. Plus, there's a canopy slide.

In the community of La Gloria, near Aguas Zarcas, is **Albergue San Juan** *(☎ 506/259-3401, www.turismoruralcr.com, $)*, a cultural cooperative with rustic cabins. You can take a tours of their protected rainforest and medicinal plant gardens, or hike to see green macaw nests. Upscale **Tilajari Resort** *(☎ 506/469-9091, www.tilajari.com, pool, tennis, restaurant, $$-$$$)* near Muelle, is a country club in the country, with spacious rooms, big pool and popular tennis courts. Lovely grounds and excellent restaurant.

The most remote of all these lodges is **La Laguna del Lagarto Lodge** *(☎ 506/289-8163, www.lagarto-lodge-costa-rica.com, restaurant, $$)*, which sits north of Pital and Boca Tapada, not far from the Nicaraguan border. It's a naturalist's and eco-tourist's paradise set way out of the way. More than 90% of owner Vinzenz Schmack's 500 hectares/1,235 acres of land are forested, with a well-maintained trail system. Several lagoons can be explored via canoe.

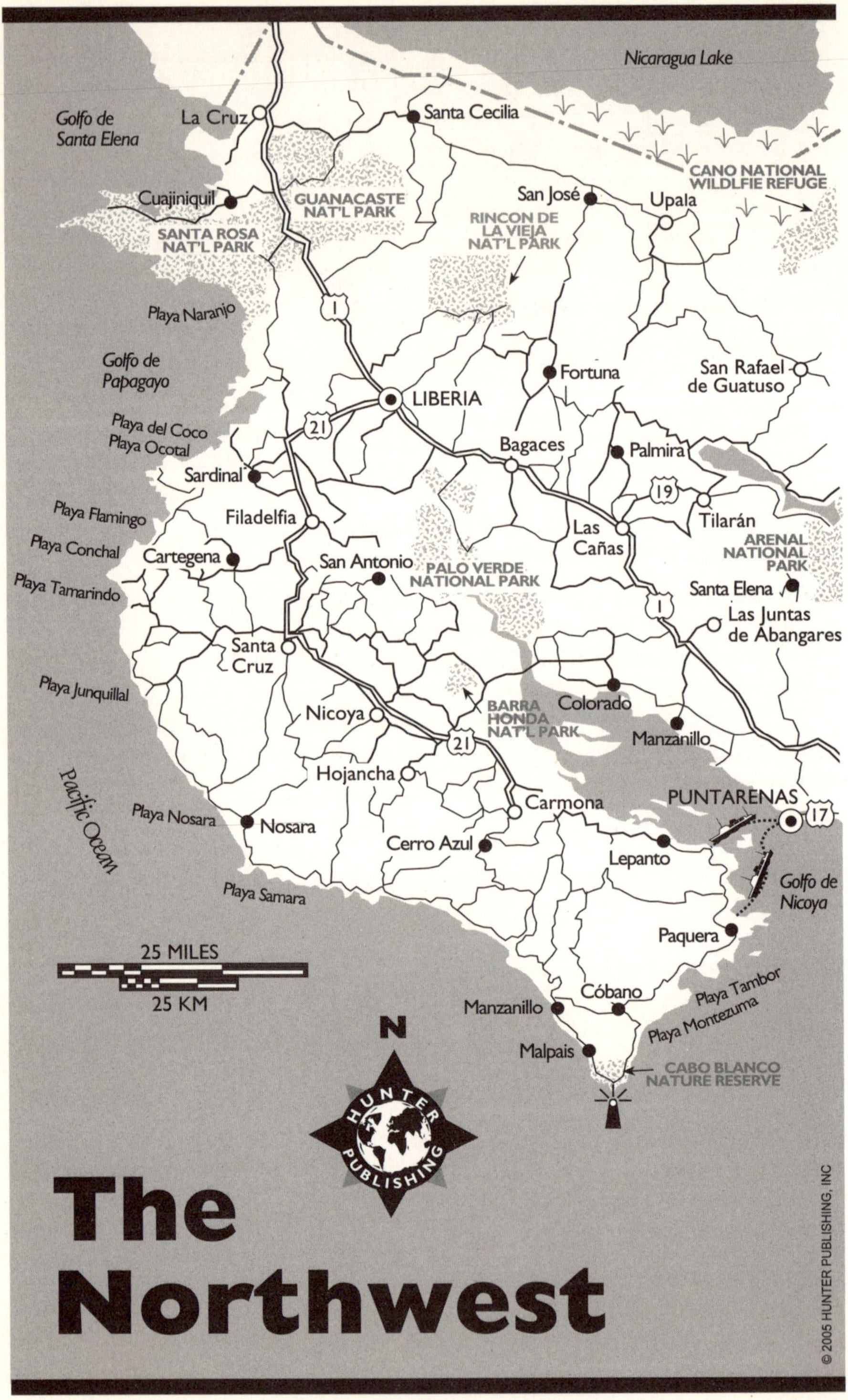
Nicaragua Lake
Golfo de Santa Elena
La Cruz
Santa Cecilia
CANO NATIONAL WILDLFIE REFUGE
Cuajiniquil
GUANACASTE NAT'L PARK
SANTA ROSA NAT'L PARK
San José
Upala
RINCON DE LA VIEJA NAT'L PARK
Playa Naranjo
1
Golfo de Papagayo
Fortuna
San Rafael de Guatuso
LIBERIA
21
Playa del Coco
Playa Ocotal
Bagaces
Palmira
Sardinal
19
Playa Flamingo
Filadelfia
Las Cañas
Tilarán
Playa Conchal
Cartegena
ARENAL NATIONAL PARK
San Antonio
PALO VERDE NATIONAL PARK
Playa Tamarindo
Santa Elena
Las Juntas de Abangares
Santa Cruz
Playa Junquillal
Colorado
Nicoya
BARRA HONDA NAT'L PARK
Manzanillo
Pacific Ocean
Hojancha
Carmona
PUNTARENAS
17
Playa Nosara
Nosara
Cerro Azul
Lepanto
Golfo de Nicoya
Playa Samara
Paquera
25 MILES
25 KM
Cóbano
Playa Tambor
Manzanillo
Playa Montezuma
N
Malpais
CABO BLANCO NATURE RESERVE
HUNTER PUBLISHING
The Northwest
© 2005 HUNTER PUBLISHING, INC

North by Northwest

Go confidently in the direction of your dreams.
Live the life you have imagined.
~ *Walden*, Henry David Thoreau, 1817-1862

IN THIS CHAPTER

This section of the country contains arid land and sprawling cattle ranches, a windswept lake, cloud-shrouded forest preserves, deep caves with thousands of bats, mangrove swamps, indigenous Indian reserves, active volcanoes, flat land, deep valleys, mountains and a shoreline pounded by waves. The territory borders Nicaragua to the north and the Central Valley to the south. For organizational purposes, the area we are covering in this section looks somewhat like a slice of pizza with one end of the crust at Santa Rosa Park on the Pacific, and the other at Los Chiles border crossing. The pointy end is above San Ramón and Zarcero. The slice covers most of Guanacaste and Alajuela provinces.

Arenal Area

■ Arenal Volcano

Nearly all men can stand adversity,
but if you want to test a man's character, give him power.
~ Abraham Lincoln, 1809-1865

Arenal is the most active volcano in the western hemisphere and its presence dominates the northern town of La Fortuna. If Poás and Irazú have only whetted your appetite for volcanoes, Arenal will make you drool. The farmland around is a deep green, primary and secondary forests blanket the midlands, and towering above it all is classically cone-shaped Arenal – smoking, rumbling, spewing car-sized boulders and spitting red-hot lava that lights up the night sky (when it's not covered in clouds and raining, that is). There are no guarantees that you'll even get to see the volcano's summit. One time we were here it rained and was cloudy for four days straight.

Arenal began its life about 4,000 years ago and grew to 1,633 meters/ 5,372 feet in height from near continuous eruptions until just before the Conquest. Then it went silent and allowed nature to cover it with forests and vegetation. So everyone forgot it was a volcano until a few small fumaroles opened at its summit in 1938, then again in 1958 and 1960. In 1967 the water temperature of the Río Tabacón, a spring-fed river than descends the slopes of Arenal, suddenly rose. It was a warning of danger, but so few people lived in the area – and no volcanologists – that it went unheeded. At 7:30 am on July 29, 1968, Arenal Volcano erupted with a pyroclastic flow that raced down the mountainside and incinerated the villages of Tabacón and Pueblo Nuevo – taking the lives of 78 townspeople. Huge incandescent boulders exploded out of the cone, halfway up the mountain, and left large craters as far as 10 km/6.2 miles away.

Things are much quieter today, but Arenal can still be deadly. During the summer of 2000, an eruption flowed down a crevice and enveloped a young American woman, her daughter and a Costa Rican guide while they were on a hike at the edge of the safety zone. All three were badly burned. Sadly, the guide died two days later; the girl a short time after.

Adventures

Hikers are now prohibited from getting as close as they did, but the element of danger, however small, is what transforms a visit to Arenal from an interesting ecological attraction to a thrilling adventure experience. Just the idea that you are so close to its power makes a stay memorable.

The town of La Fortuna is six km/four miles from the mountain, which is clearly visible from the town's main square. A road leading northwest around the volcano toward the lake is home to many lodges and hotels, and it's these places that offer the best views of Arenal. Unfortunately, clouds and rain often cover the volcano. Be patient – the wind will blow them away sooner or later. When it's quiet at night, listen for the growl and rumble.

KABOOM

Bet you didn't know that eruptions are classified by their sound and described by onomatopoeias. *Kabooms* are brief violent eruptions that throw boulders and blobs of lava into the air and shake the ground in their intensity. A *whoosh* shoots lava and ash in the sky; the sound is like a jet engine. *Chugs* are rhythmic expulsions of gas from the lava fountains.

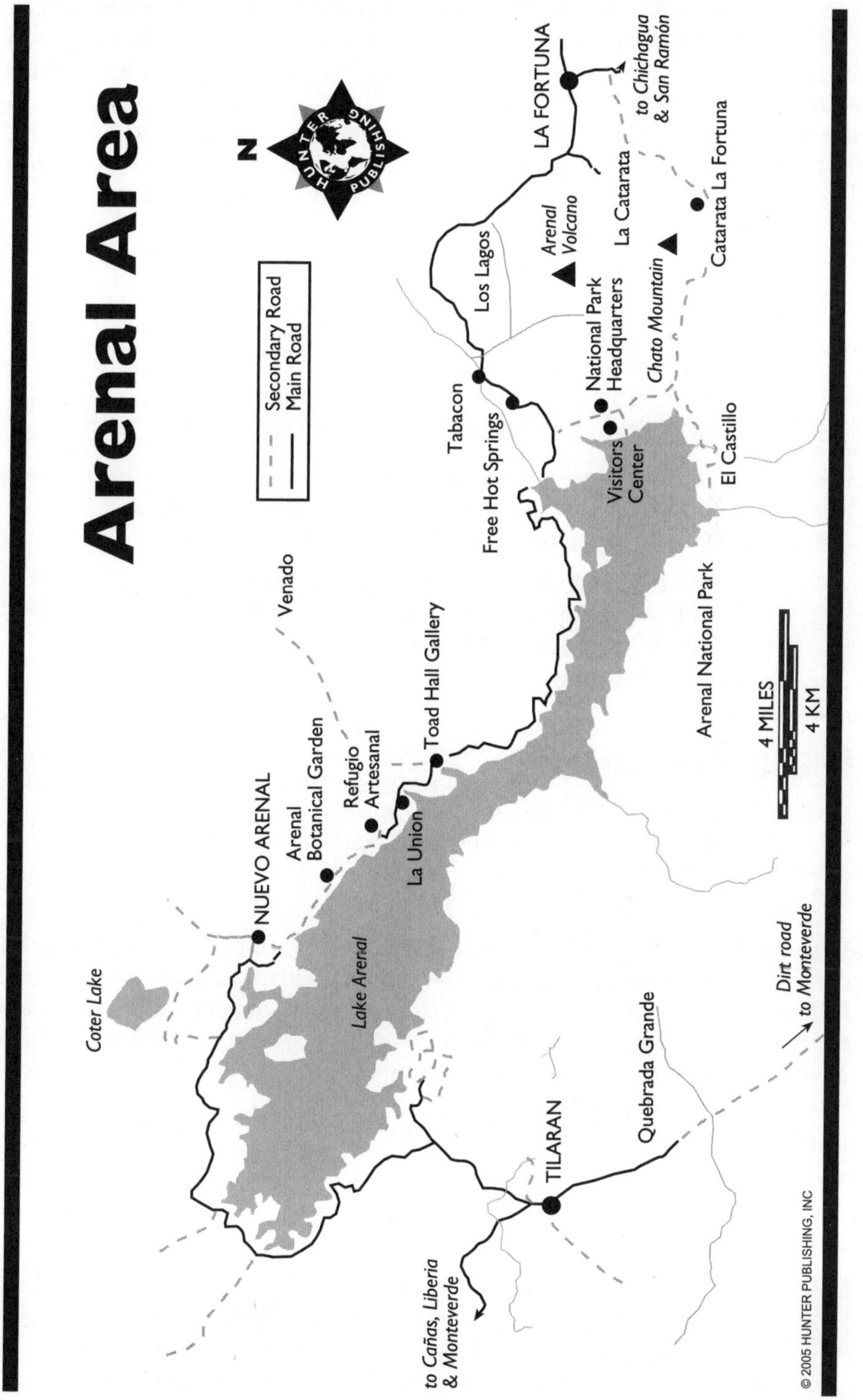
Arenal Area
N
Secondary Road
Main Road
LA FORTUNA
to Chichagua & San Ramón
Catarata La Fortuna
La Catarata
Arenal Volcano
Chato Mountain
Los Lagos
National Park Headquarters
Tabacon
Free Hot Springs
Visitors Center
El Castillo
Arenal National Park
4 MILES
4 KM
Venado
Toad Hall Gallery
Refugio Artesanal
Arenal Botanical Garden
NUEVO ARENAL
La Union
Lake Arenal
Coter Lake
Dirt road to Monteverde
Quebrada Grande
TILARAN
to Cañas, Liberia & Monteverde
© 2005 HUNTER PUBLISHING, INC

Lastly, don't miss a dip in the warm waters of the **Tabacón River**. The classiest way to enjoy them is to go to the Tabacón Resort & Spa and swim in their pools or in the stream itself. A less-expensive (under US $8) swimming pool spa, **Baldi Termae**, is on the way there. Another option is the low-cost pool, Las Fuentes, just past the Tabacón Spa entrance, which has facilities. Even farther up the road is a free spot in the stream. It's not marked, but there are usually a couple of cars on the left.

> **AUTHOR TIP:** Don't bother to enter the official **Arenal National Park** (entrance and visitor's center located up the road to Arenal Observatory Lodge). There are no facilities and not so much to do after driving up the rough dirt road. However, the hiking trails here do go closer to the lava wall on the volcano's less-active side, so it's a bit safer.

It is said that only 10 people out of three million have a chance to see a volcanic eruption, and only four out of three million see a lava flow. At Arenal, if the Guatuso god of fire wills it, you have a chance to see both.

> **WARNING:** Don't be stupid and try to hike close to the volcano without a guide. In 1988, Steven Simmler, a young American, was carbonized by an eruption while trying to climb to the top.

■ La Fortuna

Fortune knocks at everyman's door once in a life,
but in a good many cases the man is in a neighboring saloon
and does not hear her.
~ Mark Twain, 1835-1910

With clean, fresh air and beautiful countryside, the town of La Fortuna has been discovered in no small way by ecological and adventure tourists. It feels a bit like either a college or frontier town, depending on your perspective, with many happening bars and restaurants and numerous pleasant hotels in a variety of price ranges. A rash of hotel break-ins have recently plagued the town.

From San José, take the Autotransportes San José-San Carlos bus, about a 4½-hour ride, from Calle 12, between Av 7 & 9.

Adventures

There are more active adventures than bathing in the hot springs or watching the active volcano. **Whitewater rafting** is big in the area on the **Toro** and **Peñas Blancas** rivers, both of which have some excellent Class III and IV (even Class V) rapids.

Arenal Volcano, seen from La Fortuna.

La Fortuna Waterfalls, southwest of town toward Chachagua and a long walk from downtown, is a popular trip for the steady of feet. A US $6 entrance fee (bring your swimsuit) sets you out on a very steep hike down slippery steps cut into the jungle cliff. At the bottom the water pounds down from an impressive height. Don't swim at its base, where the powerful water rate can be deadly. Instead, head just around the corner, where the steam widens and you can enjoy swimming in the warm swift water. Don't leave valuables in your car here.

Guided **hikes**, such as the one to La Fortuna Waterfall, **mountain biking trips**, or a visit to explore the **Venado Cavern** can be arranged with local adventure tour agencies. **BoBo Adventures** (*in La Amistad Hotel, ☎ 506/479-9390, www.boboadventures.com*) specializes in cave tours. Contact any of the following well-established local tour agencies for more eco-adventure activities in the area: **Desafío Adventure Center** (*☎ 506/479-9464, www.desafiocostarica.com* **Aguas Bravas** (*☎ 506/479-9025, www.aguas-bravas.com*) raft the Rio Toro; **Sunset Tours** (*☎ 506/479-9800, www.sunsettourscr.com*); and **Aventuras Arenal** (*☎ 506/479-9133, www.arenaladventures.com*). Canopy tours, hanging bridges, hikes and zip lines abound in Costa Rica and this popular tourist area is no exception. Put on your adventure boots and enjoy!

Sport Gym Fortuna is located in back of Nenes Restaurant, open 5 am-9 pm weekdays and 9-6 pm on Saturday. It has weights and exercise machines.

There's also **horseback riding**, including a ride over the mountain to Monteverde (skip during the rainy season) or one to the nearby extinct cinder cone volcano, Cerro Chato (1,100 meters/3,618 feet). By car, take an educational side trip to the **Guatuso Indian Reserve** near San Rafael de Guatuso, north of La Fortuna. There isn't too much to see, but the Indians need the money tourism brings. If you're there during a traditional ceremony, you'll be lucky enough to see their colorful native costume.

If you're fortunate enough to be in La Fortuna on a weekend, take off your hiking boots, put on your dancing shoes and head for **Volcan Look** *(☎ 506/479-6961)*, a huge disco a short distance from town on the road toward Tabacón. Put 200 people on the dance floor and it still looks empty. Bar is open Wednesday through Saturday.

GIRL POWER

A nation is only as good as its women.
~ Muhammad Ali, boxer

When La Fortuna changed governing districts from San Ramón to San Carlos in 1950, it marked the first election that allowed women to vote in Costa Rica, following the passage of national suffrage legislation.

Places to Stay & Eat

On the Way In

Chachagua Rain Forest Lodge *(12 km/8 miles outside Chachagua, ☎ 506/239-6464, fax 290-6506, www.chachaguarainforesthotel.com, 22 rooms, pool, restaurant, pick-up service, conference facility, $$$)*. This is one of Costa Rica's hidden treasures. While everyone goes to hotels at Arenal or Monteverde, a short side trip leads to this really appealing private lodge just below a rainforest. Individual rustic cabins speckle the landscaped gardens where streams splash down from the mountaintop. Each good-size cabin boasts a large wooden porch for sitting and birdwatching. Inside, the bedrooms have two beds, dresser, table and chair. They are made of natural wood, with memorably big, bright and airy bathrooms, lush plants and lots of light. The wooden deck restaurant (everything they prepare is grown on the farm, except the seafood) boasts a bar. From there, it is a 20-minute walk to a five-meter/16-foot waterfall where you can swim. Heck, you can

Hotel Prices	
Prices are per night for two people	
[No $]	less than US $20
$	US $20-$40
$$	US $41-$80
$$$	US $81-$125
$$$$	US $126-$200
$$$$$	over US $200

even swim in any of the numerous cold-water streams on the property – or in the large swimming pool. It's a two-hour hike up to the mountaintop – Wellington boots supplied. One of the main activities at Chachagua (pronounced cha-CHAG-wa) is horseback riding. Their lodge stables 10 horses, including Spanish stallions, that ride in festivities all over Costa Rica. Breakfast is not included, but you should negotiate for it. A fine place to relax or take tours to the area's natural wonders.

Arenal Country Inn *(Arenal-Chachagua Road, ☎ 506/479-9670, fax 479-9433, www.costaricainn.com, air, room safe, pool, terrace, breakfast included, $$$$)*. Once a working hacienda, this appealing country inn consists of rooms in attractive duplex tropical cottages that are painted pumpkin and gold and covered with a red metal roof. Cottages are scattered around the lovely gardens in the flatlands beneath the volcano. The large reception area is open-air, under a big roof where the former cattle barn was. It has a small breakfast area, recreation area with a pool table and reading area with wicker furniture. The outdoor free-form pool is large and inviting. You are only one kilometer from La Fortuna town. Plus, a river runs through it – Robert Redford, take note.

UNDERWORLD CONNECTIONS

The Venado (deer) cave entrance was discovered in 1945 on a private farm in the rural town of Venado (about 10 km/six miles north of La Fortuna) by following a stream that flows from its mouth. The interior is undeveloped, without railings or cement walkways, and no enlarged headroom where the cave roof drops to under two meters/6.6 feet. Resident cave dwellers include a large colony of bats (not always hanging out), tarantulas, big crickets and the occasional snake. Be prepared to get dirty – rubber boots, hard hats, flashlights and face masks are supplied by tour operators who run trips here. Use them.

In Town

La Fortuna offers the best opportunity to stay close to the volcano without paying extra for the nighttime view. Well-established **Hotel San Bosco** *(☎ 506/479-9050, www.arenal-volcano.com, pool, jacuzzi, $$)* is on a side street near the square, next to Desafío Tours. Beautiful gardens and grounds surround this quiet modern hotel known for its cleanliness. Their small blue pool is inviting and the adjacent jacuzzi sits under a shady overhang. All the rooms are new, built in response to the huge increase in tourists that now flock to La Fortuna. Sand tiles, white walls, cherry trim, Italian ice-colored quilted bedspreads, porch. Across the street is **Rancho La Cascada Restaurant**, a popular eatery and bar under a huge palapa roof. Also across the way is the **Cabinas Las**

Tinajas *(☎ 506/479-9308, $)*, a tiny, four-room motel-like accommodation. It features clean rooms and baths, a patio porch and very low prices. Behind them is a fish farm that raises Tlapia. This is a quiet spot, but it comes with bragging rights that you found the best value in town.

A La Fortuna standard that fills up fast is **Luigi's Lodge & Pizzeria** *(☎ 506/479-9909, fax 479-9898, www.luigislodge.com, air, pool, spa, jacuzzi, restaurant, breakfast, $$+)*, opposite Restaurant Las Brasitas on the road to the volcano. This is the largest and best of in-town hotels, with a pizza and Italian restaurant that is always crowded. **Hotel Arenal Jireh** *(☎ 506/479-9004, www.arenalexperience.com, air, cable, $-$$)* is an appealing six-room hotel on the corner, one block closer to the square. Pool and Bible provided. Next door, **La Amistad Inn** *(☎ 506/479-9364,* $) offers basic rooms.

Sunset Tours has opened a small place, **Don Manuel Inn** *(☎ 506/479-9069, www.sunsettourscr.com, breakfast, $)*, with simple rooms, some that share a bath. Two other basic accommodations with a little extra are **Cabinas MonteReal** *(☎ 506/479-9357, cable, $)*, 50 m/150 feet west and 100 m/300 feet south of the school. A half-block away is **Mayol Lodge** *(☎ 506/479-9110, $)*, which has a pool.

For small and quiet, you won't find better than **Cabinas La Riviera** *(☎ 506/479-9048, $)*, a basic little cabana hotel several blocks east of Hotel San Bosco. Individual cabins are set in manicured gardens filled with birds and butterflies, with the river just behind the hibiscus hedges. It's very appealing, but out of the way.

Las Cabanitas Resort *(☎ 506/479-9400, www.cabinitas.com, breakfast, $$$)*, is just east of downtown. Las Cabanitas is classy. It features a large free-form pool, a restaurant called Arara (tasty food) and large lovely hardwood bungalows with tiled roofs. Dollar for quality, this is a best buy. If they're full, check out **Villa Fortuna**, also very good (and a tad cheaper), next door.

> **AUTHOR NOTE:** Be sure to ask the front desk of your hotel for a wake-up call if the sky is clear and there's a nighttime flow.

For the best value seafood restaurant in town, head to **Nene's Restaurant**, one block off the main road, east of the square park. You can always ask a local for directions – everyone eats there. There are many small "sodas" and restaurants in town. One of the most popular — for good reason — is **Lava Rocks Café**, a hip open-air restaurant under a clear canopy roof. Large menu, gift shop, green plants, and good food, it's located opposite the park square on the main drag. A favorite of the backpacking crowd is **Soda La Parada**, a clean and tasty soda opposite the bus stop and park square. Good cheap eats can be had at **Soda El Rio**, which backs against the rushing river, a block south. There is also a **Wall's Ice**

Cream shop facing the park square to the north. Heading toward the volcano a few blocks, **Las Brasitas** features wood-fire roasted chicken and Tico-Mex specialities.

At the Foot of the Mountain

Every time we come to Arenal there is a new hotel springing up on former dairy pasture and hills at the base of the volcano. If there's lodging here, you can be sure it has a view. The following hotels are not in any particular order.

Volcano Lodge *(☎ 506/460-6080, fax 460-6020, www.volcano-lodge.com, 20 rooms, air, pool, jacuzzi, restaurant, breakfast included, $$-$$$)*. Owned by the same family who own Sunset Tours and the Hotel San Bosco in town, this pretty lodge on a hill features large rooms with big picture windows and comfortable beds. Each of the 10 tile-roof duplex bungalows has a little garden, a patio and a good view of the volcano. We have stayed here twice since its opening in 2000 and enjoyed it. The restaurant serves Tico food and kids like the huge aquarium in the center of the *al fresco* dining area. Behind it, up a dirt road, is the **Cabinas Palo Verde** *(☎ 506/479-9306, $$)*, rustic cabins on a farm.

Tabacón Resort *(☎ 506/256-1500, fax 221-3075, www.tabacon.com, pool, 2 restaurants, 82 rooms & suites, air, cable, health spa, $$$$, breakfast included)* is our splurge accommodation at the base of the volcano. This luxury hotel complex is located up the hill from the warm Tabacón River, where they have built an elaborate spa. The rooms are large and attractive, with a great nighttime view of the volcano from either balconies or garden patios. New rooms were added in 2001. But the best reason of all for staying here is to use the spa (your room rate includes unlimited access). Three very warm swimming pools, one with a swim-up bar and cement tube slide, are very popular, as is the stream that features a natural waterfall with a ledge to sit under. The volcano heats the pools and river and it's great to go in the evening and see steam rise from the bathtub-warm river. Open to the public (US $18), the spa offers free changing rooms and towels (tips accepted).

A tropical garden weaves along the streambed, and visitors will delight in the colorful setting here. If you're not staying at the hotel and want to visit the spa, which has mud baths and massages by appointment, it's best to make a reservation for entrance and/or dinner, especially in high season. The restaurant overlooks the volcano (on a clear day, watch the boulders roll down) and river. It's a good choice for buffet or à la carte dining. Fifty meters/164 feet down the road on the right is their lower-cost warm-water pool, Las Fuentes, where locals go. It's more like an open garden area. There is a swimming pool and changing rooms, but most visitors sit in the waterfalls of the stream. The advantage is that it's less crowded and costs less (US $8), but it has fewer facilities. About a half a kilometer farther down the road you may find cars parked on the left. This is the entrance to a free swimming hole in the warm river.

Arenal Lodge *(200 meters / 658 feet after Arenal Lake Dam, ☎ 506 / 228-3189, fax 289-6798, www.arenallodge.com, restaurant, full breakfast included, $$-$$$).* If you ever wondered what was above the clouds, turn into the entrance of Arenal Lodge, just over the dam at the east end of the lake. Follow the serpentine brick road up through a cloud forest to the top of a mountain. Half the fun of Arenal Lodge (not to be confused with Arenal Observatory Lodge) is driving up their very good – but very steep – road through the cloud forest. Four-wheel-drive is strongly recommended. The lodge's economy rooms (*sans* view) are pleasant, with one queen and one single bed. If it's rainy season, which is actually high season around Arenal, you won't miss not having a view from your room. The popular junior suites are very large and comfortably appointed, with wooden balconies that overlook tropical gardens, with the lake and the volcano in the distance. Fruit bird feeders in the gardens attract scores of colorful song and hummingbirds. Swiss-style chalets, complete with fireplaces and balconies, sit on the very crest of the mountaintop. They cost just a little more than the junior suites. Free shuttle bus to the hot springs and La Fortuna town.

Los Lagos, Junglas y Senderos *(☎ 506 / 479-8000, $$-$$$),* which means "Lakes, Forests and Trails," is the hotel closest to the volcano's active side and lava flow. You won't believe how close until you hike up to the observatory, which you can do even if you're not staying here (just look out your window if you are). Signs abound for emergency exit routes – how exciting. Rooms in villas, restaurant, a water splash park for kids, pool.

Other properties on the road uphill include **Montaña de Fuego** *(☎ 506 / 460-1220, www.montanadefuego.com, $$$)* and **Hotel Arenal Paraíso** *(☎ 506 / 460-5333, www.arenalparaiso.decostarica.co.cr, $$).* Montaña was the first of these side-by-side hilltop *casita* hotels, both owned by the same family. They offer large simple cabins, a glass porch restaurant at Montaña and unobstructed volcano views. Arenal Paraíso is the newer and least expensive of the two, with smaller wooden cabins and decks. **Arenal Vista Lodge** *(☎ 506 / 221-0965, www.arenalvistalodge.com, $$)* has a great vista of Arenal, with simple connected rooms on a hilltop. You can't miss it.

If you get a hankering for a good steak or just a hearty meal someplace other than your hotel, try the **Mirador Arenal Steakhouse**, opposite the Lavas Tacotal Hotel. It features big views from big glass windows and fine food in a rustic chic setting that some might call romantic.

On your way around the volcano to the lake, turn left for the National Park and **Arenal Observatory Lodge** *(☎ 506 / 257-9489, www.arenal-observatory.co.cr, $$$-$$$$),* a former farm on the base of the volcano. It was converted into a lodge in 1973 for the volcanologists from the Smithsonian Institution. This well-known hotel, which has a close, bird's-eye view of Arenal, is protected from a lava flow by the Agua Caliente River gorge. Located within the national park and bordering the Monteverde Cloud Forest, it's no surprise that the lodge received a "Four Leaf" rating from Costa Rica's sustainable tourism certification program. The lodge owner, William Aspinall, former director of Monteverde, has instituted a new three-day hike through the rainforest to Monteverde. The lava flow at night no longer faces this side of the mountain, but you get to see day-time eruptions.

Just a bit farther along the dirt road and across two streams is the **Hotel Linda Vista del Norte Lodge** *(☎ 506/380-0847, www.lindavistadelnorte.com, breakfast included, $$)* on the other side of the lake. Its "linda vista" (pretty view) is of the lake and the volcano. This is the sunnier side of the mountain. Basic, comfortable accommodations in a pretty setting; rooms facing the volcano cost a bit more. The personable owner runs the farm surrounding this hilltop lodge.

■ Caño Negro Wildlife Refuge

In a remote area of the north lies the large, swampy, wildlife mecca, **Caño Negro Wildlife Refuge**. It's a long drive (124 km/77 miles north of Ciudad Quesada), but of huge interest for dedicated eco-travelers, naturalists and birders. A large lake covers much of the 9,969-hectare/24,623-acre park; most sightseeing is done by boat. Arrange your visit with La Fortuna agencies, listed on page 201. The best time to visit here is the rainy season – more water means more birds.

NICARAGUA SIDE TRIP

Just to the north of Caño Negro is **Los Chiles**, on the border of Nicaragua. If you're thinking of crossing the border, it may be possible to travel by boat from Los Chiles to San Carlos on Lake Nicaragua. Immigration is closed weekends. A much easier and safer way to get a **Nicaraguan passport stamp** is to go back south and take a boat trip to Tortuguero from Muelle, near Puerto Viejo de Sarapiquí. The boat stops at a Nicaraguan customs and guardhouse and everyone must show ID. **Costa Rica Expeditions** *(☎ 506/257-0766)* occasionally runs this route from Caño Negro. Call ahead.

■ Lake Arenal

What a woman says to her ardent lover
should be written in wind and running water.
~ Caius Valerius Catullus, Roman poet, 87-54 B.C.

If you weren't impressed by – or couldn't see – the volcano, take a ride around Lake Arenal, Costa Rica's largest lake at 33 km/20 miles long. If nothing else, the size of the potholes in many parts of the road should impress (or depress, depending on your driving skills). The view of the countryside is wonderful.

Tectonic upheavals created Lake Arenal about three million years ago and it provided fresh fish for ancient Guatusu inhabitants of its shore. In 1973 an electrical energy dam raised the lake level and drowned the old villages of Arenal and Tronado. Arenal rebuilt on higher ground as Nuevo Arenal, but lost to the water was archeological evidence of pre-historic habitation, including a native burial ground. The eastern edge of the lake has an incredible view of the volcano, although that side no longer features the lava flow.

As you drive west you pass dairy farms, secondary forests and the occasional habitation. The lake's water is warm and fine for swimming, but the most popular sports around are sailing, windsurfing and fishing for *guapote* (bass), who put up a fierce fight. Due to the lake's alignment and climate, it's famous for warm winds that whip across its western end – at 60 or more knots, especially from December to February. Outside of Tilarán you'll come to huge white windmills rotating with a steady hum on the hills above the lake. Be Don Quixote and chase windmills by turning up one of the dirt roads that rise to the huge metal dragons.

The first stop along the lake circuit should be **Los Heroes** *(☎ 506/692-8012, www.hotellosheroes.com, $$)*, a restaurant, gift-shop, hotel and dairy farm right out of *The Sound of Music*. The owners, a Swiss husband and Tica wife, offer small rooms and a mixed menu at their huge Swiss chalet.Imported embryos from Alpine bovines makes even the family's dairy herd Swiss. They're working on a fascinating narrow gauge diesel train (imported from Switzerland, of course) that will take visitors up the mountainside to the cloud forest above their farm, or carry a wedding party from their small chapel.

Arenal Botanical Gardens *(www.junglegold.com, ☎ 506-694-4305)* was founded in 1991 by Michael LeMay as a "plant reserve and nature library." This lush, lovely hillside garden also contains a butterfly farm. Open 9-5 daily, admissions about US $8. Worth a peaceful and quiet visit, and they do mail-order exotic plant seeds.

Places to Stay & Eat

Toad Hall, a famous store that offers a great variety of local and national handicrafts, is located in the tiny town of La Unión. It has a pleasing vegetarian/health food restaurant overlooking the lake. Alternately, drive two km/1.25 miles farther to the artistic **Restaurante Sabor Italiano**, a pizzeria run by a real Italian family.

Hotel Prices	
Prices are per night for two people	
[No $]	less than US $20
$	US $20-$40
$$	US $41-$80
$$$	US $81-$125
$$$$	US $126-$200
$$$$$	over US $200

La Ceiba Tree Lodge *(☎ 506/814-4004, ciebaldg@rasca.co.cr, $$)* is a captivating little lodge – not in a tree but up on a hillside with a sweeping view of the lake. A 500-year-old-plus ceiba tree is the inspiration for this wonderful B&B with five rooms and an apartment. Over 80 species of orchids are grown in the gardens surrounding the magnificent ceiba.

A short distance from the botanical gardens is **Villa Decary** *(☎ 506/383-3012, www.villadecary.com, $$$)*, another wonderful B&B, this one owned by North Americans. It has nice rooms and offers great birding.

THE TREE OF LIFE

Ceiba trees were considered holy in the Maya-influenced pre-Columbian civilizations such as the local Guatusu. They are called the "tree of life." Their roots lead to the underworld; their trunk is the world in which we live. The tree's spreading branches hold up the sky, home of the gods.

The quaint but prosperous town of **Nuevo Arenal** replaced the original town that was flooded in the late 1970s. Achtung! Not to be missed on the main drag is Ellen and Thomas' German bakery café, **Tom's Pan** *(☎ 506/694-4547)*. They serve delicious daily baked goods, good coffee and rich German /Tico lunches. Try the strudel. Perched on a rise on the side of the road, Tom's is a must stop for anyone passing through Nuevo Arenal. Tom also has a cabin available to rent. It's around back in the garden and features a sensual waterbed. Find diner-style gringo food at **Restaurant/Bar Bambu**, next to the gas station. At last, from here to Tilarán the road improves. Good eats can be found at **Willy's Caballo Negro**, three km/1.9 miles west, where vegetarian and European cooking are served side by side. If you follow the signs inland of Nuevo Arenal to Cote Lake you'll come to **Lake Coter Eco-Lodge** *(☎ 506/440-6768, www.ecolodgecostarica.com, $$)*, an inviting little eco-adventure lodge with a canopy slide and environmental programs for eco-tourists.

Back along Arenal lake road is **Chalet Nicholas** *(☎ 506/694-4041, nicholas@sol.racsa.co.cr)*, a hidden gem of a B&B run by North Americans and surrounded by woods. Bordering their property is a nature reserve; hiking and horseback riding are popular. Non-smokers and dog-lovers only.

The **Rock River Lodge** *(☎ 506/692-1180, $$, www.rockriverlodge.com)* advertises the "Best Mountain Biking in Costa Rica." But we know it best for its popularity as a windsurfing outfitter. The restaurant-bar has a welcoming fireplace for cool or rainy evenings. The **Mystica Resort, Bar & Pizza** *(☎ 506/692-1001, mystica@aol.com, $$)* never had Julia Roberts stay there (*Mystic Pizza*, her first movie... get it?) but does offer wood oven pizza in the restaurant. Their pretty ranch-style cabins have spacious rooms with porches entwined by flowers.

■ Tilarán

The roundabout-the-lake route described above leads to Tilarán and from there, access to the western beaches of Nicoya and Guanacaste or to the famous Monteverde Cloud Forest. Tilarán is a cool country town at the center of farmlands and horse and cattle ranching. It's an unpretentious place that can serve as a base for trips to several destinations, with a comfortable climate and lots of activities. Its main claim to fame, however, is the popular **Fiesta de Dias Civicas** – a rodeo, livestock show and country fair held the last weekend of April each year.

Eat at **La Carreta Restaurant** behind the cathedral; you can't miss the signs. Billie and Tom Jaffe from Florida opened it in 1994 and their food is legendary. It has an unusual gift shop and a good menu. Dine on the shady porch where plastic bags, filled with water, hang from the rafters. Apparently, the sun's rays refract in prisms that seem to keep away flies. They have opened the eight-room **La Carreta Bed & Breakfast** *(☎ 506/695-6593)*, which is a good choice. Alternatively, stay cheap at the **Hotel Naralit**, across from the modern-looking cathedral. From Tilarán you may be able to see the **Tenorio Volcano** (no facilities) to the north, which looks like any other mountain. A good new B&B has opened just south of where the road from Tilarán reaches the lake. Turn toward Tronadora and go one km/.6 miles to **Casa Manaña** *(☎ 506/693-1234, www.casamanabedandbreakfast.com, $$)* near the old Lake Arenal Country Club. Say hello to Alex and Laura.

From Tilarán there is a back way to Monteverde (see below) and a good road to the Inter-American Highway, which can take you to Guanacaste or back to San José.

■ Backcountry

A scenic alternate way to the far side of the mountains, without going around Lake Arenal, is to head north from La Fortuna to San Rafael and up to Upala (the road between San Rafael and Upala is narrow but drivable) and take the road that crosses the valley between the Tenorio and Miravalles volcanoes. The rural road reaches the Inter-American Highway about seven km/4.4 miles north of Cañas. Three km/1.9 miles from Bijagua town is a Cooprena lodge, **Albergue Heliconia** *(☎ 506/286-4203, www.agroecotourism.net)*, run by 12 local farm families growing coffee and grain.

People splashing in the hot springs.

A rustic, all-inclusive eco-adventure lodge on the slopes of Tenorio – look for a sign six km/3.7 miles north of Bijagua, after Bar Mirador – is **La Carolina Lodge** *(☎ 506/380-1656, www.lacarolinalodge.com, $$)*. This ranch-style lodge has a river and a pool for swimming and offers horseback trips or some great hikes to this little-visited area of Costa Rica. How do waterfalls, thermal springs and mineral mud pits within a national forest sound? Even better is that fact that it's only about a two-hour drive to the beach!

Once you come to the Inter-American Highway, your choice is to proceed north to Liberia and the coast, or drive south through Cañas and backtrack up to Monteverde. If you choose, there is also a gravel road **shortcut** from Tilarán to Monteverde, bumpy and rough but passable with four-wheel-drive. A public bus follows the same route. If you go this way you'll pass **El Trapiche**, an old-fashioned sugar mill. The route is "iffy" during the rainy season, so ask first. Along this route (six km/3.7 miles before Santa Elena) is another Cooprena lodge, **Albergue Ecoverde** *(☎ 506/286-4203)* run by 12 families of Los Olivos.

Monteverde

May your trails be crooked, lonesome, dangerous,
leading to the most amazing view.
May your mountains rise into, and above, the clouds.
~ *A Prayer for the Traveler*, Edward Abbey, 1927-1989

■ Monteverde Cloud Forest

A colony of Quaker farmers from Alabama, searching for a more pacifistic environment, came to Costa Rica in 1951 just two years after it abolished its army. They set up dairy farms in Monteverde ("Green Mountain"), part of the steep Tilarán mountain range. The community prospered when they developed a market for their excellent cheeses. Come here hungry – **Productores de Monteverde Cheese Factory** produces a ton of cheese a day.

The original settlers included forward-thinking pioneers who recognized the frailty of the environment and determined to preserve the Guacimal River at its source in the primary cloud forests. By preventing development in their original 554-hectare/1,368-acre **Bosque Eterno**, the area remained pristine until the threat of homesteading in the surrounding forest forced further action. Visiting scientists George and Harriet Powell teamed up with long-time resident Wilford Guindon to promote an "official" nature preserve. With the help of the **Tropical Science Center**, who now run and maintain it – it is *not* a national park – the **Monteverde Biological Cloud Forest Preserve** *(☎ 506/645-5122, www.cct.or.cr)* became official in March, 1972. The foggy forest (elevation 1,440 meters/4,662 feet) boasts six distinct ecological zones. Since its inception, the preserved region has grown rapidly – both in size and fame – and now covers 10,500 hectares and sees well over 50,000 visitors annually. Bordered by the Santa Elena and Children's rainforests, plus Arenal and San Ramón forests, the area is one of Costa Rica's most important protected ecological zones. Entrance into the park is restricted to 150 people at any one time, which may mean a wait. US $12.

The Monteverde Cloud Forest is a diverse biological wonderland that contains old-growth trees, favored habitat of the spectacular quetzal. But it is also home to 400 other kinds of birds, including 30 types of hummingbirds and such odd specimens as three-wattled bellbirds and bare-necked umbrella birds. Ground-dwelling residents include the powerful jaguar, the ocelot and the tapir, 2,500 species of plants (including 420 kinds of orchids) and 1,200 amphibian and reptile species. Because there is so much to see, it's best to have a guide who knows where to look for wildlife in the wild. Guided tours run about US $15 per person and night tours, when

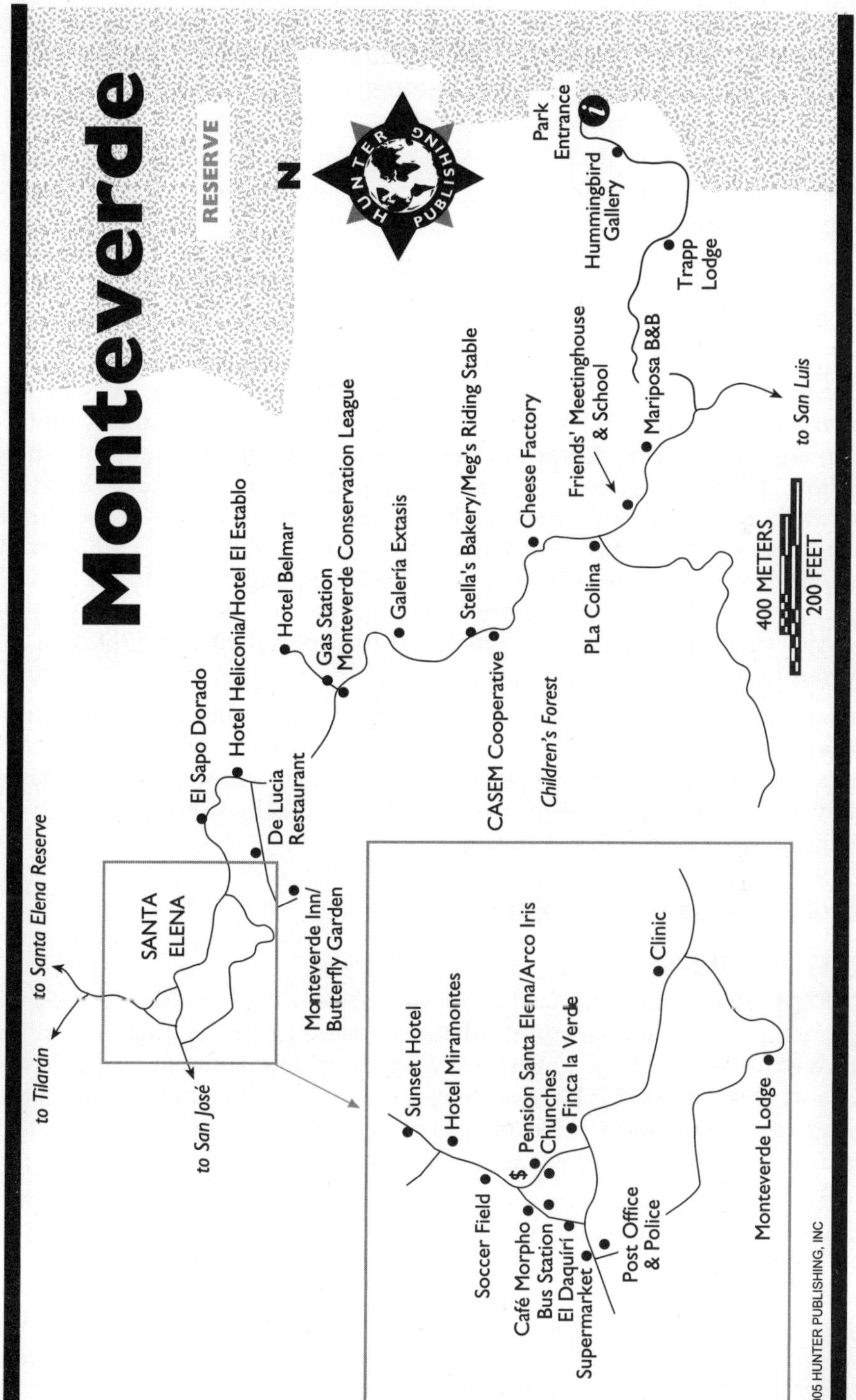
Monteverde
RESERVE
N
HUNTER PUBLISHING
Park Entrance
Hummingbird Gallery
Trapp Lodge
Friends' Meetinghouse & School
Mariposa B&B
to San Luis
Cheese Factory
Stella's Bakery/Meg's Riding Stable
Galería Extasis
Monteverde Conservation League
Gas Station
Hotel Belmar
Hotel Heliconia/Hotel El Establo
El Sapo Dorado
De Lucia Restaurant
Monteverde Inn/ Butterfly Garden
CASEM Cooperative
Children's Forest
PLa Colina
400 METERS
200 FEET
to Santa Elena Reserve
to Tilarán
SANTA ELENA
to San José
Sunset Hotel
Hotel Miramontes
Pension Santa Elena/Arco Iris
Chunches
Finca la Verde
Clinic
Soccer Field
Café Morpho
Bus Station
El Daquirí
Supermarket
Post Office & Police
Monteverde Lodge
$
© 2005 HUNTER PUBLISHING, INC

it's not raining, start at 7 pm. You can reserve a guide on-line at guide@monteverdeinfo.com. An excellent website focusing on both Monteverde and the Santa Elena area is www.monteverdeinfo.com.

If you're driving the Inter-American Highway there are two routes up the mountain. Both feature hard driving on rough roads – a conscious effort by the Monteverde community to slow development by jarring your kidneys. The **southern route** from San José is just before the Río Lagarto Bridge at Km 149; look for a tiny sign on a bar to the right. There's also a shortcut south of the Lagarto Bridge at Río Sardinal, which joins the Lagarto road, but this "shortcut" may sometimes take longer because the road is often in terrible shape. Plans are afoot to improve the main route.

The **northern route** is better marked and leaves the highway at the turn for Las Juntas. It's a good road for a third of the way up the mountain, and then it deteriorates to be as bad as, if not worse than, the other. Follow the sign in Juntas for San Rafael, not Los Dos. Monteverde is two hours from the highway and about four hours from San José. If you have time, check out Las Juntas, once the gold-mining capital of Costa Rica. It has an Eco Museo outside town with trails leading to an old gold mine. Boston – yes Boston – is a nearby town with an active gold-mining cooperative that you can visit. Call **Mina Tours** *(☎ 506/662-0753)* for an escorted tour.

As of press time, the controversy of improving the road up to Monteverde from the Inter-American Highway has renewed. The road's condition definitely restricts the number of tourists and extends the length of time they stay which, it can be argued, benefits the ecology. Four-wheel-drive vehicle is strongly recommended.

Adventures on Foot

Hut Hiking

Three rustic back country huts are available in Monteverde for just US $4 per night, plus the entrance fee for each day. It's a five-hour hike to **Eladios Hut** (Portland Audubon Center), the largest of the three. **La Leona Hut** features an enchanting setting on the Río Peñas Blancas with a cable car upstream for crossings. It takes 3½ hours to reach. The closest, **El Valle Hut**, takes 2½ hours to reach.

Huts are rustic, with dorm bunk beds, water, propane and wood stoves, cooking pots and utensils, but no sheets or blankets. Bring food, candles, sleeping bags and toilet paper. Arrange with the visitor's center.

TO GUIDE, OR NOT TO GUIDE?

No guides are needed for the following Monteverde Cloud Forest trails, although we recommend you hire one to point out the wildlife. Use your hotel's recommendation or hire a private guide through **Asociación de Guias Monteverde** *(☎ 506/645-5483)*.

Hiking Trails

Many of the trails can be combined to form loops.

MONTEVERDE HIKING TRAILS
All times given are approximate for round-trip hikes.
SENDERO BOSQUE NUBOSO (Cloud Forest Trail) is the most popular trail, winding past some fascinating strangler figs. It takes about 1½ hours and is just shy of two km long (1.2 miles), rising 65 m (213 ft). A self-guided tour booklet of this trail is available in English and Spanish at the visitor's center.
EL CAMINO (The Road) rises 45 meters/148 feet through a less forested area; the sunlight attracts butterflies and birds. Two km (1.2 miles); allow about an hour.
SENDERO PANTANOSO (Swamp Trail). This 1.6-km trail (one mile) takes a little over an hour to traverse a swamp forest. It features wild magnolias and a podocarpus, the only conifer in the preserve.
SENDERO RIO (River Trail) follows the Quebrada Cueda for nearly two km (1.2 miles) to a waterfall. Near the waterfall check out the zapote trees and their huge, buttressed roots.
SENDERO CHOMOGO rises 150 meters/492 feet to reach a height of 1,680 meters/5,510 feet above sea level. During your 1¼-hour hike you'll encounter oak, bamboo, heliconia, and hot lip plants abound.
SENDERO BOSQUE ETERNO (Eternal Forest Trail) is in the original Quaker preserve. This 20-minute hike shows good examples of strangler figs.
SENDERO GEORGE POWELL is named after one of the founders of Monteverde. It's a quick 10 minutes through a second-growth forest.
SENDERO BRILLIANTE (Brilliant Trail) leads you along the Continental Divide to La Ventana, with a wide view of the elfin forest en route. Bamboo is plentiful. You can hike this .2-mile trail in 10 minutes.
SENDERO ROBLE (Tree Trail) is an appealing narrow trail with a beautiful heliconia grove. It will take 10 minutes to hike.

Outside the reserve there is an excellent hike to the top of Cerro Amigos, along the dirt road to the Television Tower. The road begins just before the gas station on the road to Monteverde and passes the Hotel Bel Mar.

It's about 3½ hours round-trip. On a clear day, you can see forever from the 1,842-meter/6,000-foot peak.

Bosque Nuboso de Los Niños

Never doubt that a small group of thoughtful,
committed citizens can change the world.
Indeed, it's the only thing that ever has.
~ Margaret Mead, American anthropologist

The **Children's Eternal Rainforest** is good for children of all ages – in more ways than one. Its 3½-km/2.2-mile **Bajo del Tigre Trail** (Jaguar Canyon) snakes along the Pacific slope in a transition zone between wet and cloud forests. Thirty tree species within its border have been identified as previously unknown and the area is rich in diverse wildlife. The forest is managed by the Monteverde Conservation League and it borders the Monteverde Preserve.

But what is most magical about these woods is the fact that it was inspired by a fourth-grade students' Children's Rainforest Campaign in Sweden in 1987. It is now the largest private preserve in Costa Rica at over 50,000 acres.

There are two small private complimentary ecological endeavors in the area. **Finca Ecologica**, an Ecological Farm *(☎ 506/645-5363)* with a forest, waterfall and fields, is great for birding and has guided tours available. The **Reserva Senduro Tranquillo** (Quiet Path Reserve, ☎ 506/645-5010) is a 200-acre reserve belonging to the Lowther family. It has excellent bi-lingual trail guides. Up the hill from the Cheese Factory.

Santa Elena Rainforest Reserve

A culture is no better than its woods.
~ *Woods*, Wynstan Hugh Auden, 1907-1973

Much less visited, so in some ways more virginal, is the nearby **Reserva Santa Elena**, one of the first successful community-administered private reserves in Costa Rica. More than 750 acres of rainforest land was permanently leased by Santa Elena High School seven km/4.4 miles from the town of the same name. It was originally meant to be a training ground for farming, but in 1989 students and faculty decided to open it as a cloud forest preserve instead. Now it is an outdoor classroom in botany, biology, ecology and tourism. Entrance fees go towards maintenance and benefit local schools. One of the highlights in the 12 km/7.5 miles of trails is an 11-meter/36-foot viewing platform above the rainforest canopy.

A four-wheel-drive leaves from Pension Santa Elena and the Banco Nacional each morning and heads to the reserve. Make reservations for a ride and/or a guide by phoning ☎ 506/645-5014. There's also a welcome center at the north end of town.

On the way to the reserve is **Mirador San Gerardo** *(take the left fork in the road, ☎ 506/645-5354, www.monteverdesangerardolodge.com)* that offers sweeping views of Arenal Volcano and Lake – but usually only in the dry season (it's always cloudy in the rainy season). The ranch features horseback rides and hikes on their property, including lunches, hostel accommodations and a restaurant. One enticing horseback ride heads to the volcano – a unique, sore-butt, overnight excursion.

■ Santa Elena Pueblo

Peace is more precious than a piece of land.
~ Anwar Sadat, 1918-1981

There is no real town of Monteverde, but a charming little village called **Santa Elena**, six km/3.7 miles downhill. This is the center of most activity and it's where you'll go for provisions and socialization. It features shops, bus station, restaurants, supermarket, bars, post office with Internet access and even a very likeable bookstore/laundromat/coffee shop, **Chunches**, *(☎ 506/645-5147)*, owned by Jim Stanley and Wendy Rockwell. The bulletin board here is used extensively by the community and visitors.

Shopping

When not trekking around the woods, serious shoppers head for "Community of Artisans of Santa Elena and Monteverde" (**CASEM**), a women's crafts cooperative. They also feature lots of one-of-a-kind T-shirts, clothing, local coffee, weavings, pottery and souvenirs. From a modest start with eight women in 1982, the co-op has grown to over 140 women and 10 men.

The **Monteverde Park Visitor's Center** also has a well-stocked gift shop. The **Cheese Factory** shows how they make and sell cheese. Up the road is a **Ceramics Studio** and there is a **Butterfly Garden** near the Finca Ecológica.

Gift and art galleries abound. Check out the **Hummingbird Gallery**, up on the way to the park entrance, or **Galería Extasis**, also on the main road uphill. Weekly workshops in various art forms can be arranged for long-term visitors at the **Studio of the Arts** *(US ☎ 800/370-3331, CR ☎ 506/645-5434, www.mvstudios.com)*, a cooperative of local artists.

Adventures

More active travelers should contact the **Original Canopy Tour** *(☎ 506/645-5243, www.canopytour.co.cr)*, which claims to be the pioneers of canopy tours. Participants are strapped into a safety harness and proceed to zip along steel wires stretched between tall trees. We've been on several canopy tours in Costa Rica and found this one pleasing. The best part for us was climbing to a platform through a hollowed strangler fig lattice that had killed the tree it once wrapped around. This activity is great for kids from eight years old to 80. Prices start at US $45 for walk-ins.

Safely harnessed, a tourist enjoys a Monteverde canopy tour.

Higher and longer thrills can be found at **Sky Trek** *(☎ 506/645-5238, www.skytrek.co.cr)*, a complex of high suspension bridges, platforms and zip lines to shoot over and through the trees. You'll be as much as 127 meters/417 feet above the ground and travel 427 meters/1,400 feet on one of the runs. Of course, you're strapped into a safety harness. Its sister attraction, **Sky Walk**, offers a more sedate experience. Two km/1.25 miles of hiking trails lead to five long, narrow suspension bridges as much as 40 meters/132 feet above ground. All attractions are safe. Cost is about US $35 per person for the Trek and US $15 for the Walk.

An even more sedentary tour is with **Natural Wonders Train** *(☎ 506/645-5960)*. It mimics the tram with cars suspended on rails traversing 1. 5 km/.9 miles of open fields and woods. The slow-moving cars go as low as ground level and as high as 12 meters/39 feet. Near the Finca Ecológica.

Horseback riding is favorite pastime of both tourists and residents in Monteverde. **Meg's Stables** *(☎ 506/645-5052)*, at Stella's Bakery, is the area's original horse haven and is very experienced. **Sabine's Smiling Horses** *(☎ 506/645-5051)* has happy equines as well as customers. Calm steeds are the specialty of **El Palomino** *(☎ 506/645-5479)* and **La Estrella** *(☎ 506/645-5075)*. Gitty yap.

Most hotels will make your day-before arrangements for the tours listed above.

Music lovers might like to coordinate their visit with the **Monteverde Music Festival** *(☎ 506/645-5053, www.muiinstitute.org)*. It runs March through April at the Monteverde Institute and offers live performances nightly.

The main offices of **Selvatura Park** (☎ *506/645-5929, www.selvatura.com, restaurant*), a multi-recreational, eco-tourism park, are across the street from Santa Elena's Catholic Church. This private commercial park boasts canopy zip lines, treetop walkways, a reptile and amphibian exhibit and hummingbird and butterfly gardens. Also located here is the world's largest private insect collection, **Jewels of the Rain Forest**, whose worldwide specimens are presented in a way that can only be described as an "art form." The famous entomologist, Dr. Richard Whitten, who has collected insects since age five, presents this unusual and fascinating exhibition. *Ants in your pants?* If you're lucky he'll be there to expound on his passion – or play the pipe organ.

All told, there are four canopy zip line tours, including Selvatura, in Monteverde. The least expensive is **Aventura Canopy Tour**, which offers free transportation and features an exciting Tarzan Swing. The forest over which you travel is not as dense as at other zip rides, but thrill-seekers will love stepping off a two-story high platform and swinging 20 meters/66 feet or more back and forth over a ravine.

Places to Stay

If I've only got two weeks to live,
I'll take the last week in July and the first week in August.
~ Ian Baker, cartoonist.

All in all, we found Santa Elena a great place to hang out for several days or longer. There are some good hotels near town and in the area, plus more on the road uphill toward the forest. A bus leaves to Monteverde Preserve from the Banco Nacional at 6:15 am and 1 pm daily, returning at noon and 4 pm. Only a couple of hotels are less than a half-hour walk uphill to the entrance.

Hotel Prices	
Prices are per night for two people	
[No $]	less than US $20
$	US $20-$40
$$	US $41-$80
$$$	US $81-$125
$$$$	US $126-$200
$$$$$	over US $200

In Town

Hotel Finca Valverde's *(250 meters/822 feet southeast of the Banco Nacional, ☎ 506/645-5157, fax 645-5216, www.monteverde.co.cr, 22 rooms in 14 cabañas and 8 standard rooms, restaurant, $$)*. Just on the outskirts of Santa Elena, where the road inclines toward the Monteverde Cloud Forest, is this working coffee plantation that has sprouted a very pleasant hotel and bungalows in the forest below their *finca* (farm). A metal bridge suspended over a rushing stream – a free mini-Sky Walk – leads to the comfortable standard rooms housed in a large wooden building with wrap-around balcony and a good view. The only disadvantage to these quarters is the lack of a bathroom towel rack. The individual cabañas, scattered along the wooded hillside, are more luxurious, incorporating a large bathroom, individual veranda with garden chairs, main bedroom and upstairs sleeping loft with queen-size beds. The walls are all natural wood. Several trails fan out past the cabins into the coffee fields that climb the steep surrounding hills. The farm has been in Victor Valverde's family for years and his sons now work with him. His daughter-in-law manages the hotel's candle-lit restaurant, **Don Miguel**, where starched napery and curtains hand-painted with images of Costa Rican birds and flowers add to the romantic atmosphere. Excellent local cooking and very generous servings, but reservations are requested for all three meals. Private enough to make it feel like the forest, but just a two-minute walk to the town and its community attractions.

Arco Iris *(behind the Pension Santa Elena, ☎ 506/645-5067, fax 645-5022, www.arcoirislodge.com, 10 rooms in 7 cabins, $$)*. The Arco Iris is German-owned and features attractive wood cabins on a tropically landscaped hill at the edge of town. The inviting duplex cabins are spread around the hillside set back off the road to Monteverde and the rooms in them are clean and basic. Well known as a pleasing place to stay. The larger family cabin has two bedrooms and a small sitting area with futon couches that can be converted into beds. Our favorite is the Doll House, a charming miniature cabin with bunk beds, a little desk and a private bath. It's a real bargain. Full breakfast is an extra US $5 per person.

Pension Santa Elena *(☎ 506/645-5051, fax 645-6060, www.monteverdeinfo.com, $)*. This bohemian low-rent lodge is the heart and soul of Santa Elena and if it ever changes it will doom the tiny town to gentrification. It has some private rooms with bath, and some shared rooms with shared bath. The people that run it are the town's Internet gurus. The pension is a landmark, with bulletin boards for communications to meet, arrange, rent, hire, bum, beg and borrow.

Toward the Cloud Forest

Trapp Family Lodge *(☎ 506/645-5858, fax 645-5990, www.trappfam.com, 10 rooms, restaurant, TV lounge, $$)*. If they ever find a human gene for geniality, they'll find it in the Trapp family lineage. The young owner of this lodge is one of the warmest and friendliest hosts around – we really felt welcome here. Andrés Trapp Belmar grew up in Chile and came to Costa Rica to work as a guide in the rainforest. His grandmother was a

Trapp from Austria, and hotels are in his blood – his uncle owns the Bavarian-like Belmar (see below). The rooms are clean and appealing, with lots of polished tropical hardwoods. We'd like the rates to include breakfast, but that may come only with competition. This is the closest lodge to the Cloud Forest entrance, although it's still a fair hike. We would stay here again, and again.

El Establo Hotel *(on the road to Monteverde in Cerro Plano, ☎ 506/645-5110, fax 645-5033, www.hotelestablo.com, 50 rooms, 2 restaurants, jacuzzi, cable TV lounge, fireplace lounge, pool, gym, full breakfast included, $$-$$$)*. We cannot do justice to this hotel by translating its name as "The Stable," nor by describing it as built in a style that resembles a converted barn. But it has a very homey kind of American horse-loving appeal, accented by room doors that look as if they open top and bottom – Dutch or cottage doors – and high-ceilinged public areas instead of hallways. Large, pleasing quarters have queen beds. The unique country style is a reflection of the owner, Ruth Campbell, who came to Monteverde as one of the original Quaker settlers when she was a baby. Large windows around the hotel look out into the 61-hectare/150-acre gardens and fruit trees that attract many birds and butterflies. Spotlessly clean and comfortable, El Establo is simple and priced very reasonably for the excellent service provided. Family friendly atmosphere. No smoking and no alcohol served.

Hotel Belmar *(on road to Monteverde, ☎ 506/645-5201, fax 645-5135, www.hotelbelmar.com, 34 rooms, jacuzzi, restaurant, $$$)*. If it were not for the warm temperature and surrounding cloud forest, you'd swear you were in the Alps. This well-known Swiss chalet-style hotel is built into the mountainside facing extravagant gardens that sweep downward to a small lake. The common areas have a rustic feel, but the individual rooms are well appointed, with two beds, dresser, natural wood walls and small balconies overlooking the grounds. The rather expensive restaurant, where dramatic sunsets can set the mood, is renowned for its international cooking.

Hotel El Sapo Dorado *(on a side road not far uphill from the fork, ☎ 506/645-5010, fax 645-5180, www.sapodorado.com, restaurant, $$$)*. The Sapo Dorado, or Golden Frog, is often sited as one of the best places to eat in Monteverde. It is – and it helps that the owner, Hannah Lowther, is a gourmet chef and cookbook author. But it is also a luxury hotel that boasts beautiful suites with fireplaces, terraces with sweeping views and private cabins nestled in the woods. This is one of Monteverde's best.

Mariposa B&B *(opposite La Fonda Vela Hotel on the road to Monteverde, ☎ 506/645-5013, ravargas@costaricense.com, 3 rooms, breakfast included, $)*. A young Tico family rents three simple rooms set in a green building up a private driveway along the road to Monteverde. Close to the cloud forest. You get a bathtub (hot water too), queen-size

*Opposite: This rainforest flower looks other-worldly.***

mattress, lamp, a big dresser and breakfast. You may ask for other meals at an additional charge.

North of Santa Elena

Swiss Hotel Miramontes *(500 meters / 1,645 feet from the soccer field on the road from Las Juntas, ☎ 506/645-5152, fax 645-5297, www.swisshotel/miramontes.com, 6 rooms, restaurant, $$).* Set up on a sunny hill, the Miramontes is a small hotel run by a Swiss couple. The four standard rooms are off a hallway that leads to their appealing restaurant (it's open for lunch and dinner and features international cuisine, including Swiss). The best of the bedrooms is the first one, which has big bright corner windows. Each good-size room has a queen and twin bed and natural wood-paneled walls. The two rooms in the cabin (#7 and 8) are larger, with a small table and chair, shower and bathtub, all wood walls and ceiling, and a veranda. By showing even a small interest in orchids or butterflies, you may encourage part-owner Walter Faisthuber to take you on a nature walk or butterfly hunt.

Sunset Hotel *(outside town on the road to Skytrek and Santa Elena Forest, ☎ 506/645-5048, fax 645-5228, 7 rooms, restaurant, full breakfast included, $-$$).* The squeaky-clean Sunset offers motel-like rooms, all with a great view of the sunsets over the Nicoya Peninsula – if the clouds don't get in the way. It's an excellent value in Monteverde, set on a knoll with expansive open gardens and nature hiking trails. The hotel's bright and welcoming rooms contain one queen and one twin bed each, tile floors and bath, table lamp and front sitting porch. The restaurant, open for breakfast and dinner, features German/European cooking with an expansive view from its panoramic windows. Mucho birds and butterflies.

Outside of Town

Ecolodge San Luis & Biological Station *(☎ 506/645-5277, www.ecolodgesanluis.com, $$$).* This is a unique attempt to bring together researchers, their work and interested eco-tourists in a fairly luxurious setting. Located at 1,400 meters/4,600 feet near a primary cloud forest, facilities include housing and research labs for students and professionals, a forest preserve and comfortable lodgings. Choose from well-appointed *cabina* rooms with balconies overlooking the San Luis River gorge, roomy bungalows or bunkhouse facilities (real bunk beds in dorm rooms). All have shared baths). A stay here is like being back in summer camp. Food is served family-style with plenty of company. The Ecolodge attracts environmentally aware tourists curious about life and smart enough to stay here. In San Luis, a few kilometers from Santa Elena, but 40 minutes by car on a very poor road.

Nature Tours

"Customizing Your Adventures
In Our Country Since 1991"

Photo courtesy Bruce & June Conord

Mexico,
Central
& South
America

Hiking, whitewater rafting, kayacking, diving, deep-sea fishing, surfing, birdwatching.

Soft adventures as well as high adrenaline adventures.

Fly-and-drive packages, escorted guided tours, individuals and groups.

www.naturetoursinc.com
info@naturetoursinc.com
☎ 800-444-3990

Above: Tree-top canopy, Braulio Carrillo Park.
Opposite: Poás Volcano crater lake.
Below: Payaso masks.

Above: Bananas for sale at a roadside stand.
Opposite: Basilica de Nuestra Señora Los Angeles, Cartago.
Below: Boulder covered by ferns, Montezuma.

Above: Beautiful cascading flowers.
Opposite: Staircase engulfed by huge leaves.
Below: A scenic road snakes through the Costa Rican countryside.
Following page: Arenal Volcano, seen from the lake.

Places to Eat

Reservations? No, we definitely want to eat here.
~ Dave Carpenter, cartoonist.

Just when you think you'll never survive another creep-along, crawl-along bumpy road, a short way down this one provides a respite at **Restaurante De Lucía** *(on the road to the Butterfly Gardens, Cerro Plano, ☎ 506/645-5337, $-$$).* We didn't even see a sign at the intersection (where the Hotel Heliconia and El Establo are) for this surprisingly good eatery featuring international cuisine and excellent wines. Owner Lucia and her husband José (who was once a lawyer in Chile – so don't get him started about Pinochet) have built a lovely all-wood restaurant with a European ambience, not to mention a paved parking lot. Open from 11 am for lunch and dinner. Prices are reasonable and the food commendable.

Restaurant Prices

Prices are for an average entrée

$	less than US $5
$$	US $5-$10
$$$	US $11-$20
$$$$$	over US $20

Don't miss **Stella's Bakery** *(☎ 506/645-5560)*, across from CASEM store – just follow the scent of home baking and Monteverde coffee. Stella's is a hip little coffee shop, bakery and sandwich shop with eclectic artwork and earth mothers in attendance. A Monteverde institution.

Most hotels have good restaurants, and you don't need to be staying there to dine. A higher-class entry is the one in the **Sapo Dorado Hotel**, which receives the cuisine kudos from us.

In town, morph into **Café Morpho**, across the street from the bus station. This is by far the most popular place in town and they've recently expanded. You'll get a good economical meal – veggies and *batidos* are big here – and a chance to see fellow wanderers. Coffee up and hunker down at **Chunches**, a bookstore, restaurant, everything store on the side street across from Pension Santa Elena, or eat and drink at the popular **El Daquirí**. Nearby, **La Taverna Bar & Disco** heats up the crowd after sundown.

Liberia & Guanacaste

Last night as I lay on the prairie, And looked at the stars in the sky,
I wondered if ever a cowboy, Would drift to that sweet bye-and-bye.
~ song: *The Cowboy's Dream*, anonymous

Liberia

The capital of Guanacaste Province is the growing city of Liberia, now complete with an increasingly popular international airport. Overlooked as a tourist destination, the dusty colonial town is a center for the many ranches and farms in the mostly agricultural area. The sun shines more in Guanacaste especially in the dry season, but even in the rainy season when things are green, you get more sun time here. The year-round weather down on the plains is hot and hotter, but the town offers some respite in its cool, shady central square and *mercado*. Liberia is a more colonial town than any we've mentioned before. Spanish architecture survives in corner homes that have *puertas del sol* (wooden doors on each side of the corner, one to let in the morning sun, one for the afternoon); streets lined with flamboyant trees; brilliant white-washed walls; adobe and stucco; brown tiled roofs; wooden porticos; bright color trim; and bougainvillea peeking over garden walls. All these elements combine to make a pleasant stroll in the early morning or late afternoon. A good area is **Calle Central** (also called Calle Real), from the cathedral south, a street lined with historic, colonial-era buildings.

QUARTZ DELIGHT

The bright sparkles in the walls of older buildings are reflections of quartzite contained in the white coral limestone used to build most of the original Ciudad Blanca, the White City.

If you're walking around, stop at the **Casa de la Cultura**, three blocks south of the main plaza. It features a display (in Spanish) of the area's cowboy, *sabanero*, heritage. Another destination could be the **Iglesia de la Ermita de la Resurreccíon** (Church of the Hermit of the Resurrection), thankfully shortened by locals to **La Agonía**, The Agony. The simple, quiet, colonial-style church dates back to when this corner of Costa Rica succeeded from Nicaragua. Liberia holds a major festival each year on July 25 to commemorate its succession from its northern neighbor. Count on cowboys, rodeos, parades, processions, fireworks, concerts and bloodless bullfights.

Liberia

1. La Agonía
2. Casa de Cultura
3. Cathedral
4. Pharmacy
5. Hotel Liberia
6. ICE (phones)
7. Posada del Tope
8. Hotel La Siesta
9. Gas Stations
10. Best Western El Sitio
11. Best Western Las Espuelas
12. Post Office
13. Pulmitan Bus Station
14. Tracopa Bus Station
15. Central Market

500 METERS
1500 FEET

© 2005 HUNTER PUBLISHING, INC

Set along the Inter-American Highway, Liberia is the last large city before Nicaragua, 75 km/47 miles north. For sea lovers, it's a crossroads for traffic headed to the beaches on Nicoya; nature lovers use it as a gateway to the Santa Rosa, Guanacaste, Palo Verde and Rincón de la Vieja national parks, plus San Ramón (Lomas Barbudal) Nature Preserve.

Places to Stay & Eat

If you're staying, several hotels and services cluster around the busy highway intersection where you turn for Nicoya and the international airport. You'll know you're there by the big **Burger King** on the corner. There are two Best Western hotels in Liberia: **Best Western Las Espuelas** *(US ☎ 800/528-1234, CR ☎ 506/666-0144, www.bestwestern.com, $$)* is south of the main intersection and traffic light; and **Best Western El Sitio** *(US ☎ 800/528-1234, CR ☎ 506/666-1211, www.bestwestern.com, $$)* is west of the main intersection and light, toward the airport. Both are commendable hotels with big pools, air conditioning, restaurants, cable and secure parking. Farther along the road past the airport is the wonderful **Café Europa & German Bakery**, famous for sweet treats, lunches and German baked specialties. You'll encounter German Bakery goods sold or served in various hotels and restaurants on the shore. Big souvenir stores are nearby.

Hotel Prices	
Prices are per night for two people	
[No $]	less than US $20
$	US $20-$40
$$	US $41-$80
$$$	US $81-$125
$$$$	US $126-$200
$$$$$	over US $200

In Liberia's town, around the main park square, one of the prettiest motels is **Hotel La Siesta** *(Calle 4, Av 4 & 6, ☎ 506/666-0678, ropeflo@hotmail.com, pool, restaurant, air, cable, $)*, with an inviting blue pool that nudges up to the porch of its rooms. A block away on Calle 4 is **La Cocina de José**, one of the few upscale eateries in town. It serves typical local food. The inexpensive **Posada del Tope** *(Calle Central between Av 2 & 4, ☎ 506/666-3876)* offers small rooms with shared baths in a restored colonial home. They manage the hotel across the street, **Casa Real**, which boasts an astronomical observatory up top. One block north is the clean and cheap **Hotel Liberia** *(☎ 506/666-0160).*

Our favorite place to eat breakfast in town is the venerable patio restaurant **Jardín de Azucar** (Sugar Garden) near the Central Park. We also like the seafood **Restaurant Paso Real** on the corner facing the park on the south side with a balcony. East of the park and south on Calle 8 is an intriguing new place, **Restaurant El Café**, which serves light French food, cheeses and better wines in air-conditioned comfort. Right now, it's open only Monday through Friday until 7 pm.

Just 200 meters/660 feet north of the highway intersection, as you head toward Nicaragua and Rincon de la Vieja Park, look left for basic **Hotel Aserradero** *(☎ 506 / 666-1939, abalto@rasca.co.cr, $)* in a converted lumber mill, and **Jardín Cervecero La Jarra**, a chic new seafood restaurant that also serves chocolate fondue! Melts in your mouth and in your hand.

Northern National Parks

Who has seen the wind? Neither you or I,
But when the trees bow down their heads, The wind is passing by.
~ *Who Has Seen the Wind?*, Christina Georgina Rossetti, English poet

■ Palo Verde National Park

Nestled down in the Tempisque River Valley is one of the most important bird paradises in Costa Rica, the **Palo Verde National Park** *(☎ 506 / 671-1062, www.ots.ac.cr)*. The park was once a massive ranch that was used – and still is – by the Organization for Tropical Studies to investigate its unique dry-forest ecosystem. In 1977 it was named a national park, yet cattle still graze in many parts and help keep down the vegetation, reflecting a symbiotic relationship with the birds that have come to seek refuge. Plus, cattle businesses benefit the local economy.

To get to Palo Verde National Park, follow the signs one hour west from Bagaces, a small town halfway between Cañas and Liberia on the Inter-American Highway.

Birding

The best time to see birds is during the dry season when they congregate around the water holes very near the ranger station. Scarlet macaws are the big attraction here, even for non-birders. The administration will refer you to local boatmen who will provide transportation over to **Isla Pájaros**, an important nesting ground. Arrange in advance. Observation of "Bird Island" is from the boat only.

Do not scare up a flock for photos – it disturbs their nesting cycle and can result in a fine for you.

Places to Stay & Eat

If you want to stay close to Palo Verde, try the new luxury B&B, **Rancho Humo** *(☎ 506/255-2463, $$$)*, located near Puerto Humo on the west bank of the Tempisque River. Reservations required.

Eco-tourists will be happy a bit upriver at the **Coopeortega** *(www.turismoruralcr.com is their Spanish-language site)* cooperative in Bolson, accessed from Santa Cruz on the Nicoya Peninsula. The communities of Bolson and Ortega offer family lodging, boat trips on the Tempisque, Charco and Bolson rivers to see the crocodiles and birds, as well as horseback riding through the towns, farms and wetlands. They cook typical country food and have folk dancing and music.

Don't be surprised to see wandering cattle as you explore Palo Verde.

■ Lomas Barbudal

The **Reserva Agroecologica de San Ramón** is more commonly known as Lomas Barbudal, or Bearded Hills. It's just 15 km/9.3 miles southwest of Bagaces – look for a dirt road just north of Bagaces and follow it for six km/3.7 miles. Open only in the dry season, this bird and animal refuge also boasts a diverse plant life, including rare cannonball trees (*balas de canón*), upon whose pendulous fruit scarlet macaws come from Palo Verde to dine. Camping is permitted, but the park has only rudimentary facilities. Swimming and hiking. You can get a taxi from Bagaces. Admission, US $25.

A few kilometers north of the town of Cañas, on the banks of the Corobicí River, is the resort hotel **Hacienda La Pacífica** (☎ *506/669-1203, pacifica@pacificacr.com, $$)*, which began life as a research facility and now offers tranquil lodgings, a nice pool and tours on horseback. Its Restaurante Miravalles is popular if you're hungry about now.

■ Rincón de la Vieja National Park

Life is mostly froth and bubble.
Two things stand like stone,
Kindness in another's trouble,
Courage in your own.
~ *Ye Wearie Wayfarer,* Adam Lindsay Gordon, 1833-1870

Rincón de la Vieja National Park is an impressive park surrounding two volcanoes, 25 km/15.5 miles northeast of Liberia. The park's 14,090 hectares/34,802 acres have been split into two sections, Las Pailas and Santa Maria. Las Pailas (The Cauldrons) includes the active volcano, Rincón de la Vieja (Old Woman's Place), complete with fumaroles, mud pots and steam vents, plus a number of waterfalls.

The park is part of Area de Conservación Guanacaste, and you can contact their headquarters for information, ☎ 506/666-5051. Admission is US $6 per day.

Contribute to the ecology and local economy by booking and planning your visit through the local micro-business Chamber of Commerce association with the acronym, **ACATURIVIE** (☎ *506/691-8177, www.acaturivie.com*). The organization helps tourists and benefits women's cooperatives in the area's small towns. Headquartered in Curubanda Lodge and Finca Nueva Zelandia.

OLD WOMAN'S PLACE

The volcano's name comes from a Romeo and Juliet native legend about Princess Curubanda, whose lover from a feuding tribe, Prince Mixcoac, was thrown into the volcano. In order to be near him, she moved to the mountaintop and became a great healer, bestowing healing qualities to the ash and mud from the volcanic mud pots.

Rincón is a massive 600,000-year-old geological wonder with at least nine volcanic cones. The rim of the most active crater is 1,806 meters/5,940 feet high with near-vertical sides denuded of vegetation. No less than 32 rivers flow down the mighty volcano that bridges the Continental Divide, and its activity once served as a lighthouse for ships at sea. The most recent eruption of consequence was 1997 when it caused problems for some small towns on the Atlantic side.

Rincón's diverse ecosystems offer a variety of flora. Its high forested slopes feature gnarled, dwarfed trees draped in moss mats that provide the arboreal base for orchids and epiphytes. It hosts a large population of Costa Rica's national flower, the guaria morada orchid, and in the center of the brushwood is an immense 800-year-old ceiba tree – the start of one of several **canopy tours**. Rincón de la Vieja is home to 300 bird species, including crested guan, blue-crowned motmot and emerald toucanet, as well as a variety of mammals such as collared peccaries, agoutis, nine-banded armadillos and several species of monkeys. Many more people would visit here if access were a little easier, but the roads are often as bad as those at Monteverde.

> **WARNING:** Do not wander off the trail to get a closer look at the hot springs and vents. The crust is sometimes fragile.

The good news is that hiking is less difficult, and there are a number of good trails in both sections.

■ DIRECTIONS

Two routes lead to Rincón Park. From Liberia itself, head west on Av 3 through Barrio La Victoria. Once out of town this road is very rough – dramatically beautiful, but very rough. Turn left in Colonia Blanca. The more common route is to go five km/3.1 miles north of Liberia on the highway to the town of Guadalupe and follow the gravel road toward Curubandé, for a total of 20 km/12.4 miles. This road crosses the private land of Hacienda Lodge Guachipelin, which charges US $2 per vehicle as a toll. Four-wheel-drive only.

Places to Stay

Hacienda Lodge Guachipelin *(☎ 506/666-8075, www.guachipelin.com, $$-$$$ with meals)*. You might stay here and take their tours to the hot springs and other natural wonders. (If this is your first time, tours with any of the hotels are more rewarding than wandering around on your own.) The Guachipelin accommodations are a bit sparse but comfortable, with a lovely little pool. It's set amid a working 19th century-era cattle ranch. Their exciting adventure tour, the **Kazm Cañon** (www.canopytour.co.cr), involves hikes, climbs, traverses and rappelling in a lush canyon of Río Colorado. There's river swimming and diving, too. Loads of fun. You can go here on a tour from San José.

Albuergue Finca Nueva Zelandia *(☎ 506/691-8177, www.tourism.co.cr/hotels/nzelandia, $$)* and **Rincon de la Vieja Mountain Lodge** *(☎ 506/661-8198, www.rincondelaviejalodge.com, $$)* are rustic but comfortable farm lodges close to the park entrance. The new rustic luxury resort of **Buena Vista Mountain Lodge** *(☎ 506/661-8158, www.buenavistacr.com, $$)* welcomes tourists to its Adventure Center. It has cool rooms, a pool, restaurants, a spa and various activities.

Despite its being called a *posada*, the **Posada El Encuentro** *(☎ 506/382-0815, www.posadaencu.com, pool, $$ with breakfast)* is an excellent B&B in a former private home en route to the park. One of the "Charming Nature Hotels" Group, the *posada* is owned by a German couple who welcome guests. Although located farther from the park than others mentioned, it offers guided tours and pleasant rooms in which to relax.

Hotel Borinquen *(☎ 506/666-5098, fax 666-2931, www.borinquenresort.com, pool, $$$-$$$$)* is a luxurious and private mountain resort and thermal spa within range of the volcano's warm underground springs. Drive 26 km/16 miles north of Liberia, turn right through Cañas Dulce and continue to the main gate, about another 20 km/12.4 miles. In the middle of nowhere, guests are indulged in spacious luxury cabin suites with satellite TV, air conditioning and other amenities, such as spa beauty treatments, hiking, horseback riding, ATV rides, mud baths and swimming.

■ Guanacaste National Park

I travel not to go anywhere but just to go.
~ Robert Louis Stevenson, 1850-1894

The park that bears the name of the Province is a sizeable 32,500 hectares/80,275 miles of land reserved to protect animals and their migrations through the grasslands at the foot of the Orosí and Cacao volcanoes. Many rare and endangered species crisscross the park as part of their migratory routes and still others live in the area and depend upon a protected ecosystem for their survival. In the new ecological thinking, habitats – rather than individual species – are protected. This has the added benefit of helping multiple species. In the case of Guanacaste, naturalists are attempting to reclaim the original unique "dry forest" that once covered this corner of Costa Rica and the Pacific lowlands as far north as Mexico. Biologists have created a very successful program that trains local *campesinos* in identification of species, conservation and fire prevention – brush fires are the bane of reforestation efforts.

Because the park was set aside to directly benefit animals, very little in the way of facilities are available for the public. Of the four Biological Research Stations in the park, the most interesting for visitors is the Maritza Station, located in the Pedregal section at the base of Orosí Volcano. It has rustic hostel accommodations. From here you can arrange a tour of the intricate, rock **petroglyphs** dating back to around 500 A.D. Turn east at the intersection where a west turn would be for Guajiniquil.

The park is part of Area de Conservación Guanacaste, and you can contact their headquarters for information, ☎ 506/666-5051.

■ Santa Rosa National Park

Sometime they'll give a war and nobody will come.
~ *The People*, Carl Sandburg, 1936

Divided from Guanacaste National Park by the Inter-American Highway, Santa Rosa National Park *(www.acguanacaste.ac.cr; click for the English-language version)* was Costa Rica's first official park. In the quixotic world of environmental protection, the government created Santa Rosa to protect a battlefield – but wound up benefitting a multitude of microhabitats, including dry deciduous forests, mangrove swamps, oak forests, evergreen forests, dry savannahs and turtle nesting beaches.

The park is part of Area de Conservación Guanacaste. Contact their headquarters (based here) for information, ☎ 506/666-5051. Admission is US $6; pay at the booth on your way in. Open 7:30-4:30 daily.

WILLIAM WALKER'S BATTLE

This fateful battle occurred in late March, 1856, when William Walker, an American *conquistador*, invaded Costa Rica after taking Nicaragua with a mercenary force called the Filibusters. He made the Hacienda Santa Rosa (**Museo La Casona**) his headquarters and it became the scene of the battle between a hastily assembled Tico force and Walker's mercenaries. National hero, Juan Santamaria, a drummer boy from Alajuela, turned the tide by supposedly setting on fire the roof of La Casona, forcing Walker's command out into the bush. The good guys eventually won, as they did again in 1955. On January 11 of that year, a well-armed band of *calderonistas*, a self-proclaimed anti-communist group made up of exiled Costa Ricans and Nicaraguan mercenaries, invaded Costa Rica and captured the city of Quesada. One of their planes even strafed San José. The United States immediately sold Costa Rica four fighter planes – which deterred further air attacks – and the invaders were driven out 10 days later. The final pitched battle was fought at Santa Rosa.

The landscape in Santa Rosa is similar to the African plains – high temperature and a long dry season that browns everything, including your skin. The park is managed in two large sections, **Santa Rosa** to the south and **Murciélago** along the north coast of the peninsula that the park encompasses. It was recently expanded with the addition of a huge cattle ranch, complete with an airstrip used in the Iran-Contra affair by the infamous Colonel Oliver North. The gringo owner won a lawsuit over the expropriation of his land by the Costa Rican government so some of your park fees will go toward the US $16 million he was awarded.

Very close to La Casona (recently rebuilt after a fire) are camping facilities and a hiking trail, **El Sendero Indio Desnudo** (Naked Indian Path). But most people come to this park for the beaches, especially surfers and windsurfers. A rutted road leads down to the long bright-white beach of **Playa Naranjo** (camping but no drinking water) and the most famous surfing break around, **Witch's Rock**. Access to **Playa Nancite** is restricted because of its importance as a turtle nesting ground.

Beaches in the Murciélago sector can be accessed from the fishing village of **Cuajiniquil**. North, along the ocean from this little town, is an annex to the park, the **Bahía Junquillal Wildlife Refuge**, which boasts a good swimming beach and camping. Remember – because it will affect your enjoyment – this entire area is buffeted by strong, warm, westerly winds in the dry season. Two entrances on the Inter-American Highway access Santa Rosa National Park.

The next town north is **La Cruz**, 20 km/12.4 miles south of the Peñas Blancas border crossing (the border station is open daily 8 am-6 pm). There's not much to do or see here, but if you take the six-km/3.7-mile dirt road to the fishing village of Puerto Soley, you can hire a boat for a trip to the National Refuge for Forest Fauna on **Isla Bolaños**, three km/1.8 miles offshore. Look but don't touch. Visits are prohibited from December to March so as not to disturb the nesting seabird colonies. Isla Bolaños is home to some 200 pairs of brown pelicans in the scrub trees on the north side and 100 pairs of magnificent frigate birds on its southern cliffs. It's also the only place in Costa Rica where American oystercatchers have been found nesting. If you're hungry while in La Cruz, eat at **Coopetortillas** and support a women's cooperative. If you're a windsurfer, breeze on over to Salinas Bay, second only to Arenal for sailboarding.

A Place to Stay

Fourteen km/8.7 miles from La Cruz near the Nicaraguan border is the famous **Los Inocentes Lodge** *(☎ 506/679-9190, www.losinocenteslodge.com, $$$, meals included)*, a working hacienda that's more than a century old. This was one of the first eco-lodges open for guests in Costa Rica. It borders Guanacaste National Park and has a fabulous view of the Orosí Volcano. Rooms are in the main lodge or in very spacious separate cabins. The food and service are excellent and the ranch exudes old-fashioned charm. Day visitors can eat and horseback ride. Take a drive up, but only if you have an SUV.

Santa Rosa Sector
~ Santa Rosa National Park ~
N
HUNTER PUBLISHING
to La Cruz (24 km) & Nicaragua (42 km)
Inter-American Highway
Waterfalls
Tanks (1955)
to Liberia & San José
Ojo de Agua Las Pilas
Gateway of the Dogs Rd
Ojo de Agua La Penca
Río Cuajiniquil
Escondido Lake
Ojo de Agua El Copel
Río Calera
Park Admin.
Naked Indian Nature Trail
Monument to "The Heroes"
La Casona
Río Misperal
Quebrada Nance
Orange Trail
Duck Trail
Restricted Area
Cannon del Tigre
Playa Tule
Playa Nancite
Camping
Río Pozo Salada
Estero Real
Witches Rock
Playa Naranjo
2 KM
5 MILES
Camping
El Limbo Lake
Gulf of Papagayo
Carbonal Mts
Santa Rosa National Park
Murciélago Sector
Santa Rosa Sector
Park Boundary
Secondary Road
Trail
© 2005 HUNTER PUBLISHING, INC

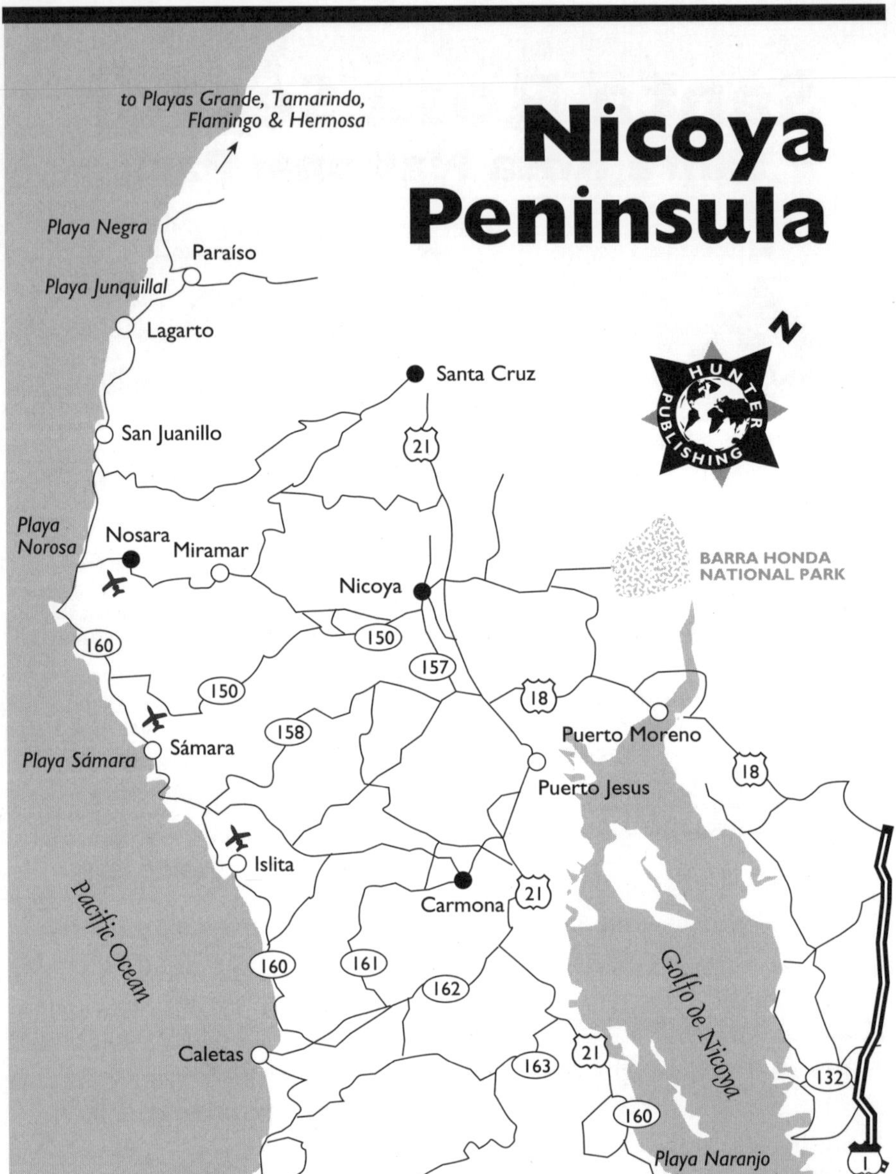
Nicoya Peninsula
to Playas Grande, Tamarindo, Flamingo & Hermosa
Playa Negra
Paraíso
Playa Junquillal
Lagarto
Santa Cruz
San Juanillo
21
Playa Norosa
Nosara
Miramar
Nicoya
BARRA HONDA NATIONAL PARK
160
150
150
157
18
Sámara
Playa Sámara
158
Puerto Moreno
Puerto Jesus
18
Islita
Carmona
21
Pacific Ocean
160
161
162
Golfo de Nicoya
Caletas
163
21
132
160
Playa Naranjo
1
Santa Teresa
Cobanó
Tambor
Malpais
160
Puntarenas
Paquera
Montezuma
Playa Tambor
CABO BLANCO ABSOLUTE NATURE RESERVE
HUNTER PUBLISHING
N
© 2005 HUNTER PUBLISHING, INC
NOT TO SCALE

Nicoya Peninsula

~ Endless Summer ~

Life is a Beach.
~ Bumper sticker slogan

IN THIS CHAPTER

If Guanacaste Province is the shoulder of Costa Rica as you look at a map, the Nicoya Peninsula is its right arm. The big attraction of Nicoya is its sandy beaches, which stretch from the end of the Santa Rosa Park down to Cabo Blanco Nature Reserve on the tip of the peninsula and around to Curú. The area is dry and hot, and gets much more sun than any other part of the country, making it doubly alluring for sun and sand lovers.

Costa Rica's beaches are not as visually appealing as Mexico's Yucatán coast or the islands of the Caribbean. They don't offer that tropical feel and bright white sand demanded by resort-goers worldwide. But that's not to say its beaches aren't beautiful in their own right – plus, they attract ecotourists because of the many turtle nesting sites and empty beaches in a natural, unspoiled state. Did we mention, "surf's up?"

Sand colors run from black or gray to brown, from gold to white, depending on its origin. Darker sand is volcanic. Some beaches are easily accessible and crowded, others hidden and magical, and still others accessible only by sea and completely deserted. A vacation to Nicoya, or as a stop in your travels, is a wonderful experience.

Beaches are grouped together in our book because of the natural terrain but also because of the road system that spiders off from Route 21, the main peninsular highway. There is no serviceable road along the coast running north and south, which, fortunately, limits development at some of the more remote locations such as Junquillal and Ostinal. We have grouped our beach destinations according to their accessibility and location. Mr. Spock would be proud of our organizational logic.

To reach the beaches of northern Nicoya from Liberia, simply head west to Guardia or Comunidad and follow signs for your destination. This road goes south to Santa Cruz, the town of Nicoya and farther. You can take it all the way to Montezuma. From the Central Valley, cross the Río Tempisque on the bridge or take a ferry from Puntarenas to Playa Naranjo (not to be confused with other places with the same name in Costa Rica) or, for the southern beaches, to Paquera. Or consider making the crossing by a *lancha*, a 25-foot launch, which you can hire from Puntarenas.

Northern Peninsula

He was my North, my South, my East and West,
My working week and Sunday rest,
My noon, my midnight, my talk, my song;
I thought that love would last forever: I was wrong.
~ *Stop All the Clocks*, W. H. Auden, 1907-1973

A direct road from Liberia through Comunidad to Playas Panama, Hermosa, Playa del Coco and Ocotal reaches the northernmost set of beaches. The road from Belén eventually forks to two other groups of beaches. The right fork takes you to Playas Pan de Azúcar, Potrero, Flamingo, Brasilito and Conchal. The left fork leads to Playas Grande, Tamarindo and Junquillal.

■ Playa Hermosa Area

Great wide, beautiful, wonderful world,
With the wonderful water round you curled,
And the wonderful grass upon your breast,
World, you are beautifully dressed.
~ *The Child's World*, William Brighty Rands (Matthew Browne), 1823-1882

The beaches of this horseshoe bay are wide and expansive, the water warm and inviting. Plus, they offer a wide range of accommodations.

Playa Hermosa

Playa Hermosa, or Beautiful Beach, has developed in the last few years, thankfully with smaller properties. It has gray sand, flat and lined with trees, and is a favorite with swimmers. If you're thinking of going under the waves, you're in luck. **Diving Safaris** (*formerly Bill Beard's, ☎ 506/672-0012, www.costaricadiving.com*) is located here. It's one of the area's oldest and largest operators. **Agua Sport** (*☎ 506/672-0050*) on the beach will rent all kinds of watersports equipment. They also have a little beachside eatery to keep your belly happy.

Places to Stay & Eat

Make the first left after your turn toward the beach and you'll encounter **Villa del Sueño** *(☎ 506/672-0026, www. villadelsueno.com, air, pool, restaurant, $$)*, one of the most appealing places to stay set just a short walk from the beach under shade trees on the dirt road. Rooms are clean and simple. There's a small pool, large restaurant, expansive gardens and condos across the street that they also rent. French-Canadian owners. Popular place at night with live music at the bar.

Hotel Prices	
Prices are per night for two people	
[No $]	less than US $20
$	US $20-$40
$$	US $41-$80
$$$	US $81-$125
$$$$	US $126-$200
$$$$$	over US $200

A good bet if you want an equipped kitchen is **Villa Huetares** *(on the second beach turn, ☎ 506/672-0052, pool, basketball court, air, $$-$$$)*. These large motel suites with kitchenettes can sleep four or more. Super value in the off-season – and you're only a block or so from the beach. Right across the street are two new indulgent hotels, the **Hotel Resort Villa Acia** *(☎ 506/672-1000, www. villaacia.com, $$$)*, which features luxury rooms and villas and a great pool, and **Hotel Mangaby** *(☎ 506/ 672-0048, www.hotelmangaby@ hotmail.com, $$-$$$), a handicap-accessible hotel that features large, clean rooms, a big pool with waterfall and a jacuzzi.*

Restaurant Prices	
Prices are for an average entrée	
$	less than US $5
$$	US $5-$10
$$$	US $11-$20
$$$$$	over US $20

On the beach our favorite is **El Velero Hotel** *(☎ 506/672-0036, elvelerocr@yahoo.com, restaurant, pool, $$)*. The bar is a happening place in the evening and the restaurant back in the shade near the beach is very pleasing for three meals. Rooms are simply furnished, but comfortable. The hotel offers a very popular sailing day-trip on their 38-foot catamaran. More monkeying around at night is at the popular local watering hole, **Monkey Bar**, set back from the road so it doesn't share the fun with neighbors, on the road to Playa Panama. Farther along from the Monkey Bar's island atmosphere is **Gingers**, a trendy restaurant-bar that serves drinks and Nuevo Cuisine appetizers at open-air tables under a winged roof, high on a deck above the road. If it's too hot, they offer air conditioning, while overhead fans cool the tables outside.

Playa Panama

Playa Panama, just north of Hermosa, is the location of some resorts that are part of the planned Cancún-like mega-resort development known as Papagayo. Expensive and expansive, the project has been delayed for years by lawsuits and other impediments, but several large and beautiful hotels are open, such as the ritzy Four Seasons.

Playa del Coco

Playa del Coco, south of Hermosa, has been a popular destination for Costa Ricans and tourists alike for many years. It has a paved road going right up to the beach, which makes for easy access. Weekends are generally boisterous and loud downtown – head here if you're looking for something to do. The beach itself has a dull volcanic gray sand (for white sand, sneak down to Ocotal). Coco isn't a surfer beach, but it is a port, which makes it very popular with divers and fishermen, who ship out from its anchorages, and surfers bound for Witch's Rock. The southern end houses a commercial fishing fleet. Mid-July sees a very big celebration – the Virgin del Mar Festival. In fact, Cocos has a reputation as a fun, year-round party town with good restaurants and shopping, which makes it the number one destination of the northern beaches. Divers will want to connect with **Rich Coast Diving** *(☎ 506/670-0176, www.richcoastdiving.com)* for some of the country's best dives. Rich Coast was one of the first dive operators in the area and is very experienced. Sports fishermen can catch **Arco Pacífico Sportfishing** *(☎ 506/670-0707)*.

The town's social activity is centered along the road to the beach, past the soccer field, with an easy-to-miss turn west for Playa Ocotal, plus a road east along the dark-sand coast. *Watch your valuables in town.*

Places to Stay & Eat

As for Coco lodgings, **Rancho Armadillo** *(☎ 506/670-0108, www.ranchoarmadillo.com, $$$)* stands out as a comfortable luxury hotel in the foothills above the beach. It was formerly owned by a Texan, hence the name, but Rick, the new owner, offers a much more laid-back quality stay. The small pool has a stunning ocean view and a covered deck to view incredible sunsets, watch for birds or to star gaze at night. The spacious rooms are wood paneled and the gardens are manicured. Turn left just before the road into

Hotel Prices	
Prices are per night for two people	
[No $]	less than US $20
$	US $20-$40
$$	US $41-$80
$$$	US $81-$125
$$$$	US $126-$200
$$$$$	over US $200

town becomes a divided boulevard; follow the driveway lined by bougainvilleas, uphill. As a former restaurant owner state-side, Rick also offers cooking lessons.

Closer to downtown, **El Pato Loco Inn** *(☎ 506/670-0145, www.costa-rica-beach-hotel-patoloco.com, $$)* offers clean, very pleasing rooms around back of their delicious Italian restaurant. The friendly owners traded in the hills of the ancient city of Romulus and Remus for the mountains and natural wonders of Costa Rica. Very appealing place. Nearby is the new supermarket, where everyone shops, sooner or later. It boasts a well-used bulletin board and sells local newspapers.

The big boy on the block is **Hotel & Casino Coco Verde** *(☎ 506/670-0544, fax 670-0555, www.bestwestern.com, pool, air, restaurant, cable, spa, casino, $$).* This 33-room Best Western property sits in the middle of all the action, just a long block from the waterfront. It offers large rooms and all the comforts of home. A safe bet. Another choice is **Hotel Villa Flores B&B** *(☎ 506/670-0269, www.villaflorescostarica.com, air, $$),* located on the sharp right turn just past Coco Verde. It has very large rooms, a swimming pool and pretty gardens. The high season restaurant, for which you should make reservations, features gourmet continental cooking. Almost opposite is **B&B Laura's House** *(☎ 506/377-3119)*, a simple but funky guest house with a tiny pool.

A right turn on the side road here leads to the town's more elegant hotel entry, **Hotel La Puerta del Sol** *(☎ 506/670-0195, www.lapuertadelsol.com, air, cable, phone, $$-$$$).* The "Doorway to the Sun" hotel features a gym, jacuzzi and pleasing tropical gardens, plus inviting junior suites for longer stays. Its *al fresco* restaurant, **Sol y Luna**, is renowned for fine Italian food and is worth the difficulty to find in the evening.

We enjoyed our stay at the **Villa del Sol B&B** *(US ☎ 866/815-8902, Canada ☎ 866-793-9523, CR ☎ 506/670-0085, www.villadelsol.com,* $$). This small, friendly Canadian-owned hotel is farther up the east road parallel to the beach. It has bright and airy rooms plus a small pool, but is not close to town.

There is an ever-growing number of dining choices as Cocos increases in popularity. The two high-quality restaurants mentioned above, in Sol y Luna and Villa Flores hotels, are more quiet and upscale than ones on the main drag, but those are a lot more fun to people watch. Try **Papagayo's Seafood**, famous for you know what, or, for a special treat, eat at **Bob's Louisiana Bar & Grill**, upstairs on a modern veranda. We had an excellent dinner, but chuckled over our choice of humus as an appetizer; they served us peanut butter. Still, we like Cajun food and Bob cooks up a good meal. On the beach and best to visit when there's a sea breeze is **Papagayo Pura Vida**, a seafood restaurant.

Night owls can head along the beach to **Disco Coco Mar** for dancing, dining and drinking until the wee hours. Facing the soccer field is a new German restaurant, **Valentin's**, located next to **Hotel Coco Palms**, a decent place to stay for budget travelers.

Playa Ocotal

Playa Ocotal is the southernmost beach in this bay. Its gentle cove is home to sailing yachts at anchor and steep hills that come right to the water at each end of the bay. For tranquility, scenery, black sand and swaying trees this was our favorite.

The **El Ocotal Beach Resort and Marina** *(☎ 506/670-0321, fax 670-0083, www.ocotalresort.com, $$$$$)* is a lovely hotel with rooms, bungalows and suites set up on a rocky bluff above the gentle bay. Fabulous views and a very good restaurant, with a large veranda for *al fresco* dining. Look for lizards that hide in the rocks as you hike by and listen to birds sing in the trees. Two pools and a lovely beach. They also operate a casual beachside restaurant that we found delightful, **Father Roosters**. With wooden plank floors, billiards and foot-tapping music from a stereo, this bar and restaurant is fun to eat in or to hang out and drink. Try the Tequila Matar, a tequila to die for.

Speaking of heaven, **Hotel Villa Casa Blanca** *(☎ 506/670-0518, fax 670-0448, www.villablanca.com, pool, restaurant, jacuzzi)* is about as close as you can be to the sand without being on the beach itself. This sumptuous Spanish-styled mansion sits amid a lovely garden on a bluff overlooking the ocean and the surrounding, steep rolling hills. The intimate rooms are decorated with formal heavy furniture and fabulous artwork. Lately, it's become a favorite for weddings and small groups looking for a unique hotel experience. *Muy romantico!*

■ Playa Flamingo Area Beaches

The shells upon the warm sands, have taken from their own lands,
the echo of their story. All I hear are low sounds as pillow
words are weaving, and willow waves are leaving,
but should I be believing, that I am only dreaming?
~ *Anywhere Is*, sung by Enya, lyrics by Roma Ryan

These beaches center around Playa Flamingo, long considered Costa Rica's premier beach resort. Located on a jutting peninsula, Flamingo's beach is pure white sand, as its pre-development name, "Playa Blanca," suggested. **Playa Conchal** is even more enticing because much of its beach is made of finely ground white seashells. But if you're not staying at the Meliá mega-resort there, your shell beach access is restricted. So do what the locals do: drive or walk along the beach the half-mile from

Brasilito town and park yourself on the sand between the hotel and the bay.

To get to the Flamingo area beaches from Playa Cocos, go back inland through Filadelfia and follow the signs for Flamingo. Or, if you have a four-wheel-drive, it's dry season, and you have good shock absorbers, follow the "Monkey Trail," 16 km/10 miles of rough road from the outskirts of Cocos to Playa Potrero. You'll pass nearby the **Congo Trail Canopy Tour** *(☎ 506/666-4422, seven km/4 miles along the road)* and the **Jardín Botánico 4 Mangos** *(☎ 506/670-0851)* along the way. Look for signs.

Playa Flamingo is fully developed with exclusive hotels and is no destination for budding naturalists – or low-budget beach bums. But the yachting crowd loves its protected harbor and marina. If you're looking for a sailor's life, search out the sailboat ***Shannon*** *(☎ 506/654-4536)*, or catch fish with **Flamingo II Sportsfishing** *(☎ 506/654-4867)*.

Playa Potrero is north of Flamingo. Its sand is not as appealing, but it is a favorite for private homes and vacation getaways of retired North Americans.

Farther north is **Playa Pan de Azucar** (Sugar Bread Beach), which relatively undeveloped and has just one resort (see below).

Playa Brasilito is the south side of the jutting peninsula of Flamingo. It features a dull gray sand and gentle surf for good swimming. Like Potrero, it is a quiet little town. The next beach south is **Playa Conchal**. Until a few years ago, Conchal was a Tico-secret beach without hotels, just camping spots. Now Meliá resorts have an exclusive all-inclusive there.

Places to Stay & Eat

At Playa Flamingo, gringo vacationers stay at the **Flamingo Marina Resort** *(☎ 506/654-4141, www.flamingomarina.com, $$$-$$$$)*, a quality resort hotel on a hill above the beach. Pools, views, restaurants, tennis – everything you expect. Lower down the hill, with a lower price scale too, mates can walk the plank to **Mariner's Inn** *(☎ 506/654-4081)*, which features small, clean, air-conditioned rooms and a seafood restaurant. Outside town, high on a bluff overlooking the harbor, you passed **Hotel Colores del Pacífico** *(☎ 506/654-4769, fax 654-4976, www.coloresdelpacifico.*

Hotel Prices

Prices are per night for two people

[No $]	less than US $20
$	US $20-$40
$$	US $41-$80
$$$	US $81-$125
$$$$	US $126-$200
$$$$$	over US $200

com, infinity pool, air, breakfast included). This stately hotel features upscale rooms with private terraces and plenty of activities for guests. Closed in the rainy season, September and October.

The best dining in Flamingo is at the popular **Marie's Restaurant**, near the Mariner's Inn. It serves delicious fresh fish and Mexican specials. Nearby is the **Hillside Café**, offering excellent and affordable sandwiches.

In **Playa Potrero** we lived like Europeans on good food and fine wine at **Monte Carlo Resort & Restaurant** *(☎ 506/654-4676, www.montecarlo-beach-resort.com, $$)*. The beachside hotel has a lovely pool next to its dining area and pretty pink individual casitas with kitchenettes. Follow your nose across the street to **Harden's Bakery Garden**, a bakery-snack bar as famous as Amos for chocolate chip cookies and cinnamon sticky buns. Near where the Monkey Trail spills out to the beach is **Hotel Bahía Esmeralda** *(☎ 506/654-4480, www.hotelbahiaesmerelda.com, air, cable, pizzeria, $-$$)*. They rent rooms and furnished apartments at reasonable prices. On the beach, look for **Bar Las Brisas**, which serves Tex-Mex and salads.

Other pleasing places include **Cabinas Cristina** *(☎ 506/654-4006, www.cabinascristina.com, $-$$)*, which has larger, older bungalows, some with kitchenettes, and a pool; or **Cabinas Isolina** *(☎ 506/654-4375, www.isolinabeach.com, $)* with small, inexpensive rooms. Camp or take a rustic room at **Myra's Camping and Cabinas** *(☎ 506/654-4213)* right on the beach. In what passes for Potrero's downtown, there are many good little restaurants and sodas. Try them all and let us know.

Playa Pan de Azucar offers the sweet **Hotel Sugar Beach** *(☎ 506/654-4242, www.sugar-beach.com, $$-$$$)*. We found this beach intriguing, with its rocky snorkeling in some spots and sandy areas otherwise. Wooded forest in back, infinity pool in front. It will be our next, long-weekend beach layover. That is, if we can face the torturous road that leads to it.

We already told you about the not-so-secret way to get to Playa Conchal from **Playa Brasilito**. You can stay in Brasilito village at the low-key, Caribbean-style wooden **Hotel Brasilito** *(☎ 506/654-4247, www.brasilito.com)*. Or maybe just stop on your way through town and eat at the memorable **Restaurant Camarónes Dorado** (*☎ 506/654-4028*), with beach table service of mouth-watering seafood. They're so popular they run a shuttle to Flamingo to pick up the *hoi polloi*. We found our culinary contentment at the corner of the soccer field and the beach at American-owned restaurant/bar **Happy Snapper**. It has excellent seafood and good prices. Nighttime dancing and loud music attract locals and tourists in droves. Several Italians have opened pizza/pasta places on the road into town. Another choice for lodgings is **Apartotel Nany**

(☎ *506/654-4320, pool, cable*), recently renovated to offer cold-water apartments. There is a **gas station** just outside town on the way to Flamingo. If you're looking to snorkel, fish or just go for a boat ride in his 24-foot launch, ask in town for Eugenio "Chino" Amey, or phone him at ☎ *506/654-4907.*

Playa Conchal is the home of the enormous, all-suite **Meliá Playa Conchal** *(US ☎ 888/336-3542, in CR ☎ 506/654-4123, www.melia-playaconchal.com, $$$$$)*, which is a village unto itself. **Paradisus Playa Conchal Beach and Golf Resort** *(US ☎ 888/336-3542, CR ☎ 506/654-4123, www.solmelia.com)*, is another self-container all-inclusive. Play tennis, 18 holes of golf, or swim in what is purported to be Central America's largest swimming pool. Not counting the ocean, of course.

Eight kilometers south of Conchal is **Playa Real**, accessed only by another very rough road. Two bungalow-style resorts there are **Condor Lodge** *(☎ 506/654-4050, www.condorlodge.com)* and **Hotel Bahía de los Piratas** *(☎ 506/654-4651, www.bahiadelospiratas.com)*.

■ Playa Tamarindo

Surf City, here I come.
~ Jan and Dean

Partly because of paved roads that lead almost all the way there (plus a local air strip, with service from San José) and partly because of its word-of-mouth reputation among surfers, Tamarindo is one of the most popu-

This surf shop in Tamarindo gets plenty of business.

lar beach destinations on Nicoya. It was once a sleepy little fishing village, but the hillsides up from the dusty road that runs parallel to the beach are slowly filling up with small hotels and private houses. The main drag is home to restaurants and artsy-crafty gift stores. At the end of the downtown road, where restaurants and the beach square a circle, street vendors sell jewelry and art in a tiny park. It's a central meeting place, made more popular by its bank of pay phones. Some of these vendors are North Americans trying to eke out enough money to stay in Costa Rica longer – a difficult task if you're looking for inexpensive hotel rooms, and Tamarindo isn't nearly as cheap as its laid-back surfer reputation might lead you to think. On our last visit, one young woman plied the streets selling her rich, homemade jelly donuts. Yum.

The express bus ride from San José, which arrives in Tamarindo in the evening, takes you out of the valley and over dramatic mountains down to the rolling plains of Guanacaste and the Nicoya cattle country. Like the American West, the land is gritty, especially in the dry season, when light brown dust, *polvo*, coats the landscape along unpaved roads. In Tamarindo, most restaurants have learned to set their tables well back from the road to avoid coating your spaghetti sauce with grit. But no one seems to mind the windy weather when the surf is up in the warm Pacific. Surfing season at Tamarindo winds down at the end of July with a major professional/amateur competition in both long and short boards, dude. Robert August, the California surfboard manufacturer and star of Bruce Brown's classic surfer film, *Endless Summer*, has a home on the hill overlooking the beach.

Even though it's a surfer area, swimming at Tamarindo isn't bad at all. The beach is long, wide and shallow, and is protected by an offshore reef. The whole of the *playa* is framed to the north and south by two river estuaries that empty into the Pacific.

> **WARNING:** Be careful wading or swimming across the easily accessed estuary to the north between Tamarindo and Playa Grande. The tidal current can be very strong. Wait for low tide.

Locals already know that swimming in the ocean in front of the **Barceló Playa Langosta** hotel, where the Río San Francisco estuary flows into the Pacific, can be very dangerous because of rip tides. According to the former sheriff of Tamarindo, the hotel refuses to post danger warnings and access to the beach is blocked for all vehicles – even for emergencies! Although it may be an attractive-looking resort, it was built under a cloud of controversy concerning alleged damage to the environmentally sensitive estuary. The hotel's lawyers even pressured Hunter Publishing to tone down the *Adventure Guide's* criticism. So, we won't criticize – but we won't stay there and we don't recommend it to our readers.

CRIMINAL CONCERN

Recently, a rash of petty thefts in town has kept beach visitors on the defensive. One honeymoon couple lost the luggage they left locked overnight in their rental car, right in front of their hotel's reception entrance. The thieves popped the lock, so be aware. Read all about it in *The Howler*, a fun local paper.

Despite these minor cautions, Tamarindo is a welcoming beach destination, offering an eclectic social and nightlife, an appealing beach, interesting ecological trips nearby and a variety of hotels and restaurants to please any taste. Get there by plane or express bus from San José. If you're driving, turn west off the Liberia-to-Santa-Cruz road, in the town of Belén, and follow your nose. All but the last few kilometers are paved.

Adventures & Attractions

ADVENTURES ON WATER: The big sport in Tamarindo is **surfing**, and **Iguana Surf** *(☎ 506/653-0148, www.tamarindo.com/iguana)* is the largest supplier of everything in watersports in Costa Rica. They have two locations, a small surf shop right in the downtown and their large four-story palapa-roofed headquarters and restaurant on the road to Playa Langosta and the Río San Francisco estuary. They rent and sell boogie boards, beach chairs, beach umbrellas, snorkel gear, sea kayaks and both long and short surfboards. Iguana Surf's guided **sea kayak nature tour** on the Río San Francisco estuary, is excellent.

Guided tour on the San Francisco River.

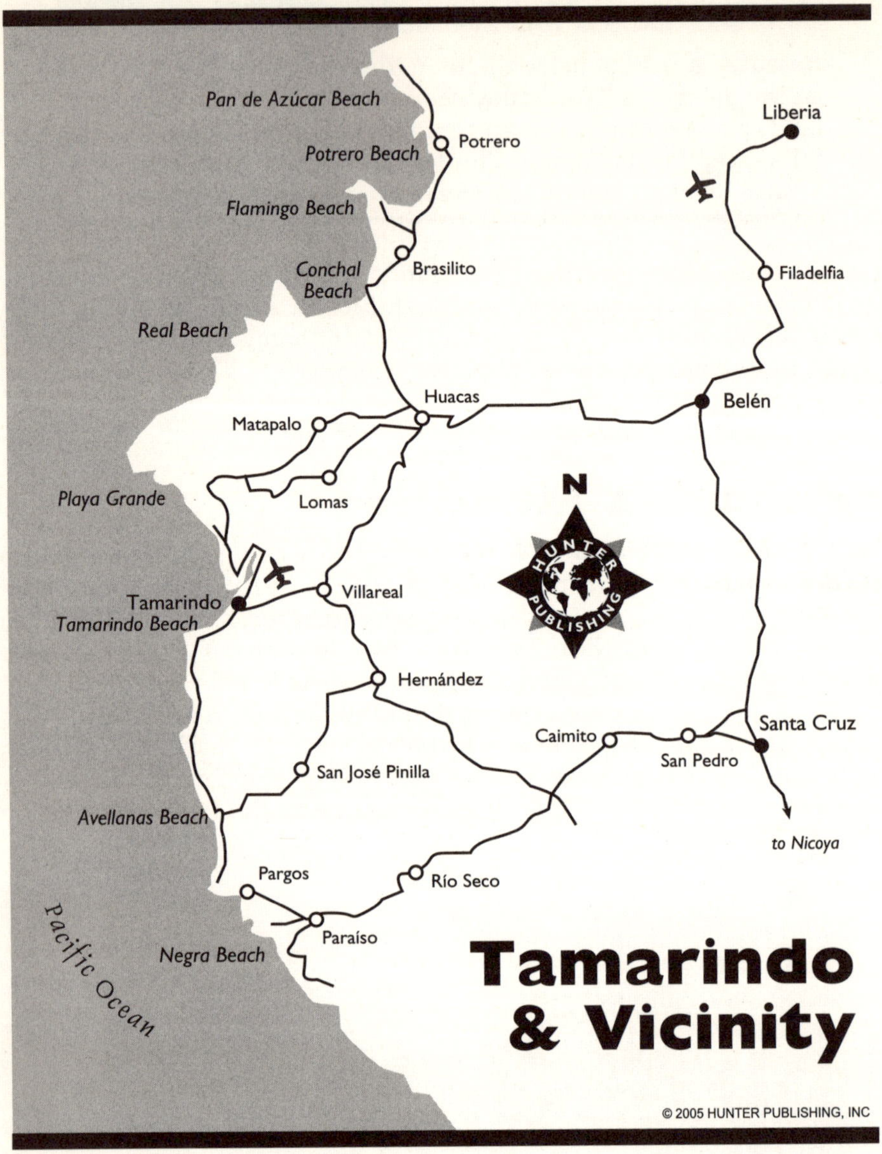

Anyone going deeper can check into **Agua Rica Dive Shop** *(☎ 506/653-0094, www.tamarindo.com/agua)* on the main street for organized dive tours, equipment and certification. Day and half-day **deep-sea fishing** trips are the specialty of **Papagayo Excursions** *(☎ 506/653-0254, papagayo@sol.racsa.co.cr)* at the Hotel Diría. They can arrange land tours as well. **Tamarindo Sportfishing** *(☎ 506/653-0090, www.tamarindosportfishing.com)* also offers day and half-day fishing trips out to the rich offshore Pacific grounds.

ADVENTURES ON LAND: **ATV tours** can be taken from **Iguana Surf** (see above) on private trails, or with **Tamarindo Adventures** *(☎ 506/653-0640)*. If you want a horseback ride, horses are often corralled up the estuary on the beach in back of **Tamarindo Tours** *(☎ 506/653-0078)*. Hourly rates are inexpensive. A number of places rent **mountain bikes**, including **Iguana Surf**.

ADVENTURES IN NATURE: A great place to rent horses and get a nature tour of Guanacaste's disappearing dry forests is **Casagua Horses – Cantina Tours** *(☎ 506/653-8041, kaydodge@racsa.co.cr)*, located near Portegolpe on the main road into the beaches from Belén. Kay Dodge, Ph.D. and her husband Esteban Peraza, are extremely ecological minded and deliver some of the most professional, and most fun, nature tours in Guanacaste and Nicoya.

To see **turtles nest** in season at Playa Grande – an activity that requires some stamina, a bit of luck, and lots of patience – ask in the **Coopetamarindo** office on the main drag, where guides have set up a locally based organization. If you're over on **Playa Grande**, try the **Matapalo Conservation Association**, which has a small office next to Las Tortugas hotel. This co-op benefits the local community. Sign up at 6 pm for their participant-limited turtle tours.

The world of the turtle is explained at Playa Grande's unique small museum, **El Mundo de La Tortuga** *(☎ 506/653-0471)* through a self-guided cassette player presentation in Spanish, English, Italian and German. Admission is about US $6.

OXCART TRAILS

One of the best ways to see waterfowl, or a rural Guanacaste/Nicoya that is fast disappearing, is to follow one of the old oxcart trails that connects towns. An old, rutted oxcart trail, still in use, runs from Categena to Santa Cruz through seasonal ponds, swamps and rivers. Proceed by car (best in four-wheel-drive in dry season only) about a half-mile west of the turn for Cartegena. Better yet, arrange for a guided trip on horseback with **Casagua Horses** *(☎ 506/653-8041)*.

Places to Stay

Surf House *(northern end of the beach, call at Hotel Santo Tomás ☎ 506/255-0448, fax 222-3950, www.hotelsantotomas.com, 3 bedrooms, 2 full baths, washing machine, telephone, $$$-$$$$)*. The Surf House is owned by the same owner as the Hotel Santo Tomás in San José, but is much more casual. The two-story rental is a good value for a family or group looking for a laid back vacation in surf-

Hotel Prices	
Prices are per night for two people	
[No $]	less than US $20
$	US $20-$40
$$	US $41-$80
$$$	US $81-$125
$$$$	US $126-$200
$$$$$	over US $200

crazy and growing-more-pricey Tamarindo. It's located right on the beach and offers privacy, full kitchen, bed space for 10 in five queen beds and maid service. It's truly a home away from home. There are a number of more appealing condos/houses available to rent in a higher price range as developments burgeon, but the Surf House sits right where the estuary runs into the Pacific and a group of monkeys hang out in the trees in the front yard. There's a three-night minimum.

Capitan Suizo *(southern end of beach, ☎ 506/653-0075, fax 653-0292, www.hotelcapitansuizo.com, 8 bungalows, 22 rooms, a/c & fans, restaurant, room safe, bidet, refrigerator, pool, $$$$$).* Capitan Suizo has emerged as an outstanding small luxury hotel right on the beach in Tamarindo. Its gracious gardens host comfortable tropical bungalows with large bathrooms and *al fresco* showers. Standard rooms are in a two-story building, downstairs rooms feature air conditioning and terraces, and upstairs rooms have breezy balconies. The modern décor features terra cotta tile floors and wood-beamed ceilings. Splashes of tropical color and touches such as bidets give it a decidedly European flair. The centerpiece of the gorgeous gardens is the large free-form pool. The bar/restaurant overlooking the pool and white beach serves international cuisine. A four-bedroom apartment over the lobby is available for large families.

Pasatiempo *(1 block off the beach, ☎ 506/653-0096, fax 653-0275, www.hotelpasatiempo.com, 11 cabins, 2 suites, air, continental breakfast, restaurant, $$).* This is the first hotel we stayed at in Tamarindo and it's still a popular spot despite the plethora of hotels opened since then. Pasatiempo means "spending time," and it's a good characterization of the laid-back atmosphere. Rooms are in duplex thatched cabañas, each with a tiny patio with hammock and hand-painted murals inside. Large closets. The rooms were remodeled in mid-2000 to keep up with the competition. Their small tiled pool is next to Pacho's, an open-air bar/restaurant featuring pastas and fish and a festive happy hour. Suites have a king-size bed and sofa bed, plus cable TV. Discount rate for cash instead of credit cards.

Cabinas Marielos *(in town, along the beach road, ☎/fax 506/653-0141, 17 rooms, in-room safe, cold water only, air or fan, $).* Not to be confused with the Hotel Marielos farther south down the road, Cabinas Marielos is a budget hotel in the downtown. Its plain but pleasant rooms are set in two buildings that feature blue Adirondack chairs on long porches in front. The upstairs room #10 is the best. The gardens between buildings provide for a cool and colorful atmosphere with trellis-covered walkways. There's a communal kitchen, as well as bicycle and boogie board rentals. It's the best value in the budget category.

Downtown Tamarindo boasts several other boutique hotels, although not much exists in the lower end of the budget category. The well-established and well-known **Hotel Diría** *(☎/fax 506/653-0031, www.tamarindo-diria.co.cr, air, cable, breakfast included, $$$-$$$$)* went through two recent remodelings and an expansion in 2000. It's located on the beach in the center of everything and offers a quality continental restaurant under a big palapa roof. If you want a safe bet, check in at the **Best Western Vista Villas** *(☎ 506/653-0119, fax 653-0115, in US ☎ 800/528-1234, www.tamarindovistavillas.com, $$$-$$$$)*, a lovely hotel across from the beach with a view of the ocean from flower-covered balconies.

Villa Alegre *(Playa Langosta, ☎ 506/653-0270, fax 653-0287, www.villaalegrecostarica.com, 4 rooms, 3 villas, pool, air, parking, breakfast included, $$$$)*. Villa Alegre, or Happy House, was specifically designed as a bed and breakfast and built on the beach by North Americans Suzye and Barry Lawson. The rambling California ranch hides behind huge wooden doors set in a tall, bougainvillea-covered white garden wall. Two of the four rooms inside are configured to convert into a suite and all the bedrooms have a private terrace. The travel-happy Lawsons have decorated in themes with typical objects they collected in Mexico, Guatemala, the Caribbean islands and the good old USA. Guests have access to a huge kitchen, but the generous and delicious home-cooked breakfasts should fill you up for the day. The three "villas" are in small, unattached cabañas just outside the comfy communal library/living room. One is teeny tiny and very cute with a smiling gecko tile in the shower. The much larger villas are next to the pool, behind hand-carved wooden doors. The Russian villa/suite sleeps five and has a kitchenette; the simple Japanese suite is its mirror image except for the welcoming outdoor patio. Suzye has a soft touch for dogs, so there are always one or two nosing around.

In a more economical mood? Try **Cabinas Zullymar** *(☎ 506/653-0140, fax 653-0028, $-$$)*, a basic, clean, colonial-style hotel owned by the same local family as the favorite Restaurante Zullymar. Backpackers can crash at **Tito's Camping**, near Capitan Suizo's hotel.

On the other end of the price scale, one of the most luxurious hotel properties is **Cala Luna** *(☎ 506/653-0214, fax 653-0213, www.calaluna.com, 21 rooms & villas, air, pool, cable, restaurant, $$$$)*, south toward Punta Langosta. Each individual villa and hotel room is elegantly decorated and exudes comfort. Inviting free-form pool with wet bar. And to top it all, each villa boasts an individual pool and private terrace. We found this hotel particularly lovely in its gorgeous garden setting. A path leads through the woods to a rocky beach. The poolside restaurant, **Cala Moresca**, serves romantic dinners. (Reservations at ☎ 506/653-0214.)

If you don't stay at **Hotel El Jardín del Edén** *(☎ 506/653-0137, www.jardin-eden.com, $$$$)*, a boutique luxury hotel on the hill overlooking the bay, then splurge for a romantic dinner at its Italian-French palapa-roofed restaurant.

A captivatingly, dreamy place is **Sueño del Mar** *(☎/fax 506/653-0284, www.tamarindo.com/sdmar, pool, restaurant, breakfast included, $$$$)*. Despite having only three large rooms, a honeymoon suite and beach-front bungalow, the "Dream of the Sea" B&B is one of the most appealing and intimate places in Playa Langosta, the southern section of Tamarindo.

Places to Eat

A gourmet is just a glutton with brains.
~ Philip Haberman, Jr., *Vogue* magazine

Informality is the rule in Tamarindo, so restaurant reservations are not needed.

Restaurant Prices	
Prices are for an average entrée	
$	less than US $5
$$	US $5-$10
$$$	US $11-$20
$$$$$	over US $20

As you enter Tamarindo, on the right is the **Paris Bakery**, often known by its former owner's name, Johan's. For more substantial fare, dine in the restaurant in back, next to the canal, where several crocodiles also lounge. Turn uphill at the crowded El Milagro Hotel & Restaurant and go right a short way for the hotel and restaurant **El Jardín del Edén**. The "Garden of Eden" offers a fine menu on the lush hillside above the beach. For less formal Tico and gringo food, at the north end of town on the main road is **Frutas Tropicales**. Its offerings are basic, but good.

Diría (☎ *506/653-0031)* has a new chef who serves a tasty international meal poolside at their hotel on the main drag. On the beach at the circle where downtown ends, head to **Zullymar**, which has been serving good seafood and Tico dishes for a lot of years. Alternatively, try **Arrecifes**, also overlooking the beach. Walk south from there a short way and enjoy **Pedro's**, a fish shack restaurant favored by locals and adventurers looking for fresh fish on the cheap. This end of town has a number of night-club and/or restaurants.

For the very best meal downtown, diners rave over **Lazy Wave**, stuck in back of a strip mall on the left as you turn inland (toward Hotel Pasatiempo) in the middle of town. With outdoor seating around a couple of huge trees, this restaurant fills up with tourists searching for a special meal. It's not cheap, but the food is good. For sandwich cravings, walk across the street and inland a half-block to the **Shark Bite Deli**, where you might need an articulated jaw to eat their build-your-own sandwiches. Free coffee refills!

On the same side of the road, a half-block from the turn, is the **Pachanga Restaurant**, set in the leafy, vine-covered hotel of the same name. Open for dinners only, this popular place features Mediterranean dining from Chef Sholmy Koren. Nearby is the landmark, **Stella's**, which serves pricey but delicious Nuevo American cuisine, al fresco.

Heading toward the Playa Langosta end of the beach, pass the thatched roof sports store called **Iguana Surf**. Their front end, known as **Gil's**, serves Mexican food for breakfast and lunch. During the evenings the name changes to **Gecko's** and they serve seafood. Try the chocolate cake.

We already mentioned the poolside restaurant **Cala Moresca**, at Playa Langosta's Cala Luna Hotel, but our favorite place to have dinner is nearby at **Maria Bonita**, on a side street about 200 m/660 feet east of the Barceló. This small hacienda/garden-style eatery features Latin-Caribbean cuisine in a romantic atmosphere.

■ Playa Grande

A stone's throw north of Playa Tamarindo, but far by dirt road, is Playa Grande (Big Beach), featuring the Baula National Marine Park, created in 1991. The marine park expanded in 1995 to include the Tamarindo Estuary Wildlife Refuge. Playa Grande is known for the giant leatherback turtles who nest there from October through March. The area also offers tours into the estuary to see crocodiles and mangroves. The leatherbacks, however, are the main attraction. Leatherback turtles get their name because they have a leathery hide in place of a shell. These enormous sea creatures typically weigh more than 275 kilos (over 600 pounds) but can get as big as two meters (six feet) and weigh in at 680 kilos (1,500 pounds). They lumber ashore mostly at night to lay their eggs. One estimate of the world leatherback population was down to 35,000 turtles, 900 of which nest here at Playa Grande. In season, turtle watchers mount platforms along the beach so they don't interfere with the egg-laying process as they watch. Other area activities include birdwatching and ocean kayaking in the estuary wildlife preserves, and saltwater fishing, surfing and scuba diving in the ocean – your hotel can make arrangements. If you are arriving by air or bus, head to Tamarindo and get a fisherman to ferry you across the first estuary to the Playa Grande Beach.

WARNING: Be cautious swimming off Playa Grande and especially near the estuary, where people have drowned. Try Playa Ventanas, a bit farther north, for swimming, or explore Playa Carbon, a black sand beach with lots of tide pools.

Places to Stay

Hotel Prices	
Prices are per night for two people	
[No $]	less than US $20
$	US $20-$40
$$	US $41-$80
$$$	US $81-$125
$$$$	US $126-$200
$$$$$	over US $200

Villa Baula *(Playa Grande, ☎ 506/653-0493, fax 653-0459, www.villabaula.com, 20 rooms, 5 bungalows, fans, pool, restaurant, kids pool & playground, TV, $$$)*. The Villa Baula bills itself as an "ecological" hotel and it certainly looks the part. Rustic wooden buildings with barn siding are scattered among the trees in a garden along the beach. Rooms are in two module buildings. Individual two-bedroom bungalows feature upstairs porches with *al fresco* kitchenettes. There's a very large pool and a huge thatched-roof restaurant that serves German/Indonesian food. The ecological label is acquired from the efforts by the hotel to filter and recycle its black water for irrigation – plus its location right next to the Baula National Marine Park and the wildlife refuge.

Hotel Las Tortugas *(Playa Grande, ☎ 506/653-0423, fax 653-0458, www.tamarindo.com/tortugas, air, pool, jacuzzi, restaurant, $$)*. The ecologically oriented Las Tortugas features a friendly staff, turtle-shaped pool, canoe rental and horseback riding, all in a laid-back tropical atmosphere along the beach.

Other hotels well worth considering include the charming **Hotel Bule Bula** *(☎ 506/653-0975, breakfast included, $$)*, which features bright comfortable air-conditioned rooms, each named for an indigenous Central American animal. This was formerly called Hotel Cantarana. It has a pleasing pool and a very good restaurant. A relatively short walk off the beach is **Casa y Casitas Linda Vista** *(☎ 506/653-0474, www.tamarindo.com/kai, $$$)* a 1.2-hectare/three-acre hilltop estate in the dry forest overlooking the ocean. Its beautiful secluded houses are excellent values for a family or friends. The smallest, Casita Linda Vista, sleeps three.

Surfer types in need of cheap lodging can try the **Rancho Diablo Surfer Camp** *(☎ 506/653-0490, rdiablo@sol.racsa.co.cr, restaurant, $)* for a room with a private bath or dormitory-style sleeping.

■ Playa Junquillal

Playa Junquillal (hoon-key-YAL) is a broad beach, hard against wild grasslands, in a remote part of the Nicoya Peninsula. It's remote enough that the long beach is usually deserted and there is no town to speak of nearby. The surf is what brings most visitors here, but when it's not too strong there's good swimming. This is a beachcomber's paradise. Plus, it's cheaper than Tamarindo.

Places to Stay

The oldest hotel, with colonial-style buildings on a hill facing the sea, is **Hotel Antumalal** *(☎ 506/653-0425, $$$)*, a venerable standard for good accommodations. It is named after the Chilean god of the sun. If you're still feeling South American, dine at a Peruvian restaurant, **Lak'Ampu**.

If it's true that you pay for what you get, it's worth paying to stay at **Iguanazul** *(☎ 506/653-0123, www.iguanazul.com, $$$)*. Set on a breezy ridge by the ocean. This Canadian-owned hotel is the cream of the crop in Junquillal. High-ceilinged rooms in shaded cottages, red tile floors, Mexican tile bathrooms, and a huge freeform pool. It's a pleasure to stay here. Lots of activities.

Camp at low-rent **Camping El Malinche**, a short distance from the Iguanazul. It offers a little grocery store, showers, bathrooms and manicured garden campsites at the beach. Get a roof over your head at **Hotel Hibiscus** *(☎ 506/653-0437, including breakfast, $-$$)*. Pretty tile work and personal service. Across the street is **La Puesta del Sol**, an Italian restaurant.

Central Peninsula

Skins tanned to the consistency of well-traveled alligator suitcases.
~ *NY Times*, on fashionable tans, 1986

The series of beaches south of Tamarindo lack the denser development of the northern coast, hence have an attraction all their own. They vary from black sand to gray to sandy white and, as in the north, some are better for swimming, while others are known for surfing. Please note we start our coverage from Sámara, despite Nosara being geographically more north, because the road to Sámara is paved all the way.

■ Santa Cruz & Nicoya

For tourists, the cities of Santa Cruz and Nicoya are central gateways to the beaches of western and southern Nicoya Peninsula. But for Costa Ricans, these cities are traditional farming and cattle ranching centers for the surrounding agricultural areas. Santa Cruz is a sunny and hot small town known euphemistically as the "National Folklore City," although that seems to be a somewhat optimistic label. In the center of town, the *zocaló* is a shady, sleepy park; great for watching the slow Guanacaste life pass by. Across the street, a colonial-era tower houses a large clock from the original church destroyed in an earthquake.

Santa Cruz is a favorite marketplace for **Chorotega pottery**, Costa Rica's polished, hand-painted glazed pottery made by the indigenous pre-Columbian Chorotega peoples. After the Conquest, pottery-making in Guanacaste died out, perhaps because it no longer had the same religious significance. However, in recent years it has made a comeback and now high-quality replica pottery from Nicoya/Guanacaste is sold throughout the country. One of the most significant centers is the nearby artisan village of **Guaitil**, 10 km/6.2 miles to the east, near San Vicente. Here you can watch pottery being made and decorated by the indigenous Chorotega craftsman. The town has a festive annual celebration on January 15th. You can drive to Guaitil on a scenic but bumpy road.

A unique eating experience in Santa Cruz is the **Coopetortilla**, a few blocks south of the central plaza under a cavernous, tin-roofed former airport hanger with screened walls. Enjoy handmade tortillas, rice and beans, and provincial dishes prepared over an open fire by a local women's cooperative – they're inexpensive and tasty. For more local eco-tourism see the **Coopeortega**, 14 km/8.7 miles away near Palo Verde. See page 228.

Hotels and *pensíons* in town are relatively inexpensive, but also very basic. Try the **Hotel Sharatoga** *(☎ 506/680-0011, $)*, a half-block or so from the *zocaló*, which offers a miniature pool plus air-conditioners in plain rooms. On the highway is the **Hotel La Calle de Alcala** *(☎ 506/680-0000, $$)*, a modern appealing hotel with a large pool, courtyard gardens and air-conditioned rooms and suites. Can a place with a swim-up bar be all bad?

Nicoya is a much larger city, 25 km/15.5 miles south of Santa Cruz, in the hot, dry flatlands. It is much more commercial and consequently less interesting to tourists – most simply crawl through its traffic on the way to the beach. The in-town place to hire guides to **Barra Honda Park** is at the park office *(☎ 506/686-6760)*, across from Nicoya's historic **San Blas Church**, built in 1644. We can also say that the gasoline station just before town, opposite the hospital, has very clean bathrooms, a pleasant surprise at a bad moment. Another pleasant surprise is the **Ho-**

tel Curime *(☎ 506/685-5238, $$)*, named after a famous Chorotega chief. It has a cool deep swimming pool, open-air restaurant, sports fields and rooms with or without air conditioning in duplex bungalows. You'll find the Curime just west of town on the way to Sámara and the beach. If you're in the area in December, the 12th marks the **Fiesta de la Yeguita**, when a religious parade takes place.

On the road to Monte Romo, four km/2.5 miles south of Hojancha, a farming town south and east of Nicoya, is **Monte Alto Lodge** *(☎ 506/659-9394, montealto92@terra.es, $)*, a recent addition to the Cooprena list of eco-lodges. Set up to benefit the local community as well as protect the watershed of the Río Nosara, Monte Alto offers rustic overnight accommodations or an opportunity for a day-trip, perhaps a stop on your way to the beach. Typical food is served, plus visitors can enjoy five km/three miles of walking trails, a huge natural orchid garden, a view of Nicoya Bay from a mountaintop *mirador*, plus the local organizers will take you to visit their farms if you wish.

Barra Honda National Park

Barra Honda National Park is unique in Costa Rica – its 2,295 hectares (5,671 acres) were set aside to protect not the land, but the geological wonders underground. Huge limestone caves are natural phenomena of the park, 14 km/8.7 miles east of Nicoya. Barra Honda Mountain, rising 575 meters/1,891 feet above the dry plain, was thought to be a volcano as late as 1967. Foul odors and strange whooshing sounds coming from the craters that pockmark its slopes convinced local farmers of underground activity. Finally, in 1973, the caves were officially discovered by scientists from the US-based Cave Research Foundation. The strange sounds turned out to be the sounds of millions of bats, and the strong odor came from their thick guano.

The spectacular formations inside – descriptively named soda straws, cave grapes, organ pipes, fried eggs – date to the Paleocene epoch, 70 million years ago, and were formed by rainwater dripping through limestone. In the Nicoya cave, speleologists have discovered prehistoric remains and ritual offerings. Terciopelo cave has the best formations, and is quite deep at over 55 meters/181 feet. Each cave is accessed by a vertical descent, so if you want to explore them, make arrangements the day before with the **Park Service** *(☎ 506/686-6760)*. Get a tour with **Turnisa Tours** *(☎ 506/221-9185)* in San José. Even if you don't get into spelunking, the white limestone, tabletop mesa at the top of Barra Honda offers breathtaking views of Nicoya, especially from the south edge. Hike up marked trails, but bring plenty of water and stay on the path. With a guide, you can take a six-km/3.7-mile hike on Sendero al Ceibo trail, which leads to a waterfall accented by wispy calcium carbonate formations.

■ Playa Sámara

For memory has painted this perfect day.
With colors that never fade,
And we find at the end of a perfect day
The soul of a friend we've made.
~ *A Perfect Day*, Carrie Jacobs Bond, 1862-1946

This beachfront community is growing in popularity – for very good reasons. Its wide white beach is protected by rock promontories at either end and an offshore reef, which makes it good for snorkeling. It offers safe swimming in warm shallow water without rough surf. During the dry season, Ticos who own summer homes flock here, although it never gets crowded. And during the off-season, the beach feels deserted, despite the fact that the town is right there behind the palms. Another attractive reason for staying near Sámara (pronounced SAH-mara) is that the road is paved all the way into town. Good, reasonably priced hotels also help. There are a few small grocery and liquor stores in town.

Playa Sámara offers great snorkeling.

South from town the highway becomes a dirt road and snakes back to the beach where numerous hotels hide on the right. The next beach south again is the wide picturesque Playa Carrillo, lined by rows of palms.

Visit online at www.samarabeach.com.

■ Directions

By Air - The airport is at Playa Carrillo, 15 minutes south of Sámara. Daily flights by **Sansa** (☎ 506/221-9414, in Sámara 656-

0131) leave from Juan Santamaría Airport and **Nature Air** (☎ 506/220-3054) has a daily flight from Tobías Bolaños in Pavas. Flight times are subject to change. Most Sámara/Nosara hotels can arrange your pick-up.

By Bus - There is an **Empresa Alfaro** (☎ 506/222-2666) express bus from San José leaving at 12:30 pm from between Calle 14, Av 3 & 5. A six-hour trip for about $6. The bus to Nosara leaves earlier. The return bus leaves at 4 am, so don't stay up too late the night before. Call for schedules.

By Car - The road is paved all the way to Sámara and you can reach it from the mainland by crossing on the Tempisque bridge. Follow the signs to the city of Nicoya, make a left into town, and then go straight for Sámara, 36 km/22.3 miles.

Places to Stay & Eat

In Town

Hotel Belvedere *(Sámara, ☎/fax 506/656-0213, www.samarabeach.com, 10 rooms, 2 apartments, jacuzzi, air, full breakfast included, restaurant, $$)*. Cheerful and charming, the Belvedere sits high up on the hill above town, about four blocks from the beach. The cozy double-bedded rooms have balconies or terraces, mosquito netting over the bed, colorful drapes, and are very clean. The dining room is on a wooden deck with a view and cooling breeze. The best part of a stay is the chance to relax in the intimate jacuzzi, set in the middle of a lush garden. Owners Michaela and Manfred have been here since 1992. Excellent value, and great monthly rates.

Hotel Prices	
Prices are per night for two people	
[No $]	less than US $20
$	US $20-$40
$$	US $41-$80
$$$	US $81-$125
$$$$	US $126-$200
$$$$$	over US $200

Hotel Casa del Mar *(Sámara, ☎ 506/656-0264, fax 656-0129, 7 rooms, jacuzzi, air, refrigerator, restaurant, pool, includes breakfast, $$)*. This hotel, owned by French Canadians, has both shared or private baths in clean, comfortable rooms around a courtyard. One block from the beach. We think it would benefit from a friendlier front desk staff.

Hotel Giada *(Sámara, ☎ 506/656-0131, fax 656-0132, www.hotelgiada.net, 13 rooms, pool, pizza, fan, parking, breakfast included, $$)*. This Italian-owned hotel sits on your left as you enter Sámara, 150 meters/493 feet from the beach. The palapa overhang that allowed us to park our car out of the sun was what first caught our attention. Then the large attractive rooms in their pleasant two-story building wrapped around a small, kidney-shaped pool clinched the deal. Clean, at-home atmosphere and reasonable prices.

Acuario Apartments (*Sámara, opposite El Ancla restaurant, ☎ 506/656-0036, www.samarabeach.com, 2 apartments, 4 rooms, 2 houses, cable, air, kitchenettes, $$-$$$*). This small homey complex of accommodations is only 10 meters/33 feet from the beach and very reasonably priced. It's quiet and comfortable. Angela, the owner, is a Spanish teacher and is available for some free lessons. The houses have four bedrooms each, the apartments, two bedrooms, and the rooms boast coffee-makers. Very clean.

El Ancla Restaurant & Bar (*Sámara, on the beach, ☎ 506/656-0254*). There are a number of pleasant local eateries and watering holes in Sámara, but the venerable "Anchor" does it best. Fresh seafood and Tico cooking are served under a palapa roof with a cool breeze and plenty of beach in front. After we order, we swim out to catch a few waves bodysurfing and come back in time for fried fish and a cold beer. Who could ask for more? El Ancla is a favorite of locals and tourists alike.

Casa Paraiso Hotel & Restaurant (*Sámara, 150 m/500 feet east of supermarket, ☎ 506/656-0741, www.samarabeach.com, $-$$*). This piece of paradise is a clean, basic nine-room hotel with a very good restaurant attached. Chef Ana, one of the few female chefs around, features an excellent selection of fresh fish and lobster, seafood salad, pasta and meat meals. Other hotel owners eat here.

Villas Playa Sámara (*Playa Sámara, ☎ 506/656-0100, fax 656-0109, htlvilla@sol.racsa.co.cr, 62 rooms, air, kitchen, fan, pool, restaurant, breakfast included, $$$*). The villas here weave back from the wide beach and are stitched along the paths with rows of flowering hibiscus hedges. The red tile-roof villas all feature a kitchenette and sitting area, large bedroom with ample closet space and sliding glass doors that open on a terrace. They come in three sizes, from duplex to single unit with two bedrooms. The pool is wonderfully large (perhaps the best we found) and has a jacuzzi as well as a wet bar with TV. The palapa-covered restaurant is very appealing and serves excellent food. Ask to see several villas first (they'll ride you around in a golf cart). Four km/2.5 miles south of town.

Shake Joe's (*Sámara, 50 meters/160 feet east of beach entrance*). This Dutch-owned bar and restaurant on the beach serves laid-back meals (sandwiches and salads at lunch) under palapa umbrellas. Evening meals feature pepper steak, seafood and pastas, all served with French fries. And then there's the bar. Not surprisingly, it's famous for cocktails and fruit shakes.

A few kilometers south of Sámara is **Playa Carrillo**, a gently curved, broad white sand beach lined with palm trees. Very tropical and completely deserted.

■ Playa Nosara

Take Notice: When this sign is underwater, this road is impassable.
~ Sign on an Athi River Highway

A favorite area of retired North Americans, Nosara (no-SARA) has bad roads that get worse in the rainy season. From Sámara, you have to cross three narrow rivers and navigate 28 km/17.4 miles of potholes to a series of gentle beaches you reach by individual entryways. None are much different from the other, but we found some of the best accommodations near **Playa Guiones**. The farming town of Nosara itself lies inland about five km/three miles. As if the road wasn't bad enough, armed robbers have been holding up cars heading to Nosara and it's getting scary.

"NOSARA"

The name comes from folklore. Nosara was an indigenous woman who slit her wrists rather than reveal her tribe's treasure to a rival tribe. Her blood formed the river that bears her name.

The drive north from Sámara, despite hairy river crossings and a bumpy road, is very pleasant in long stretches as you motor through farm country in the land between the mountains and sea. The first big beach you come to before Nosara is **Playa Garza**, a fishing community.

■ Directions

By Air - Nosara has an airport inland (near the town) with a daily flight by **Sansa** *(☎ 506/221-9414, in Nosara 682-0168).* Flights leave from Juan Santamaría Airport at 11:30 am, returning at 12:20 pm. **Nature Air** *(☎ 506/220-3054, www.natureair.com)* flies twice a week on Tuesdays and Saturdays. Flight times are subject to change, so check first. Most Nosara hotels can arrange pick-up.

By Bus - There is an **Empresa Alfaro** (☎ 506/222-2666) express bus from San José leaving at 6 am from Calle 14, Av 3 & 5. A 6½-hour trip costs about US $7. The return bus is scheduled to leave at 12:45 pm. Check first.

By Car - Follow the signs to the city of Nicoya and, once in town, follow signs for Sámara. From Sámara, turn north on the dirt road that runs along the beach. There is an earlier turn, at the bottom of a curving hill a few kilometers before town, which joins this road.

Places to Stay & Eat

Lagarta Lodge *(Nosara, ☎ 506/682-0035, fax 682-0135, www.lagarta.com, 7 rooms, breakfast included, fan, parking, pool, $$).* Lagarta Lodge is a paradise for naturalists, birdwatchers, photographers and adventurous travelers. It's perched on a 40-meter/132-foot cliff (it's a steep ride uphill to the top), with a sweeping vista of the Nosara Biological Reserve and its black beach Pacific shoreline. The lodge's common area, which boasts a magnificent dining table, is on a porch that offers a panoramic view and a powerful telescope to see it close up. Several trails are marked for hiking into the lodge's private reserve area bordering the reserve and down to the weaving Río Nosara. Its rooms are plainer than the common areas and beautiful gardens, but the upstairs rooms (#3 & #4) have a balcony and those downstairs boast a little terrace. The mountaintop gets a welcome cooling breeze. Rooms #5 & #6 are very large. Inviting freeform pool.

Casi Paraíso *(Nosara, ☎/fax 506/682-0173, 5 rooms, breakfast included, restaurant, $$)*. This funky tropical wood hotel, whose name translates as "Almost Paradise," clings to the mountainside along the steep road, about a five-minute walk from the beach. A laid-back Caribbean atmosphere prevails inside and out. The rooms open onto a large shared porch with big Adirondack chairs and swinging hammocks under its metal roof. Bright artwork on the painted tongue and groove walls. The restaurant is a popular and inexpensive eatery perched on a wooden porch; it serves seafood. Casi Paraíso, also known as "Almost Paradise," now offers a sports fishing boat available for charter.

Casa Romántica *(Playa Guiones, Nosara, ☎/fax 506/682-0019, www.hotelcasaromantica.com, 7 rooms, pool, restaurant, fan, breakfast included, $$)*. This "romantic Swiss house" is just 20 meters/66 feet off the beach in a large white building. The restaurant, specializing in continental cuisine and seafood, is open for breakfast and dinner only. Guest rooms are big and cool and appealingly decorated. The white-tile floors and natural woods are used throughout. Very popular with Europeans.

The Gilded Iguana (*Playa Guiones, Nosara, ☎/fax 506/682-0450, www.guilded iguana.com, 4 rooms, restaurant, $$)*. Very pleasant, large rooms in a hotel building set back a little from the well-known Gilded Iguana restaurant. Owned by North Americans. Upstairs bedrooms – reached by climbing a metal spiral staircase – are breezy and white, with refrigerators and extra sleeping areas. The popular restaurant is open 10 to 10, more or less, and is closed Mondays except during football season, when they serve dinner and drinks for Monday night football. Every Saturday at 10 am there is an open but cutthroat bridge game where you can hone your skills with razor-sharp wits. This is also the headquarters of **Iguana Expeditions** (call the hotel and ask for Joe), which runs great sea kayak trips into the reserve.

Hotel Villa Taype *(Playa Guiones, Nosara, ☎ 506/682-0333, fax 682-0187, www.villataype.com, 22 rooms, two pools, restaurant, tennis, air or fan, breakfast included, $$)*. Owned and managed by a local Costa Rican family, the Taype is a large horseshoe-shaped hotel with some bungalows. Accommodations sit in lovely gardens about 100 meters/329 feet from a beach that offers a good surf break. There's a long, open public area with a TV and bar, and large kidney-shaped pool with a swim-up bar. The rooms are relatively unadorned, but modern and very clean, with hardwood ceilings and white tile floors. Garden bungalows (numbered in the upper teens to early 20s) have small patios under thatched roofs. See a few rooms before you decide. The bulletin board announces many local activities and tours. Budget travelers should ask about a smaller room without air for about half-price. Anarchists can pay in cash and avoid the tax.

Café de Paris *(Playa Guiones, Nosara, ☎ 506/682-0087, fax 682-0207, www.cafedeparis.net, 18 rooms plus bungalow/suite, air or fan, restaurant, pool, $$).* Only in Costa Rica can you have a French café that is a bakery, restaurant and hotel rolled into one big croissant. Everything opens at 7 am – the bakery for fresh bread, pastries and baguettes, the pool's swim-up bar for early-risers, and the restaurant for the first of its three meals a day. New rooms built in mid-2000 are across the road.

Doña Olga's *(Playa Pelada, Nosara, no phone, $-$$).* Olga serves mainly fish and typical Costa Rican food in her beachside open-air restaurant under a big palapa. Very popular with locals and vacationers alike. Open for breakfast and until late at night.

Other dining spots in and around Nosara include **La Luna Bar & Grill**, with rustic natural wood furniture, blue fabric tablecloths and a small disco. It's affiliated with the recently remodeled **Nosara Beach Hotel** *(☎ 506/682-0121, www.nosarabeach.com)*, which isn't a bad choice for accommodations. At **La Dolce Vita** an Italian chef makes good pizza and homemade pasta puddings. Fresh seafood is offered everywhere.

Southern Peninsula

A chieftain to the Highlands bound
Cries, 'Boatman, do not tarry!
And I'll give thee a silver pound
To row us o'er the ferry.
~ *Lord Ullin's Daughter,* Thomas Campbell, 1777-1844

■ Playa Naranjo

The scenery changes dramatically as the dry hot plains give way to hot lowland forests in the south of the peninsula. The road south to Playa Naranjo ferry is good most of the way, although all the roads in this countryside vary in condition, especially in the rainy season. It's almost a straight line, as the crow flies, from San José to here, and you can even make a trip to this area for a long weekend adventure. If you're coming to the southern Nicoya Peninsula, it's more logical to take one of two ferries across the gulf from Puntarenas. It's now less common for vacationers to use ferry across to Playa Naranjo since a new service runs from Puntarenas to Paquera, a town closer to the southern tip. The shoreline around Playa Naranjo is fairly dramatic, with steep green hills right up to the beach area. The sand isn't great, but you'll find some great driftwood here. The road south to Pacquera from here is passable; unpaved but graded in some stretches.

For an overnight in sleepy Naranjo (which is not as much a town as a ferry dock), a good choice is the **Oasis del Pacifico** *(☎ 506/661-1555, wilhow@racsa.co.cr, tennis, pools, fishing, restaurant, $-$$)*. The enchanting Oasis is set on the beach with a sprawling 4.9-hectare/12-acre garden. The simple rooms here have porches with hammocks for lazy living. An authentic oasis of the Pacific.

If you're in the area for world-class fishing, head a half-hour north of Naranjo for **Hotel Bahía Luminosa** *(☎ 506/641-0386, www.bahialuminosa.com, pool, restaurant, includes breakfast, $$-$$$)*, an American-owned, low-key resort that specializes in yachting and sailing adventures – they even offer their own yachts. The bay is a favorite protected anchorage for sailboats. Luminosa offers deep-sea fishing, diving trips and local nature tours. Tree-shaded rooms have kitchenettes and are spread out on 71 hectares/175 acres of wooded grounds (including primary rainforest) on a private bay.

The more popular ferry drops you at the **Pacquera** dock, five km from Pacquera village, which is about an hour south of Naranjo on a very rough road. It has little to offer, except as a transit point to other attractions. **Cabinas Ginana** *(☎ 506/641-0119)* is clean, economical and has a very pleasant restaurant. From Pacquera, the road is paved or graded most of the way, off and on, to Montezuma.

Karen Morgenson Reserve

The "godparents" of the Costa Rican park movement were aspiring fruit farmers, Karen Morgenson and Olof "Nick" Wessberg, who landed in the Nicoya Peninsula on Costa Rica's northwest Pacific coast in 1955. They settled in Montezuma because of its quiet natural beauty and the fact that they were living next to one of the last wilderness areas on the peninsula, **Cabo Blanco** (White Cape). When it became endangered, Olof and Karen's worldwide campaign provided funding for Cabo Blanco's purchase and donation as a park to Costa Rica. Tragically, Olof was senselessly murdered on the Osa Peninsula where the couple worked on the environmental effort that would create Corcovado National Park. Karen continued as a tireless advocate for the ecology until her death in

1994. Her personal legacy is expressed in a 600-hectare/1,482-acre private reserve that bears her name. Stay at a primitive lodge or camp. Contact the grass roots environmental group to learn more *(☎ 506/650-0607, www.asepaleco.org)*.

Curú National Wildlife Refuge

This is an anomaly among Costa Rica's protected areas. The core 84-hectare/207-acre wildlife refuge is surrounded by a 1,200-hectare/2,964-acre farm and forest under cooperative protection with the Park Service and the farm's proprietors. The story of **Curú's** *(☎ 506/200-5020)* preservation is admirable. Federico Schutt de la Croix established a large plantation here in 1933 and made it clear to his family that he wanted the area to always remain pristine, even after his death. They follow his philosophy of ecological preservation and sustainable farming. There are silent beaches along the seafront and several hiking trails in the seven habitats: deciduous, semi-deciduous, hill forest, beach, littoral woodland, mangrove swamp, and lagoon – the only one left on the peninsula. A rich diversity of wildlife includes troops of white-faced capuchin monkeys, howler monkeys, over 222 species of birds, and three types of turtles that use the shore for nesting. Located about seven km/4.4 miles south of Pacquera. Look for the farmhouse up on stilts on your left. Cover yourself with insect repellent before stepping out here.

■ Playa Tambor

Any work of architecture that does not express serenity is a mistake.
~ Luis Barragán, *Time* magazine

Playa Tambor is a destination somewhat like Papagayo, fingered as a suitable area for large, all-inclusive beach resorts. The paved airport runway and a ferry dock nearby make access easy. Yet, despite the building of several large resorts outside the little village, that type of exclusive vacation hasn't entirely captured the heart of the average Costa Rican tourist. The largest mega-resort constructed is the **Barceló Tambor Beach Resort**. The controversy over its environmental impact has died down, but it's still one of the ugliest properties we have ever come across. It may have all the wonderful luxuries of its type for guests, but the architect must have designed army camps before building these row-upon-row of ticky-tacky little boxes.

In the little town of Playa Tambor, however, there are some tempting small hotels that allow you to include the Tambor experience without staying all-inclusive.

■ Directions

The most traveled routes to Tambor and the southern beaches depend on how you're traveling and where you're coming from. The drive from Nicoya follows the book through Naranjo. However, to drive (4x4 strongly recommended) from San José, go to Puntarenas for the Paquera Ferry (departs at 5 am, 8:45 am, and 12:30, 2, 5 and 8:15 pm) or the longer Naranjo Ferry. For mass transit, take an express bus from Avenida 10 in San José to Puntarenas, then walk or take a taxi to the ferry dock. Unfortunately, the faster launch service that once left from behind the Central Market is no longer running – if service is reinstated by the time you are here, take this launch instead. At the Paquera dock, board the direct "bumpy" bus to Montezuma or take a taxi to Tambor or Montezuma ($35). The whole trip takes between five and eight hours.

Turn left onto the main street leading to the beach, then left again and 20 meters/66 feet along the shore find **Tambor Tropical** *(☎ 506/683-0011, www.tambortropical.com, restaurant, breakfast included, $$$$)*, a beachside B&B that says tropical luxury in every way. Its modern two-story hexangular *casitas* are made of highly polished hardwood and huge sliding glass windows facing the shore. Upstairs rooms have skylights and wrap-around verandas, while those downstairs open onto garden patios. The grounds are beautifully manicured and incorporate a free-form pool in the center. The dark gray beach is lined with palm trees. Tours and activities available, but no children under 16.

A very attractive alternative is **Costa Coral** *(☎ 506/683-0105, www.costacoral.com, air, cable TV, kitchenette, restaurant, safe, $$$)*, on a corner in the "downtown" area. Although it's a long block or two from the beach, this very private hotel is so well apportioned that it's worth the walk. The six upstairs/downstairs villas are hidden next to a well-stocked gift shop and store complex, with an open-air veranda restaurant above that serves delicious, reasonably-priced meals. Each villa offers a veranda or porch, sitting room with kitchenette, Mexican tile bathroom and queen-size bed. There's a cool little pool and a groovy gift shop across the street, **Salsa Gifts**.

If you find these properties are a bit rich for your blood, don't fret. It's possible to stay in Tambor cheaply at **Cabinas El Bosque** *(no phone, $)*, just north of town. Safe and secure, it features very basic rooms in small cabins with fridges and fans. The beach is just a short walk through their heavily wooded grounds. The gracious and friendly owners are a family with relatives in New Jersey. One of their guests was staying long-term, with her young daughter, in one of the two apartments that can sleep up to six.

You can stay in luxury and golf, surf, sail and swim at the original high-standard Costa Rican resort, **Tango Mar Beach Resort & Country**

Club *(☎ 506/683-0001, www.tangomar.com, restaurant, pool, golf, tennis, air, including breakfast, $$$$)*, south of town. The resort is reached through a gated entrance that leads into its beautiful golf course. The bewitching hotel and villas are on the rise overlooking a pretty, palm tree-lined beach. Rooms have large picture windows facing the sea, while the secluded villas are scattered in the tropical gardens (not all have sea views). The grounds are large enough and privacy great enough that golf carts are used to get around on the paths. Each villa – one with as many as five bedrooms – is unique; including "Tiki" suites with a bewitching Polynesian feel to them.

Who can resist eating in a yellow Caribbean-style cabin called **Scruffy's Restaurant**? We couldn't, and liked it. Open for dinner (it doesn't matter whether you look scruffy or not). It on Tambor's main drag.

■ Montezuma

I couldn't help it. I can resist everything but temptation.
~ *Lady Windemere's Fan*, Oscar Wilde, 1891

We were predisposed to love Montezuma. Everything we heard, both good and bad, appealed to us. Long a haven for hippies, new-age seekers, tree-huggers, free thinkers, nature lovers, beachcombers and entrepreneurs, Montezuma is one of the last outposts of undeveloped, beautiful beach towns. But get there soon because word is out – Montezuma is a tantalizing and enchanting ecological beach destination. Long-time local residents welcome visitors who respect the environment and the culture – all others can stay home.

■ Directions

See Tambor section for directions when driving a car or arriving by public bus. The way into Montezuma is dramatic. From Cóbano, you arrive at the edge of a plateau, with the Pacific Ocean in all its glory before you, and then descend the steep road (paved by the hotel owners so people could get out of town in the rainy season). Beyond Montezuma the road south snakes along the sea past hidden rock coves, shady nooks and secluded sand beaches, through a couple of tiny villages, until it reaches Cabuya, the last civilization before the Cabo Blanco Reserve.

Adventures

North of town are expanses of rich white sand beaches between large, dramatic rock outcroppings, deserted except for the occasional pelican. The best swimming is in front of the campgrounds – for a long stretch north of that the currents are strong and tricky and swimming is dangerous. Walk two hours north, passed Playa

Grande, to where ocean swimming is safe again and a rainforest **waterfall** cascades into the sea. Bring water and be sure to wear a hat and sunscreen for this hike. Tourists come to the magnificent waterfall for its wild beauty, but thrill-seekers dive nearly 15 meters into the pool at its base. Do not do this! Five young people have died in the last three years when they struck rocks underwater. There is no warning sign. The waterfall is also a popular destination for horseback riders, who come to swim or watch the diving pelicans. We recommend you booking rides through **El Saño Banano** restaurant *(☎ 506/642-0638)* in town. They use Gerardo Rosales Flores, who has the only real stable, with horses in better condition than the poor maltreated wretches some guides use. By the time you're here, rides may also be offered through the **Bakery Café** (see below). Or arrange them with **Montezuma Expeditions** *(☎ 506/642-0482, www.montezumaexpedtitions.com),* who do a good job with various local excursions. Serious riders should go back up the hill to Finca Los Caballos (see below). Serious dancers and party people can leave the bar scene and head one km/.6 miles south of town to **Las Manchas Disco**, open only weekends in low season.

A skilled horseman "rides" his horses.

A second waterfall makes for s strenuous but rewarding rainforest hike. Around Pension Lucy and La Casada, take the small path on the right just after the bridge. It follows the stream to a few tiny falls where you cross to follow the path on the right bank. The impressive waterfall is five to 10 minutes away. A path uphill from there is a strenuous hike to the mountaintop. Take plenty of water and insect repellent.

Learn Spanish while you're down here at **Horizontes de Montezuma** *(☎ 506/642-0534, www.horizontes-montezuma.com, $)*, a German-owned hotel up on the bluff before town. You don't have to be a guest to take classes. Four weeks of bed, board and lessons runs about US $640.

Places to Stay & Eat

Funky little Montezuma town is home to several small restaurants, some good hotels, a couple of loud bars and an eclectic mix of visitors and residents. It must be doing something right because it fills up solid during the dry season, and quite a bit in the green time as well, when prices plummet. *Bienvenido al paraíso.*

Hotel Prices	
Prices are per night for two people	
[No $]	less than US $20
$	US $20-$40
$$	US $41-$80
$$$	US $81-$125
$$$$	US $126-$200
$$$$$	over US $200

Camping is now restricted to the **Rincón de los Monos** (Monkey's Corner) public campground at the north end of the beach, which is generally full in high season. ☎ *506/642-0048.*

Nature Lodge Finca Los Caballos *(☎ 506/642-0124, www.naturelodge.net, pool, restaurant, $$).* Finca los Caballos is set above a river and secondary forest, a few kilometers before Montezuma. It's perfect for a natural country vacation, beach lovers who want a ranch stay close to the ocean, or anyone who can't relate to the bohemian travelers in town. But it's especially appealing for horse riders. Guided tours put you back in the saddle again. Rooms in Santa Fe-style cinder block buildings are fairly basic but comfortable. The restaurant, pool and view are fantastic. The ranch is high up on the crest of a hill with cooling breezes and a view of the Pacific. A separate two-bedroom bungalow is available for rent.

Ylang Ylang Beach Resort - *El Saño Banano* *(☎ 506/642-0638, fax 642-0068, www.elbanano.com, pool, coffeemakers, restaurant, $$$).* Patricia and Len have been around Montezuma since 1980 and, little by little, they built the fantastic Ylang Ylang resort just north of town. Facing the beach, but hidden in tropical gardens, are eight private bungalows. There are six rooms and suites in a separate building, plus some accommodations in geodesic domes with kitchenettes and outdoor showers. The most private bungalow is Coco Joe's, a large upstairs-downstairs cabin with a porch outside and a Mexican interior décor. The bungalows are spread out on the grounds to give you a feeling of being an your own island getaway. Follow cement paths through lovely heliconia and verdant jungle flora to the enticing grand pool, complete with a gushing waterfall at one end. Dive in! This is the kind of special hotel that makes a one-of-a-kind vacation. Reservations recommended. Even if you don't stay here, try the new casual gourmet restaurant. The setting is just beautiful, surrounded by lush flower gardens overlooking the ocean. Romantic candle and Tiki lights provide ambiance in the evening. Arrive at El Saño

Banano Restaurant and the hotel will come and pick you up in their beach buggy (no other vehicles are permitted on the beach).

Hotel El Jardín *(☎ 506/642-0548, www.hoteljardin.com,16 rooms, refrigerators, air or fan, $$)* sits up a steep driveway on a hillside with a commanding view of the main drag downtown and the sea beyond. Spacious clean rooms, with tile floors and broad balconies.

Hotel La Aurora *(☎ 506/642-0051, www.playamontezuma.net/aurora.swf, 10 rooms, shared bath and kitchen, coffee and tea all day, $)*. The activist couple that own the Aurora have created the closest thing to a youth hostel as there is in town. The airy Caribbean-style building has a huge upstairs veranda that's crowded with Europeans in hammocks reading Proust and Kierkegaard and other non-Oprah selections. Drying towels hang over railings like multinational flags. The wood walls and floors are highly polished, but the spartan rooms are a little dark. So get out of bed and make friends. Mosquito netting provided.

Cabinas El Tucán *(no phone, shared bath, taxi service, $)* is owned by an original Montezuma family. The simple Caribbean-style wood cottage is just at the edge of the village heading south along the coast. Austere rooms, but excellent prices.

El Saño Banano Village Café *(☎ 506/642-0638, $$)* is a landmark in the heart of Montezuma village. Open from 7 am to 9:30 pm daily, El Saño offers the widest selection of food around, including natural foods and lots of fresh fruit dishes and ice cream. Try our favorite breakfast: homemade yogurt and granola. You know you're in the sticks when the big event of the evening is to join with townsfolk at El Saño and watch a movie (in English) projected on a pull-down screen. Dinner and a movie! El Sano restaurant now rents soundproof rooms around back, complete with direct TV. Down on the corner, check out **La Esquina Dulce**, with its homemade ice cream stand named Iguana. Before the restaurant is an **express laundry**.

Turn into town toward the sea, past El Saño Banano, and make a left onto the short street north that ends at the campground beach. Along this stretch you'll find the following hotels.

Cabinas Mar y Cielo *(no phone, 6 rooms, fan, $)*. If you can take the nighttime exuberance of the bar almost next door, this beachside little hotel, hidden behind a gift shop downtown, is a super choice. Six very big, very clean rooms with patio or balcony overlooking a grass lawn garden in front of the crashing surf.

Pargo Feliz *(no phone, 8 rooms, $)*. A big ugly white building that features large clean, basic rooms with tile floors. A connecting porch or upstairs veranda face a lush tropical garden lawn. Walk to the beach. Next door, almost as a continuation of the Pargo's tropical garden, is a tempting little oceanside eatery called **Cocolores** *($$)*. Very good seafood, pasta, salads, ceviche and more are served in a rustic romantic candlelit setting. Open for lunch and dinner. At the end of the road you'll discover the English-language bookstore and gift shop, **Librería Topsy Bookstore**.

Follow your nose across the street to the sweet **Bakery Café** and its small dining porch. Very popular, and for good reason. The cute couple that owns it have a horseback riding tour company (ask at the bakery for details).

Luz de Mono *(☎ 506/642-0010, fax 642-0090, luzdemono@racsa.co.cr, coffeemaker, refrigerator, fan, restaurant, $$-$$$)* is a very ambitious hotel project in the woods behind the Bakery Café and the campgrounds. New in 1998. Luxurious large rooms are set alongside a stream, which forms a moat in front of each one. Cross a bridge to enter the enticing quarters with Spanish colonial décor and high ceilings. The management also owns a ranch and farm five km/three miles away with grape vines and dairy cows. They offer a tour on horseback, free for guests. The restaurant opens for lunch and dinner in a large, multi-layered round building covered by a tile roof.

If you continue south along the sea from Montezuma, the road will twist and turn a couple of times to a group of noteworthy establishments within a 10- or 15-minute walk of town.

Playa de los Artistas Restaurant *(no phone, $$)*, across from Los Mangos, may be the best place to eat in Montezuma, if only for its beachside ambiance and crowd of admirers. Open only for dinner, the "Artists' Beach" has just a few tables in back of a small house on the beach. As it gets busy, diners can take tables on the sand, lit by candles flickering in the wind. The romantic setting matches the artistic food. Specialties are seafood, grilled fish and Italian dishes, with daily specials always on offer. Meals are served in large wooden bowls with home-baked bread. This is not the kind of place to be fashionably late; you'll have to wait for a seat. But it's worthwhile.

Restaurant Prices	
Prices are for an average entrée	
$	less than US $5
$$	US $5-$10
$$$	US $11-$20
$$$$$	over US $20

Los Mangos *(☎/fax 506/642-0259, www.hotellosmangos.com, 24 rooms, pool, restaurant, $$)* is one of our favorite Montezuma hotels. Small cabins are arranged in an old mango orchard across the street from the beach. Mango trees are not like apple or orange trees; they grow tall, like oaks, and have large heavy leaves. The also have heavy fruit that is a favorite of the howler monkeys who wander down from the nearby rainforest. (If you annoy them, they'll drop fruit on your roof! Good morning Costa Rica!) These comfortable Tahitian-style cabins with porches and hammocks are terraced up the steep hillside. Ten new rooms have been built in a single building near the pool (plus, for low-budget travelers, four small economy rooms with a shared bath). The pool is large and refreshing and has a waterfall, while the restaurant above it is worth a visit. The bright blue birds that may come for scraps are white-throated magpie jays (*Ureca copetona*). The convivial host of all these guests is Nick Costas, who would love it if you brought him some Feta cheese or Greek olives.

Hotel Lucy *(☎ 506/642-0273, across from Los Mangos, $)*. Backpackers rejoice! Lucy's has not been torn down or swept away by the sea – at least not yet. Directly on the rocky beach (not for swimming) and across the street from Los Mangos, little Hotel Lucy clings to the shore with its rustic rooms, with or without fans, and good view from the top story. It has a tiny soda serving Tico fare.

Las Cascadas *(☎ 506/642-0057, $)*, next to Los Mangos, is a basic hotel with an inviting thatched-roof restaurant alongside a cascading waterfall of the Río Montezuma that flows down to the sea. It's a popular place to sit and relax in the positive ions, thinking positive thoughts, while munching on fresh seafood and nursing a cold brewsky.

Amor de Mar *(☎ 506/642-0262, www.amordemar.com, restaurant, breakfast included, $$-$$$)*. The "Love of the Sea" hotel is one of those treasures that you sometimes stumble upon when fate allows. If you're lucky enough to stumble this way, stop and stay at this lovely two-story Caribbean-style *casa*. It's built of highly varnished tongue-and-groove wood and features a large lawn on a promontory, all next to Río Montezuma. No beach to swim in, but fascinating tide pools to splash around in. A big upstairs porch and large windows make small rooms of all wood walls and ceilings brighter. Room #5 is the largest, with two queen beds and a big attractive bathroom.

■ Leaving Town

After this hotel, the road runs south, past nooks and crannies of coast, secluded sandy beaches, freshwater streams, rocky shores, a little fishing village and lots of forest. It leads to the town of **Cabuya**, and two km/1.25 miles farther brings you to the entrance of Cabo Blanco Absolute Reserve (no relation to the vodka). This is a great drive, even though the road is a bit bumpy, and the last time we did it we came upon a pig-slaughtering affair in Cabuya – preparation for the next day's big soccer match against a neighboring village. We returned to see the match and pig out. These kinds of events make for tales and memories for years to come. Also extraordinary is the Cabuya **town cemetery**, located on a small, wooded, rocky island about 100 meters/329 feet offshore, not far past Hotel Celaje.

You can walk to it when the tide is out, and local funerals are scheduled to coincide with the tides. A metaphor of life's continual ebb and flow.

Hotel Celaje *(☎ 506/642-0374, celaje@sol.racsa.co.cr)* is an isolated but charming little hotel on the shady road just before Cabuya. Unusual A-frame, two-story cabins have a bedroom on the second floor and are at least twice as wide as they are deep. One of the hotel's best features is its small, pristine pool where we swam on a brutally hot day after enjoying a delicious lunch. The Belgin owner/cook serves outstanding food. A-frames are scattered in a coconut palm woods and a path follows a stream to the rocky beach. From here you can see the island cemetery of Cabuya, which locals claim has been used as a graveyard since pre-Columbian times.

■ Cabo Blanco Absolute Nature Reserve

To do nothing is sometimes a good remedy.
~ Hippocrates 460-400 B.C.

This is the largest single area in Costa Rica set aside as an "Absolute Reserve," which absolutely restricts the impact of humans. But that doesn't mean you can't enjoy it. To aid their conservation goal, Cabo Blanco limits entrance and provides minimal facilities. You can't camp in the park and the only thing to do is hike in and hike back out along marked trails. The ranger station, two km/1.25 miles from Cabuya, has good maps of the various trails.

What's so great about so few people entering the 1,172-hectare/2,895-acre reserve is that wildlife abounds – variegated squirrel, mantled howler monkeys, white-throated capuchin monkeys, agouti, white-nosed coati, even jaguarundi. Off the tip of the shore, Isla Cabo Blanco is a bird breeding preserve, home to 400 pair of brown boobies. Cabo Blanco got its name from the Spanish explorers who saw the thick layers of white bird excrement on the island's rocks. Expatriates Karen Morgenson and Olof Nicolas Wessberg arranged the purchase of the area in 1963 and donated it to the Costa Rican government – an act that spurred the creation of the National Park Service. When they bought it, the core of Cabo Blanco (15%) was all that was left of primary rainforest; the rest of the land around it had been cut and cleared for pasture. One of the reasons this is now an "absolute reserve" is to see how well and how quickly the rainforest can heal itself. Using the core as a genetic seed bank, the remaining 85% of the reserve is now secondary forest, an encouraging sign for the world's rainforests. See page 49 for a brief history of the park that started the park system in Costa Rica.

Cabo Blanco Absolute Nature Reserve

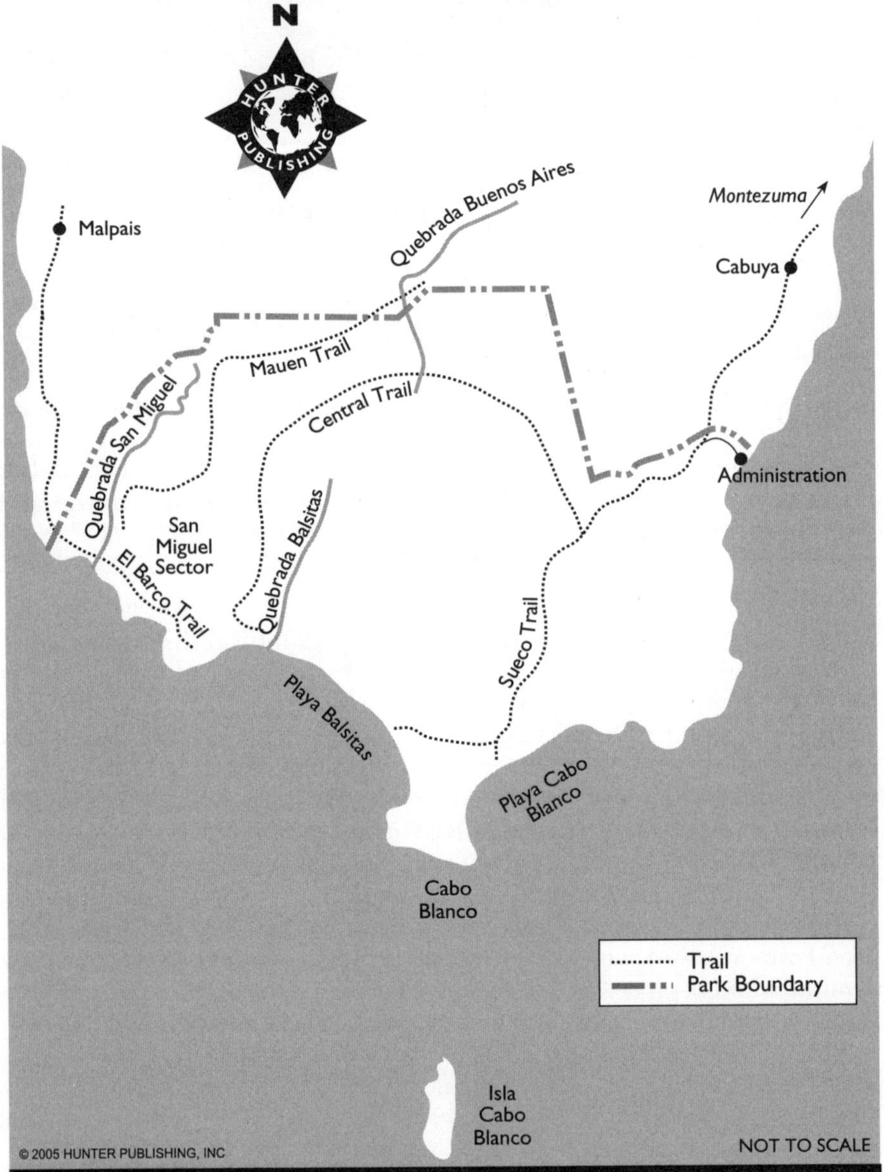

■ Malpais & Santa Teresa

Bad's the best of us.
~ John Fletcher, 1579-1625

Malpais and Santa Teresa are two beaches on the west side of Cabo Blanco with their isolation – and perhaps attraction – enhanced by their remoteness. The road that cuts across the peninsula tip from Cóbano, the town before Montezuma, is usually in terrible condition. But that seems to suit everyone just fine. Combined with the fact that Malpais means "Bad Country," you've got tourists staying away in droves. What they're missing is miles of deserted shore, scattered little beach hotels and rough surf that thrills experienced surfers. We've heard it described as "super surfing." Santa Teresa has the better sunbathing/swimming beach.

Places to Stay & Eat

Frank's Place *(☎ 506/640-0071, $)*, at the crossroads into town, is a popular place to stay. Rest in one of Frank's *cabinas* or eat at his little restaurant.

Hotel Prices	
Prices are per night for two people	
[No $]	less than US $20
$	US $20-$40
$$	US $41-$80
$$$	US $81-$125
$$$$	US $126-$200
$$$$$	over US $200

Turning south, surfer dudes ride to the **Mal País Surf and Sport Camp** *(☎ 506/640-0061, pool, restaurant, www.malpaissurfcamp.com, $-$$)*, a Malpais institution that has a loyal following. Accommodations run the gamut from open air *ranchos* (hut-like accommodations) and bunk-bed dorms to attractive individual bungalows. The bar in the main lodge is the happening place.

Eat Italian at **Albinat Dulce Magia**, opposite Cabinas Bosque Mar. Turn there for **Mar Azul** *(☎ 506/640-0098)*, a pleasant beachside bungalow hotel with a small restaurant on the sand. **Piedra Mar**, 100 meters/329 feet west of Cabinas Laura Mar and right on the beach, serves wonderful fresh seafood. Perched on a rocky promontory at the south end of the rugged Malpais coast road is the **Sunset Reef Marine Lodge** *(☎ 506/282-4160, www.sunsetreefhotel.com, $$$)* is the area's top hotel. It features large

Restaurant Prices	
Prices are for an average entrée	
$	less than US $5
$$	US $5-$10
$$$	US $11-$20
$$$$$	over US $20

rooms with air conditioning, a nice pool, good restaurant and bar, and offers a myriad of activities.

North from Frank's is **Santa Teresa** and the **Trópico Latino Lodge** *(☎ 506/640-0062, www.hoteltropicolatino.com, $$)*, a charming little cottage hotel that boasts huge rooms relatively close to the beach. Pool, jacuzzi and Italian restaurant round out the good value.

A bit more upscale is the French-owned **Milarepa** *(☎ 506/640-0023, www.milarepahotel.com, $$$)*, which reflects a wispy Indochine ambiance and is absolutely gorgeous. Lovely open-air feel, tropical gardens, rocky beach and a gourmet French restaurant to die for. Have a glass of wine; it's good for your heart.

A bit out of Santa Teresa, at nearby Playa Coyote, Tico Resorts have built a small luxury hotel called **Casa Caletas** *(☎ 506/289-6060, www.casacaletas.com, air, pool)*, with a view that makes the long hard trip to get there worthwhile. The same folks own Lake Coter Eco-Lodge, near Arenal, and Fuego del Sol Hotel in Jacó.

Pacifica

Then I felt like some watcher of the skies
When a new planet swims into his ken;
Or like stout Cortez when with eagle eyes
He star'd at the Pacific – and all his men
Look'd at each other with a wild surmise –
Silent, upon a peak in Darien.
~ *On First Looking into Chapman's Home*, John Keats, 1795-1821

IN THIS CHAPTER

The area we're calling "Pacifica" encompasses the Central Pacific coast from Puntarenas, a major shipping port and gateway to the Nicoya Peninsula, down to Dominical, a surfer's paradise that eco-tourists are beginning to discover. Inland, the soaring Talamanca mountains are the playgrounds of the cloud forest-dwelling quetzal birds, and the valleys are the fruit and coffee basket of Costa Rica. The country's size means that people in San José and the Central Valley can easily reach the fine sandy beaches here for quick weekend getaways. The social scene of Jacó, the eco-delights of Manuel Antonio, Central America's most-visited national park, the surfing haven of Dominical – and all points in between – are at the heart of Costa Rican tourism's fine line between development and conservation needs. Adventurous drivers can follow the Inter-American Highway along mountain ridges offering views that go on forever. So get on a bus, rent a car, hop a plane and get down to yet another part of Tico paradise.

To reach the shore at Puntarenas from San José follow the Inter-American (General Cañas) Highway north and watch for the signs near Barranca at Km 101. For beaches south, including Jacó and Manuel Antonio, follow the Coastal Highway from Alajuela through Atenas to Orotina and follow the signs for Jacó. This is a winding mountain road that makes for an exciting drive – the bus drivers who run this route are amazing. If you prefer a less dramatic way, go to Puntarenas and turn south past Puerto Caldera back toward Orotina, follow the sign for Jacó on a modern highway. It's an extra half-hour or so drive time, but it avoids the high mountains.

Puntarenas Province

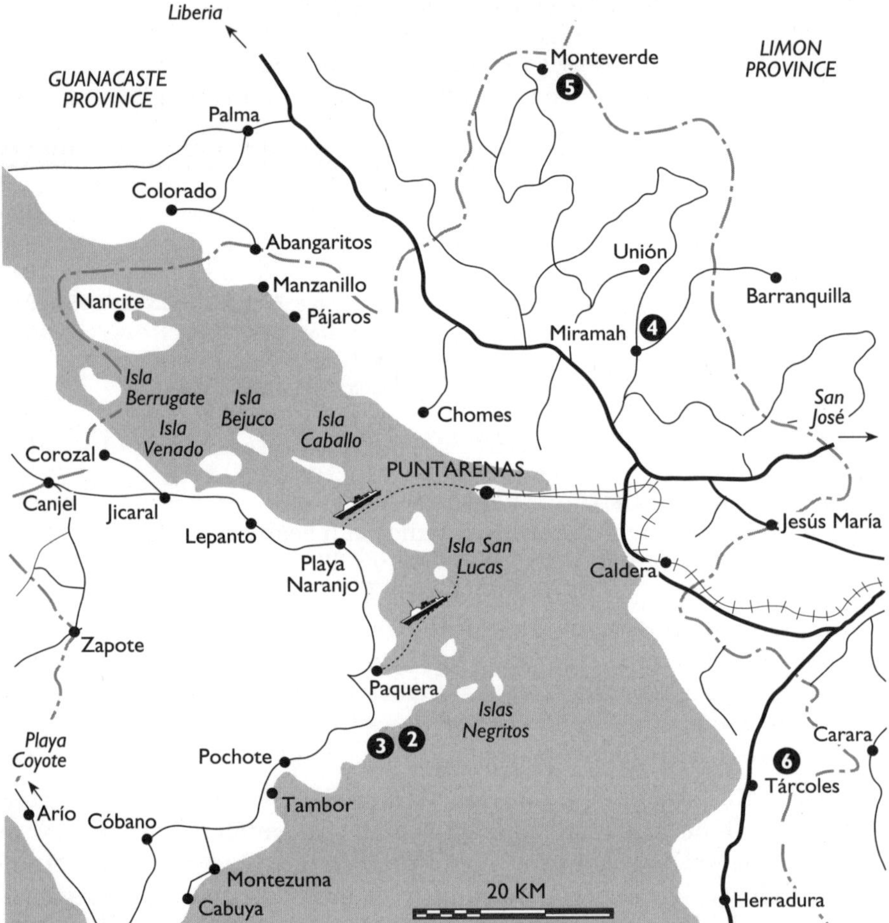

Puntarenas

They'll tell thee, sailors, when away,
In every port a mistress find.
~ John Gay, 1685-1732

Puntarenas began life as Costa Rica's premier Pacific deep-sea port and popular seaside destination for Costa Ricans. A short ferry ride across the gulf now brings much better beaches within reach, so most foreign tourists find themselves just passing through. But the Spanish *Conquistadors* who settled here in 1522 found the protected Gulf of Nicoya ideal as a port city. It was to here that, on Christmas Day, 1843, the English captain, William Le Lacheur, sailed looking for cargo. His success at transporting coffee beans to Europe fueled the country's huge – and prosperous – coffee boom. Good times for Puntarenas and Costa Rica.

Named after the long six-kilometer/3.7-mile peninsular strip of sand that houses the city, Puntarenas (Sandy Point), grew rapidly after 1846, when an improved overland road to the Central Valley was constructed for coffee exports, hauled down in oxcarts. The harbor lost much of its importance in 1890 when work crews finished the Atlantic Railroad, opening the closer-to-Europe city of Puerto Limón for coffee exports. In 1910, a railroad was completed to Puntarenas, which relieved some of the economic sting. As the closest town to the central highlands, it soon became a popular beach resort, nicknamed the "Pacific Pearl."

Over the years, wealthy businessmen, prostitutes, workmen, families, vacationers, fortune seekers, travelers, as well as seasoned sailors of the town's fishing, freight, and mother-of-pearl fleets, gave Puntarenas a unique rough-and-tumble seaport ambiance – and some vestiges remain today. In the early 1990s Puntarenas took an economic hit after the opening of Puerto Caldera, a port better suited to larger vessels and cruise ships, about 20 km/12.4 miles south. It's still trying to recover. The long beachfront was cleaned up in 1999, and *Puntarenenses* (what locals call themselves), as well as *Josefinos* are returning to the downtown again. Travelers used to dread getting "stuck" in Puntarenas overnight, but with the revitalization effort, reasonable prices and few tourists, it now makes a pleasant stop for savvy travelers, just a two-hour drive west of San José.

The long **Paseo de Turistas** *malecón* is the city's pride and joy – and so it should be. Seaside, the sandy beach stretches for several kilometers. Lined with palms, *helado* (ice cream) vendors, playgrounds, picnic areas, soccer players, swimming dogs and frolicking families, the beach and the *paseo* feature a broad sidewalk that attracts joggers, strollers and lovers walking hand in hand. Along the way there are some seaside restaurants and crowded gift shops, a favorite of passengers from small cruise ships, which stop at the refurbished waterfront pier. Across the boulevard are notable restaurants and some surprisingly good hotels. At the very tip of the sandy strip there is a public park and private yacht club.

The Catholic **cathedral** on Av Central, between Calle 5 and 7, has porthole-style windows. Built in 1902, after a fire destroyed the original 1850 structure, its walls are fortress thick. Nearby, the **Casa de Cultura** houses an art gallery and museum, plus an interesting separate maritime museum, **Museo de Historia Marina**.

Carnavale is a huge event here in Puntarenas. It's usually held at the beginning of February. There are fishing tournaments, parades, bullfights, theater, dances, rock concerts, sport tournaments such as beach volleyball, fireworks and the election of the Queen of the Carnival. We were impressed.

On the Saturday closest to July 16th the town comes alive again for the **Fiesta del Virgin del Mar**, a celebration rooted in Costa Rican folklore. According to the legend, four fishermen were caught in a bad storm at sea in 1913. In their prayers for deliverance they promised to organize a feast and boat procession in honor of their patron saint, St. Carmen. Since then all the boats in the harbor string decorative lights and colorful banners on their masts during festival time. Sailing regattas and bike races give way to evening dances, fiestas and lots of drinking.

■ Adventures on Water

The docks of Puntarenas are the embarkation points for several entertaining boat tours into the **Gulf of Nicoya**. The cheapest way to get a boat ride is to take one of the ferries as a walk-on passenger. But more fun are day-trips to **Isla Tortuga**, a private island where passengers have lunch, swim, snorkel or explore. Tour firms arrange transportation from San José if you're just going for a day-trip. This includes round-trip passage by bus, breakfast,and lunch on the island or on a boat. Try **Bay Island Cruises** *(☎ 506/258-3566)*, **Tortuga Island Tours** *(☎ 506/661-2508)*, or **Calypso Tours** *(☎ 506/256-8787, www.calypsotours.com)*, which also offers a trip to Punta Coral private reserve. Prices vary.

Tours to the seabird reserves on **Guayabo**, **Negritos** and **Los Párajos** islands are available from **Cata Tours** *(☎ 506/296-2133)*. For extra thrills, hire a launch with a guide and get permission from the Costa Guard *(☎ 506/663-5952)* to visit the abandoned **San Lucas Island**, Costa Rica's equivalent of Devil's Island. Last but not least, **Cruise West** ships (see page 136) start and end in Puntarenas.

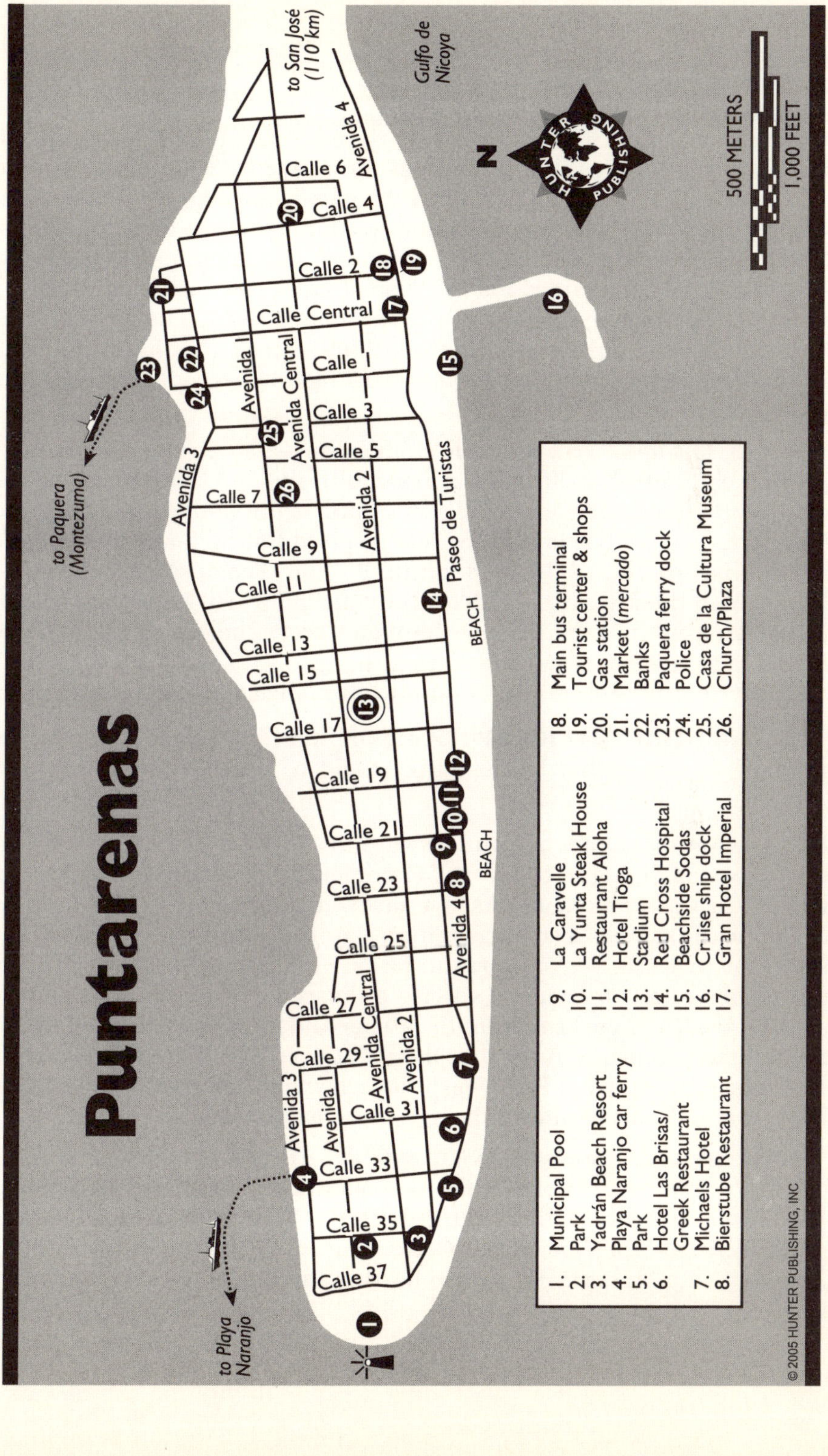
Puntarenas
to San José (110 km)
Gulfo de Nicoya
to Paquera (Montezuma)
to Playa Naranjo
Calle 6
Calle 4
Calle 2
Calle Central
Calle 1
Calle 3
Calle 5
Calle 7
Calle 9
Calle 11
Calle 13
Calle 15
Calle 17
Calle 19
Calle 21
Calle 23
Calle 25
Calle 27
Calle 29
Calle 31
Calle 33
Calle 35
Calle 37
Avenida 4
Avenida 3
Avenida 1
Avenida Central
Avenida 2
Paseo de Turistas
BEACH
N
HUNTER PUBLISHING
500 METERS
1,000 FEET
1. Municipal Pool
2. Park
3. Yadrán Beach Resort
4. Playa Naranjo car ferry
5. Park
6. Hotel Las Brisas/ Greek Restaurant
7. Michaels Hotel
8. Bierstube Restaurant
9. La Caravelle
10. La Yunta Steak House
11. Restaurant Aloha
12. Hotel Tioga
13. Stadium
14. Red Cross Hospital
15. Beachside Sodas
16. Cruise ship dock
17. Gran Hotel Imperial
18. Main bus terminal
19. Tourist center & shops
20. Gas station
21. Market (mercado)
22. Banks
23. Paquera ferry dock
24. Police
25. Casa de la Cultura Museum
26. Church/Plaza
© 2005 HUNTER PUBLISHING, INC

Pacifica

Isla del Cocos

It is, it is a glorious thing
To be a Pirate King.
~ *Pirates of Penzance*, Gilbert and Sullivan, 1879

It is from Puntarenas that the dive adventure ships head to sea for Isla del Cocos, 535 km (335 miles) out in the Pacific. Besides being a natural wonder and fabulous dive spot, its modern fame came as the supposed island that housed Steven Spielberg's dinosaur playground, Jurassic Park. The island was misnamed for the type of palm trees growing along its shores; the palm is the *Rooseveltia frankliana*, named in honor of Franklin D. Roosevelt, who visited the island three times. Chatham, one of its two protected bays, has rock-carved graffiti left by ancient mariners who stopped here.

Years ago Isla de Cocos was a safe offshore haven for pirates and corsairs preying on Spanish galleons. Sir Francis Drake and the Portuguese pirate Benito Bonito are two of the many buccaneers said to have anchored here. The isolated isle is said to have inspired Robert Louis Stevenson's novel, *Treasure Island*. In 1684, pirate William Thompson is believed to have buried the wealth looted from the coffers of Peru, known as the fabulous "Treasure of Lima," somewhere on Cocos. The entire island is rumored to be riddled with buried treasure – gold bars and pieces of eight – although centuries of treasure hunters have failed to unearth the secrets.

DID YOU KNOW? "Pieces of Eight" coins were called that because they could be broken into eight pieces.

But the secret is out about its underwater treasures – Cocos may offer the best deep-sea diving in the world. Hammerhead sharks are as plentiful in its clear waters as hammers in a hardware store. On the island, sheer cliffs line the shore, sliced by cascading waterfalls from the evergreen forests that cover its rugged terrain. Hiking trails cross the inland rainforest, home to two endemic species of lizards and three endemic bird species: the Cocos cuckoo, Cocos flycatcher and the Cocos finch, related to the famous Galapagos finch that lives off the coast of Ecuador. It is common to see the unusual Holy Spirit dove, a white bird that comes to nest here. It stands out because it hovers over visitors' heads.

The **Okeanos Aggressor** *(US ☎ 800/348-2628, CR ☎ 506/289-3333, www.aggressor.com)* is one of the two leading dive groups; the other is **Undersea Hunter** *(US ☎ 800/203-2120, CR ☎ 506/228-6535, www.underseahunter.com)*. A trip to Cocos usually lasts nine days, with six days of diving. Best times seem to be May through November, the rainy season.

■ Places to Stay & Eat

If you're staying here, the hotels along the *paseo* are the most pleasant, with prices around US $50-$80. Choose from the **Yadrán Beach Resort** *(☎ 506/661-2662, www.puntarenas.com/yadran, $$-$$$)*, **Hotel Las Brisas** *(☎ 506/661-4040, includes breakfast, $$)*, **Hotel Tioga** *(☎ 506/661-0271, www.hoteltioga.com, breakfast included, $$)* or **Michaels Hotel** *(☎ 506/661-4646)*, a small, clean hotel owned by a personable Greek-Canadian. Cheaper places nearer the ferry dock could include **Cabinas Amuebladas Arguedas** *($)*, a tiny street-front hotel, or **Casa Dulia** *($)*, a clean B&B opposite the tiny triangular park at the end of the *paseo*.

Hotel Prices	
Prices are per night for two people	
[No $]	less than US $20
$	US $20-$40
$$	US $41-$80
$$$	US $81-$125
$$$$	US $126-$200
$$$$$	over US $200

Restaurant Prices	
Prices are for an average entrée	
$	less than US $5
$$	US $5-$10
$$$	US $11-$20
$$$$$	over US $20

La Yunta Steak House *(☎ 506/661-3216, $$-$$$)* is our favorite eatery and a required stop when we're in town. It has a large wood veranda on two levels overlooking the *paseo*, with tables covered by kelly-green tablecloths and wine-color overlays. The menu features thick juicy steaks and fresh seafood. Prices are quite reasonable and the place has an old-fashioned formality that makes every meal feel like a Sunday dinner. Say hello and have a snack at the **Restaurant Aloha**, seaside on the *paseo*. For delicious traditional French cuisine in a bistro atmosphere, head for **La Caravelle** *(☎ 506/661-2262, $$-$$$)*, between 21 and 23 on the *paseo*. Around since the early 1980s, a carved carousel horse gives the restaurant its name. **Papi's Pizza**, also on the *paseo*, is American owned and makes good sandwiches. **Bierstube** *(☎ 506/661-0330, $$)* is an out-of-place German and seafood restaurant. Meat and fresh fish dishes are pretty good, and certainly above average for a beer garden.

■ Carara Biological Reserve

The way down to the beaches of Pacifica south of Orotina, one of the biggest fruit growing areas of Costa Rica, passes the important 5,242-hectare (12,952-acre) Carara Biological Reserve. This reserve straddles the transition zone between tropical moist forest and tropical wet lowland forest, with a diverse flora mix that attracts a wide variety of wildlife. The famous and colorful harlequin poison dart frog – solid black with stripes of fluorescent green – can be seen often in daytime. But the most spectacular resident is the scarlet macaw, whose brightly colored feathers are most easily seen from the Tarcole River bridge, early morning or late afternoon when the birds flock to and from the nearby mangroves and the reserve. You can't miss the bridge; it always has cars parked along its side. Gaping people look down, not for macaws, but for American crocodiles. The big (up to four meters/13 feet!) ugly prehistoric beasts often sun themselves on the muddy banks below the bridge. This is a required stop for tours and should be for you too. Remember to lock your car, or take turns staying with it; luggage theft occurs here. The ranger station and entrance is three km/1.9 miles south of the bridge.

See those crocs and other wildlife up close and personal on a Tropical Mangrove River Adventure Tour to nearby Guacalillo estuary. Groups leave from San José led by **Grupo Mawamba** *(☎ 506/223-2421, 888/246-8513, www.ranchonaturalista.com),* the operators of Mawamba Lodge over in Tortuguero. Tours are led by a naturalist guide and may be booked through a local travel agent. Fees start at $850 for a four-night excursion and go up from there.

> **DID YOU KNOW?** Over 135 species of neotropical birds migrate between North America and Central America each year.

Carara is a bright spot in a part of the country that has been pretty well deforested for agriculture. It was once part of a huge ranch, known as El Coyolar – a ranch so vast that the owner never needed to cut down the tropical wet forest that later became the biological reserve. Before that, evidence points to settlements as early as 300 B.C. Lomas de Entierro is a partially excavated village with funeral zones, dating back to somewhere between 800 and 1500 A.D. on a hilltop facing the Tarcole River. If you are driving the dramatically winding back road from Atenas to Orotina to get to Jacó, don't pass without stopping at **El Mirador del Cafetal Restaurant**, perched on the edge of a steep cliff, about half way along the route. This eatery, affiliated with El Cafetal Hotel, features spectacular views and the best soups served in all of Central America – especially the sopa Azteca. Along with good food and coffee, there is an interesting small gift shop.

Jacó

She sells seashells on the seashore.
~ English tongue-twister

The town of Jacó (hah-CO) is a weekend hotspot for beach-going Ticos and sun-loving Canadians, who come by the charter plane full. It's growing by leaps and bounds. Because of its proximity to San José, Jacó is one of the country's most developed beach towns – but developed in Tico style, rather than resort style. The fun-loving, happy-go-lucky beach town is essentially both sides of a main street that parallels the *playa*. It spreads out in a valley along the horseshoe bay and features low-rise hotels, funky restaurants, surfboard shops, bars, a disco and an unimpressive gray beach. Rough surf and strong currents limit safe swimming to calm days, so Jacó is also popular with surfers – and beach-goers who hang out in town. (If it's rough surf, swimmers should head to the public pool behind Internet café, Il Girasol, next to the Mas por Menos supermarket.) Do not leave valuables on the beach if you're swimming; petty theft is common.

Jacó is not an eco-destination but, for a social scene, it's hard to beat. And it's a popular weekend retreat for students in San José Spanish-language schools, as we once were. The two-hour bus journey is offered frequently from the Coca-Cola in San José (Calle 16 and Av 143). Once you're in town, stop in **Books & Stuff**, across from the bakery, to load up on used paperbacks for your reading pleasure – or CDs for your personal stereo.

■ Places to Stay

There are a plethora of places from which to choose; we list the most pleasing hotels, plus a few backpack hotels that still cater to the extremely poor of pocket. Campers can find primitive bliss at **Camping el Hicaco** *(☎ 506/643-3329)*, which has a seafood restaurant facing the beach. It can be a bit raucous on weekends as it's across the street from the back of the cavernous La Central Disco, in the middle of town. **Camping Madrigal** *(☎ 506/643-3329)* is quieter and shadier, located at the south end of the beach.

In contrast, a few kilometers north of town is the very ritzy Marriott Los Sueños and the stunning Villa Caletas (one of our all-time favorite hotels).

North of Jacó

Villa Caletas *(☎ 506/637-0606, fax 222-2059, www.hotelvillacaletas.com, pool, 2 restaurants, cabin-suites, air, cable, $$$$-$$$$$)* bills itself as "close to heaven," and it is – in more ways than one. The collection of villas are some of the most elegant accommodations and grounds in Costa Rica. Part of its heavenly atmosphere comes from the location, capping a high cliff overlooking the Pacific – with sunsets to die for. The main tropical Victorian mansion features eight beautiful guest rooms and antique-filled public areas. In the manicured gardens are 20 secluded individual villas, a master suite with personal pool, and a grand overhanging infinity pool. Perfect for honeymooners, perfect for us. Excellent food and big bar drinks. It's very hard to leave this hilltop sanctuary. Even if you don't stay here, enjoy the old world charm of this wonderful hotel by joining guests and locals to watch the sunset at the bar or restaurant. The elegant restaurant features a new chef, Vincent Jean Pierre Bountinaud, formerly of Le Monastere in Escazú. Look for the private road on the right, about two km/1.25 miles south of Punta Leona.

Hotel Prices	
Prices are per night for two people	
[No $]	less than US $20
$	US $20-$40
$$	US $41-$80
$$$	US $81-$125
$$$$	US $126-$200
$$$$$	over US $200

Restaurant Prices	
Prices are for an average entrée	
$	less than US $5
$$	US $5-$10
$$$	US $11-$20
$$$$$	over US $20

Los Sueños Marriott Resort *(Playa Herradura, 2 km / 1.25 miles north of Jacó beach, in US, ☎ 800 / 228-9290, in CR ☎ 506 / 630-9000, fax 630-9090, www.marriott.com, 201 rooms & suites, pool, lighted tennis courts, conference rooms, spa, marina, air, golf course, minibar, 4 restaurants, air, phone, bar, valet parking, $$$$-$$$$$).* Los Sueños, or The Dreams, and its sister Marriott hotel outside San José are two of the most luxurious and upscale large hotels in Costa Rica. The Jacó property is related to the surrounding par 72, 18-hole golf course designed by Ted Robinson. Covering land that had previously been cleared for cattle, the flat, shade-less course is nevertheless manicured. A new yacht marina has been added to attract the international sailing crowd and the protected harbor already lures small cruise ships. The huge Spanish mission-style hotel is painted an adobe brick color, with arched hallways and brick Catalan vaulted ceilings in the downstairs lounge. The public areas feature hand-set, made-to-order, aged colonial tile floors in the shape of oriental rugs. Though it is located on a dark sand Pacific beach, no one bothers to swim in the sea. Instead, everyone chooses the long, canal-like swimming pool that wanders among waterfalls and islands of gardens with rows of sun chairs to the swim-up bar. The pool's clever design reminds us of the Aztec floating gardens and the canals of Xochimilco, Mexico. Everything you would expect in a luxury hotel.

In Jacó

Hotels and restaurants line the main street that parallels the beach.

At the north end of beach, which once was the center of town, **Restaurant Santimar** is a sentimental favorite that still delivers tasty Tico delights and fresh seafood in a small, appealing open-air dining room next to the beach. Stop by their Caribbean colonial store for ice cream when it's hot out. The convivial hostess from Holland also rents very basic rooms for about US $15.

The **Copacabana Hotel** *(☎ 506/643-3131, pool, $$)* is a popular beachfront hotel, with an even more popular restaurant and bar with microbrewery on a wooden deck upon the sand. Try their blackened tuna.

La Cometa *(☎ 506/643-3615, $)* is a very good motel for clean, large, but basic rooms, with private or shared bath. Set back in gardens (not beachside), on the main road.

Gilligan's Bed and Breakfast *(☎ 506/643-3240, air, $-$$)* turned into a dream come true for us. After weeks of *gallo pinto* and other Tico specialties, we stopped in Gilligan's restaurant for their special – homemade meatloaf. What a treat! Huge servings (diner size) and deadly desserts. And why not? The owner is from "Joisey," just like us! The downstairs is enclosed by glass and the bedrooms upstairs are very appealing. Say hello to Frank for us! Also on the main road, **Cabinas La Cometa** *(☎ 506-643-3615, $)* is a motel with large, clean, basic rooms with private or shared bath. Garden setting.

The Copacabana Hotel *(US ☎ 800/788-7857, CR ☎ 506/256-3222, www.crica.com/hotels/copa.html, air, studios and suites, pool, cable, restaurant, $$-$$$)* is a popular beachfront hotel facing the beach. Three stories tall with a swim up bar and terrace restaurant, the Copa is a class operation. **Cabinas Restaurant Alice** *(☎ 506/643-3061, pool, cable, $)*, on a side street to beach, features new attached garden rooms – tiled outside and plain inside – with a big *típico* restaurant (order the butterfly shrimp). Family run, super clean.

Head down Calle Hidalgo to the beach and turn left to **La Paloma Blanca** *(☎ 506/643-1893, iwann@mailcity.com, $$)*, a duplex house that offers garden rooms and beachside suites. It's also available as a half-house for weekly or monthly rentals. Very large rooms, verandas and a unique partitioned pool (one half for each side of the house). Absolutely gorgeous décor inside and a bucolic butterfly garden outside.

A long-time favorite that keeps getting better is the stunning **Hotel Club del Mar** *(☎ 506/643-3194, www.clubdelmarcostarica.com, pool, restau-*

rant, $$), which has a shady beachside location on the south end. It has a very good restaurant (reservations recommended), and now they have added condos.

There's an established and dependable **Best Western** *(in US ☎ 800/528-1234, in CR ☎ 506/643-1000, www.bestwestern.co.cr, $$)* on the beach, but one of the best and most interesting boutique hotels is off the beach – the **Mar de Luz** *(☎ 506/643-3259, www.mardeluz.com, kitchenettes, pool, children's pool, $$)*. In the main building of this Dutch-owned hotel, the upstairs rooms are junior suites (although they lack kitchens, as are offered in lower-level accommodations). From porches or patios are views of either the mangrove field or the hotel's inviting tropical gardens. Rooms 10 through 20 are fascinating, rounded river stone lodges, with exposed stonewalls inside and out. They have smaller rooms, but include kitchenettes and a sitting area. Numbers 8 and 9 are really cute and right by the pool.

If you follow the main road south, along the beach and make a right after the bridge, you'll come to more hotels as the bay curves around. The **Arenal Pacifico** *(☎ 506/643-3419, pool, jacuzzi, restaurant, including breakfast, $$)* is owned by the same people who own the Arenal Lodge near the volcano. Their beachfront property impressed us a great deal. Big rooms and master suites face the beach, with lovely riverside gardens, cool pool and a restaurant. A good value.

Other hotels worth considering include the basic **Hotel El Jardin** (*☎ 506/643-3050, pool, bar, air*), and **Paraíso del Sol** (*☎ 506/643-3250, www.paraisodelsolcr.com, pool cable, air, kitchenettes*), a very good value, located on the road behind Pops.

■ Places to Eat

The best sushi in town is at **Tsunami Sushi**, upstairs in the Il Galeone Shopping Center, next to the bank in the center of town. It's a very popular dining spot, despite its unfortunate name. If fish isn't your favorite, head to **Nacho Daddy's**, a Mexican food restaurant in the same locale. Tico tacos. And we've already raved about **Gilligan's** diner food above. Worth a stop!

For quieter dining, **Restaurant Emily** is well recommended for typical Tico food and *mariscos* (seafood). It's a small, open-air restaurant on the main street, behind some flowering bushes and side gardens. **Monica's Pasta**, near the beach, features tasty pasta, salads and brushette. Very popular. Good drinking can be found nearby at **Bohio's Bar**.

Río Oasis has a wood-burning oven that churns out great pizza. For more typical Tico fare, find **Soda El Tucan** on the right before the second bridge. Whatever your palate, you won't lack for dining choices in Jacó, although finding cheap can be a challenge. On the main road there are several popular-with-surfer bars and eateries that have a party atmosphere – no need to name them, you can't miss 'em. *Buen Provecho.*

■ Adventures

Where there are people, there are things to do. Learn to surf at the **Academy of Surfing** (☎ *506/643-1948*) or any of the many surf shops downtown. These include **Chosita del Surf**, **Jacó Surf**, **El Surf**, **Surf City** (is there a theme here?), **Mango Surf** and **Mother of Fear Surf**. Experienced or beginner sea kayakers can go to a secluded bay or an inland river trip with **Kayak Jacó** *(☎ 506/643-1233, www.kayakjaco.com).*

Ride horses to the mountaintop and ride along the beach for 2½ hours with American-owned **Equestrian Center** (☎ *506/643-1569, www.horsetours.com*), located opposite the side road to the Marriott Los Sueños hotel.

Latin Outdoors *(☎ 506/383-1708, outdoors@latinmail.com)* feature mountain biking adventures in the woods where it's a little cooler. Ride 10 km/6.2 miles at night with helmet flashlights to a moonlit waterfall. Experienced bikers can ride to a protected wildlife sanctuary with miradors, then hike to a mountaintop. The ride back down to the beach begins at sunset.

We are not big fans of the local canopy tour on the way to Playa Hermosa. The best nearby experience for that is the **Original Canopy Tour** *(www.canopytour.com/mahogany.html)*, located at a private reserve called Mahogany Park 7½ km/three miles from the town of San Mateo on the highway back toward Orotina. It includes **Iguana Park** *(☎ 506/240-6712)*, part of a research and preservation project for green iguanas. In the reserve you can learn about the prehistoric beasts in their education center or eat the tasty beasts in their restaurant. Lots and lots of iguanas and a great photo op. Combine it with a zip through the trees on the adjacent canopy tour.

If you're into *bailando* (dancing) the cavernous **La Central Disco** is the long established place to go late at night. We offer a word of caution, however, against bringing valuables with you or indulging in getting high or drunk here. Tourists out of control have been targeted for theft and, in the case of young women, sexual attack. In general, Jacó has had more than its share of B&Es, automobile smash and grabs, and other serious problems thanks to the introduction of crack cocaine. On a lighter side, beautiful *objects de arte* can be found at the high-quality souvenir and art shop, **La Heliconia Art Gallery**, on the main street.

Playa Hermosa

The next beach south might be a good alternative for travelers who like the area but don't fancy Jacó. It's a long, surfer-favored coast that's less-developed than Jacó – as are all the endless beaches that stretch south from here to Quepos. **Hotel Villa Hermosa** *(☎ 506/643-3373, $$)* is a surfer hangout with a pool, kitchen privileges and mountain bike rentals. A solid choice for surfers or tourists like us is **Hotel Fuego del Sol** *(☎ 506/643-3737, www.fuegodelsolhotel.com, pool, restaurant $$)*, on the oceanfront. A really cool lodge is **Cabinas Las Arenas** (*☎ 506/643-3508, www.cabinaslasarenas.com, restaurant, $-$$*), a surf hotel right on the beach, near the soccer pitch. They also offer a secure camping area and a good restaurant. Cheap eats can be had at the **Goola Café** or **Jungle Surf Café**, while cheap digs can be found at **Cabinas Rancho Grande** *(☎ 506/643-3529)*. Apartments that sleep five are available at at **Casa Pura Vida** *(☎ 506/643-3422, $$$)*, which has air conditioning and a pool. On the next beach, Estrella Este (one of the two "star" beaches), seek out the charming **Auberge Du Pelican** (*☎ 506/778-8105, www.aubergepelican.com, pool, restaurant*) for a pleasant stay.

Quepos & Manuel Antonio National Park

When you collect experiences, you collect wood for your old-age fires.
~ Doun Swainpoel, Kruger National Park Director, South Africa

Unlike the hot and dry beaches of Guanacaste, the country south of Jacó transits to an increasingly humid and hot climate, creating a more tropical landscape. Humid forests spill right down to palm tree-lined sandy beaches.

Manuel Antonio Park and its many hotels sits on the other side of a mountain ridge from the small town of Quepos, about two hours south of Jacó. Before the tourism boom, Quepos survived on fishing and agriculture. It began its life as a company town, built in the 1930s to support the United Fruit banana plantations that lined the coastal plain. By the 1950s, production of African palm oil became the main source of income, after Panama disease destroyed the banana crop. On the long drive in, you'll pass rows and rows of tightly packed palm oil trees, as well as some surviving banana plantations. A few small company towns live on, distinguished by the geometrical arrangement of the turquoise houses, some built on stilts, usually around a soggy soccer field. Other agro-industries have sprung up to break the mono-crop dependency: Teak trees are harvested for furniture, gmelina trees for paper pulp, and green fields of rice dot the landscape. The recently improved coastal road follows the route of the old banana train and river crossings are still made over rickety, single lane, former train trestles. Riding south from Playa Estrillos Oeste the road curves inland from the coast through a medium-sized town (where you'll probably wait in line for the bridge) of Parrita, 25 km/15.5 miles from your destination. You reach the sea again at Quepos.

For general information – from weather and news headlines to adventure resources – log on to **www.pueretoquepos.com**.

LOST TREASURE OF QUEPOS

Didn't find any pirate treasure on Coco's Island or Isla Caño? Then keep your eyes open in the new area of Manuel Antonio Park, near the river. Just before the ruthless pirate Henry Morgan sacked Panama City, the church loaded 700 tons of gold and silver and sent it away on three ships; the paperwork was destroyed so that the pirates could never know its location. The treasure has never been found and its whereabouts remains a mystery to this day. Some think it is near the park.

■ Quepos

Three o'clock is always too late or too early for anything you want to do.
~ Jean Paul Sartre, 1905-1980

Quepos (KEH-pos), a funky little town just over a small mountain ridge from Manuel Antonio National Park, is filled with a variety of hotels and B&Bs, plus local and *extranjero* (foreigner)-owned restaurants. You'll know you're here when you see the wide *malecón*, sea wall, where the road splits; check out the weekend farmer's market there. One route runs along the wall's top, the other goes below and serves as a town street. In the evening, along the front row of buildings facing the *malecón*, street vendors hawk *parejos* (colorful beach wraps), jewelry and fancy carved hash pipes – big sellers in this bohemian seaside harbor village. Apart from young people looking to save *colones*, it also attracts North American fishermen – this area of the Pacific has taken some prize-winning catches.

As you might guess, Quepos offers lower rates for hotels and cheaper eateries than the Manuel Antonio area, which begins on the twisty road that runs over the mountain and down to the park. Turn at the last left in the downtown area and follow the street past the soccer field. After a sharp, hairpin right turn you're on your way uphill. We once walked the seven-km/4.4-mile trip to the park. It was great at the time, but automobile traffic has increased enough to make it less than bucolic now. You can also catch a half-hourly-ish bus (about 50¢)to the park entrance from the bus station next to the market. Taxis to the park are around US $5 from town.

> **AUTHOR TIP:** If you've come in by bus from San José, buy your return tickets well in advance; seats sell out, especially for Sunday afternoons and evenings.

If you're in Quepos on **Mother's Day**, August 15, you'll witness the whole town turn out for a big soccer match of sons, the Flacas (skinnys) against the Gordos (fats). Kick-off is at noon, but there are celebrations all day.

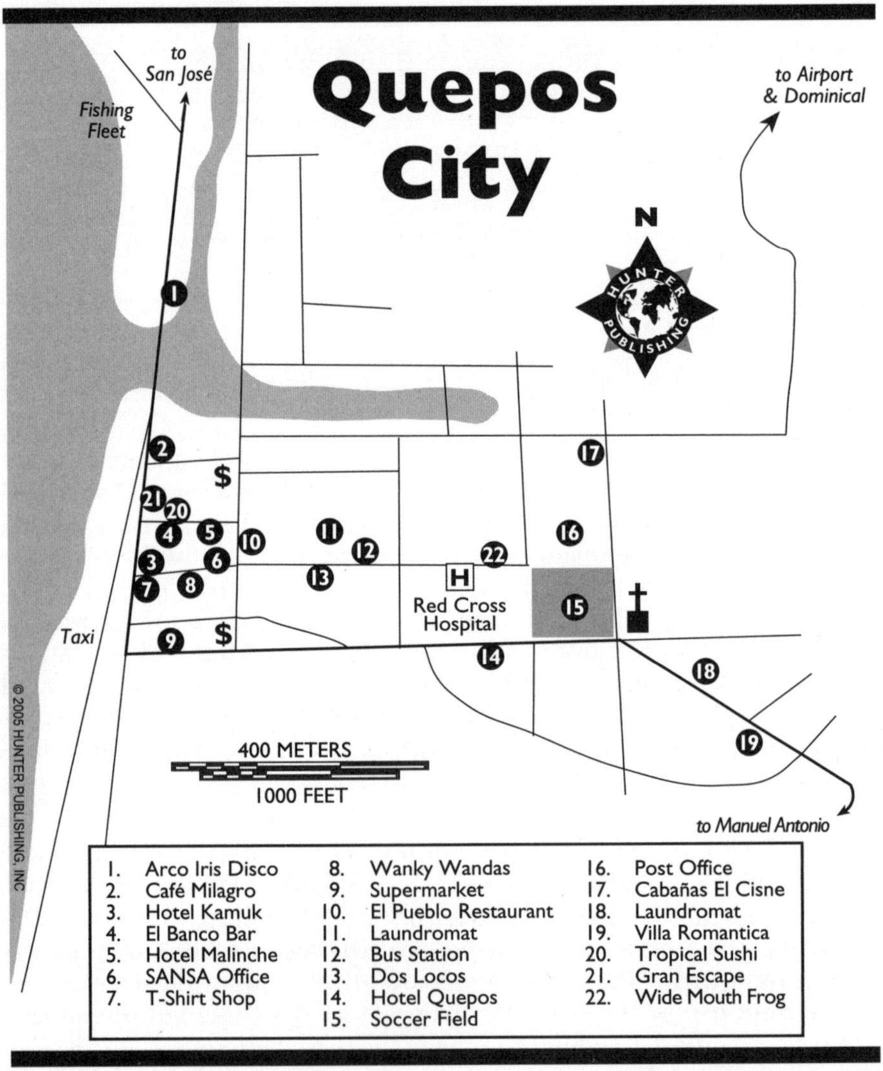

■ Manuel Antonio National Park

Everybody needs beauty as well as bread,
places to play in and pray in,
where nature may heal and give strength to body and soul alike.
~ John Muir, American naturalist

Once Manuel Antonio Park opened, tourists "discovered" the area's natural beauty – but they weren't the first. Ponce de Leon put this area on a map in 1519 while on his way to find the fountain of youth. Juan Vásquez de Coronado explored it in 1563 and encountered the Quepoa, an indigenous tribe of fishermen and divers who brought up unique pink pearls from local seas. Long dead from disease and warfare, all that remains of these people is their name.

■ Directions

To get to Manuel from San José, follow the coastal highway through Jacó and continue south. By bus, take the direct express service from Coca-Cola. By plane, jump on a 30-minute flight, offered by **Nature Air** – previously called Travel Air – (☎ *506/220-3054)* or **Sansa** *(☎ 506/221-9414)*, to the Quepos airport.

Apparently, you can't say you've visited Costa Rica unless you visit Manuel Antonio National Park *(www.manuelantoniopark.com)*. So it's no wonder the smallest national park in the country is also one of the most visited – and there would be even more people within the park grounds if the Park Service hadn't enforced a limit of approximately 600 people per day (sometimes more on weekends). Its big attractions are well-marked hiking trials through the rainforest, some incredibly idyllic beaches and abundant wildlife – especially monkeys – including 350 types of birds, plus agoutis, sloths and large iguanas.

The area became a national park in 1972, just in time to prevent a large developer from building a resort. At that time it had just 687 hectares (1,700 acres) surrounding a three-fingered spit of land that divides three bays. As you come down to the end of the road from Quepos, low-rent motels and eateries crowd the roadside. The beach on your right is **Playa Espadilla**, which gets crowded with sunbathing tourists. Be careful swimming here; the surf can be very rough. At the south end is a large rock formation and beyond that is the sand bar wash of a shallow estuary. It is across this narrow stream you need to wade to enter the park. At high tide it can be waist deep, and at those times entrepreneurial locals offer crossings in their small boats. Walk quietly on the path after the park entrance; monkeys hang out in the shady trees overhead here and agoutis often forage for tender roots in the underbrush. The long beach on the right is **Playa Espadilla Sur**, a broad band of white sand lined with shade trees. Be aware of the manchineel tree, which grows prolifically in the Manuel Antonio beach areas.

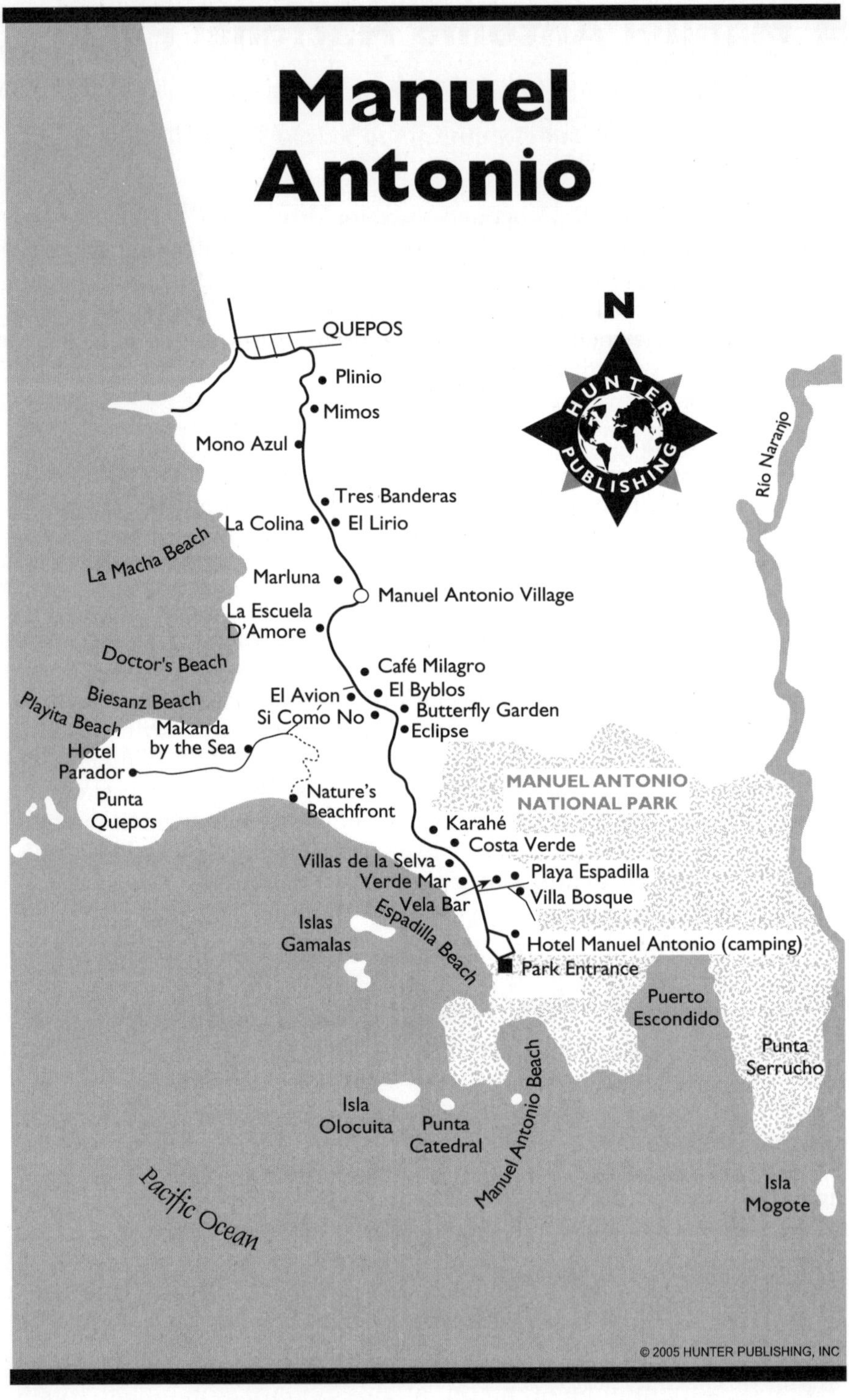
Manuel Antonio
N
HUNTER PUBLISHING
QUEPOS
Plinio
Mimos
Mono Azul
Tres Banderas
La Colina
El Lirio
La Macha Beach
Marluna
Manuel Antonio Village
La Escuela D'Amore
Río Naranjo
Doctor's Beach
Café Milagro
Biesanz Beach
El Avion
El Byblos
Playita Beach
Si Como No
Butterfly Garden
Makanda by the Sea
Eclipse
Hotel Parador
MANUEL ANTONIO NATIONAL PARK
Punta Quepos
Nature's Beachfront
Karahé
Costa Verde
Villas de la Selva
Playa Espadilla
Verde Mar
Villa Bosque
Vela Bar
Islas Gamalas
Espadilla Beach
Hotel Manuel Antonio (camping)
Park Entrance
Puerto Escondido
Punta Serrucho
Isla Olocuita
Punta Catedral
Manuel Antonio Beach
Pacific Ocean
Isla Mogote
© 2005 HUNTER PUBLISHING, INC

MANCHINEEL MENACE

The manchineel has many branches, a short trunk and elliptical leaves with a bright green sheen. Found on many Caribbean islands and throughout Central and South America, the dangerous manchineel secretes an acidic poison irritant. Its fruit is said to be the original apple in the Garden of Eden and should not be eaten under any circumstances.

Thought of the Day:

If the Garden of Eden's apple introduced deceit to humanity,isn't it possible that Adam took the first bite, then lied about it?

At the end of the beach, **Punta Catedral** (Cathedral Point) is the heavily-wooded rocky promontory that looks like an offshore island. And that's what it once was. Thousands of years of shifting sand and sediment formed a geological phenomenon known as a *tombolo*, the sandy finger that now connects it permanently to the mainland. A hiking trail runs around the *punta*.

The best beach in the park – perhaps, when it's not crowded, the best beach in all Costa Rica – is south-facing **Playa Manuel Antonio**, just past Punta Catedral. Gentle waves, celadon green water, soft sand, forest trees topped by flowering vines – this is paradise.

Hiking trails lead uphill from the beach; the one parallel to the shore connects with hidden beaches, the last of which is actually named "hidden beach," Playa Escondido. It's a strenuous hike to a completely uncrowded destination. Stop and listen to a cacophony of forest sounds on the trails. Alert ears will pick up layer upon layer of nature's music.

Just as Manuel Antonio Park represents all the best and beauty of the national park system in Costa Rica, it has also come to symbolize its failings. Although it gained its status in 1972, the government has so far not compensated all the owners for expropriated land. In 1995, a 1.7-hectare/ 4.3-acre area on property that was declared a part of the park, but not paid for in over 22 years, was clear-cut. Unresolved compensation claims (perhaps as much as 47% of the country's parkland) are an albatross around the national park system. But the good news is that Manuel Antonio was expanded by 600 to 700 hectares (1,482-1,729 acres) in 2001, making it nearly twice its original size by buying land around the Naranjo River. This "new" section is unpublicized, but it's accessible by trails (not well marked; ask a park ranger).

Unfortunately, the government failed to regulate much of the explosive growth *outside* of Manuel Antonio. In February 1992, Park Director José Antonio Salazar declared to the *Tico Times*, much to the embarrassment of the Park Service, "We have a park that's dying." According to the *Times*, the small park "cannot sustain a big enough population of tití monkeys to avoid inbreeding." One of many ecological concerns is that access of park-based monkey troops to other neighboring troops outside the park is restricted by excessive construction and deforestation.

MONKEY SEE, MONKEY DO

Softly, softly, catchee monkey.
~ African proverb

An alarming indicator of the degradation of the environment is the rapid decline in numbers of the tití monkey, also known as the squirrel monkey. This subspecies of monkey (*So citrinellus*) is indigenous to the Manuel Antonio area, and a recent survey finds only 1,500 left of the subspecies and only about 4,000 of the entire species itself, down from 200,000 in 1983. The principal cause of local deaths is electrocution on the overhead wires along the road to the park.

Kids Saving the Rainforest *(☎ 506/777-2592, www.kidssavingtherainforest.org)* is a unique organization set up in 1999 by then 11-year-old Janine Andrews and her friend Aislin Livingstone at the Mono Azul Hotel. It collects funds to pay to have rope bridges built for monkey crossings. At the Amazing Arts Gallery, the girls sell local handicrafts, jewelry and other art items, plus signed copies of their book, *The Legend of the Blue Monkey*, written by Janine's mother and illustrated by Janine, Aislin and a friend Carrie Fedor. The simple, moral story for young children introduces Coco, a blue monkey, who rises above negative peer pressure and remains true to his beliefs. The gallery includes the newest project, "Jugo Mono Rainforest Internet Café," a full-service coffee shop where you can pick up e-mail, scan photos or documents, make copies, fax, or just relax and enjoy a fruit smoothie, coffee, or herbal tea with friends. The café and gallery are open seven days a week from 8 am to 9 pm and offer a friendly welcome to all comers. All profit goes directly to the cause. Janine's environmental efforts were featured in National Geographic for Children in 2002. See our review of the group's hotel, Mono Azul, on page 308.

An influential group of hotel owners have organized a program to insulate all of the wires in an attempt to protect the tití. Visit the **Association for the Conservation of the Tití Monkey** *(www.ascomoti.org)*.

Although most of the hotel properties along the Manuel Antonio road were built by Europeans, Canadians or Americans – generally to high standards – not all the sewage systems, for example, are up to snuff. And, no matter how careful hoteliers are (some are super-sensitive to the local ecology, while for others the only "green" they relate to is the greenback) all development has an environmental impact. Eco-tourism is a matter of balance: Manuel Antonio needs to find its own. By the way, we're ignoring a big hotel project here and hope you do, too.

The best time to see Manuel Antonio Park is early in the morning on a weekday before sun worshippers hit the sands. The best time of year to visit is during the rainy season; at other times, it can be crowded, especially on weekends. You may be subject to the occasional monsoon-like downpours – usually in the afternoon and evening – but they're wonderful to fall asleep by. The advantages of coming here at this time of year far outweigh the inconveniences: the vegetation is particularly lush, far fewer people are present, prices are lower, and the park is at its most verdant. Magic. Entrance fee $6 for tourists. Closed Monday.

■ Adventures on Water

There is more to do around Quepos than visit Manuel Antonio Park. You can rent **surf** and **boogie boards** at Playa Espadilla, just outside the park. Paddle a sea kayak or take a Zodiac trip to watch dolphins. Contact **Iguana Tours** *(☎ 506/777-1262, www.iguanatours.com)*. One of their best trips is the three-hour excursion to **Damas Island**, a pristine habitat where caimans, boas, monkeys and hundreds of birds can be seen meandering in the mangrove canals. Another dolphin trip can be had with **Planet Dolphin** *(☎ 506/777-1647, www.planetdolphin.com)*.

Sunset Sails *(☎ 506/777-1304)* offers a sailing adventure with professional crew aboard a classic two-masted yacht, a romantic overnight special for honeymooners (or people who claim to be: nudge-nudge, wink-wink). Divers can go down with **Costa Rica Adventure Divers** *(US ☎ 800/317-0333, CR ☎ 506/236-5637, www.costaricadiving.com),* who also have an operation and lodge in Drake Bay, on the Osa Peninsula.

For sport fishing, look to: **Blue Fin Sportfishing** *(☎ 506/777-0674, www.bluefinsportfishing.com)*; **Costa Rican Dreams** *(☎ 506/777-0593)*; **High Tec Sportfishing** *(☎ 506/777-3465)*; or **Sport Fishing Costa Rica** *(☎ 506/777-1060, www.costaricabestfishing.com).*

Whitewater rafting in the area is on the Upper Savegre and Naranjo Rivers, both with Class III-IV rapids. A trip on the Upper Savegre takes a full day with **Amigos del Río** *(☎ 506/777-0082, www.amigosdelrio.com)*. Bewitching scenery mixed with exploding rapids. The Naranjo offers seven miles of challenging rapids only a half-hour from Quepos. A half-day tour runs between US $65 and US $100 per person.

Kayak and river trips can be had with **Still Water Tours** *(☎ 506/777-1928).*

■ Adventures on Land

One of the most refreshing adventures around is with **Canyoning Tours** *(☎ 506/777-1924, fattan@sol.racsa.co.cr)*, which features exciting descents of warm-water rainforest waterfalls, attached to a harness and rappel rope. This is a tremendous – and very active – adventure with an option of a rappel jump.

One of Rainmaker's suspension bridges.

Take a high-wire tour with **Rainmaker** *(☎ 506/375-7700, www.rainmakercostarica.com)*, which has a series of connected trails and suspension pedestrian bridges strung across a 607-hectare/1,500-acre virgin rainforest, part of the watershed area and biological corridor for Manuel Antonio. It has deep valleys and great views. Rainmaker's guided tours end at a swimming hole. Jump in! The fee (about US $65) includes transportation and lunch.

Opposite Si Como No and next to El Byblos is the new, innovative **Butterfly Garden** run by the local nature guide cooperative. It also features trails and is preserving an undisturbed land corridor at the park's edge.

The **Canopy Safari** (*☎ 506/777-0100, www.canopysafari.com*) is a half-day zip line adventure, a fun trip into the jungle about an hour out of Quepos. Swimming and lunch are included for the fee of $65. For adventure touring with a hot twist, the **rioSPAraiso** *(☎ 506/293-5465, www.gaialinktours.com)* trip to a biological reserve features a 45-minute (each way) horseback ride, canopy zip line rides, lunch and a dip in the hot springs (39°C/102°F). You'll see lots of birds and animals.

The release point for much of the rehabilitated wildlife in the area is **Albergue El Silencio** *(☎ 506/779-8250, www.turismoruralcr.com)*, a community cooperative lodge next to a palm plantation in El Silencio, a town six km/3.7 miles inland and 35 km/20 miles south of Quepos. Green and red macaws released here have joined local flocks. See *Dominical* for accommodation details.

Equus Stables *(☎ 506/777-0001, havefun@sol.racsa.co.cr)* offers local horse tours with well-treated horses. Horseback riding at its best is at **Brisas del Nara** *(☎ 506/779-1235, www.tourbrisasnara.com)*, 13 km/eight miles inland from Quepos in Londres. The family-run operation offers full- or half-day tours. Half-day trips include exploring the Río Naranjo's chasms, the Cerro Nara mountains, and several small waterfalls. The full-day trip (three hours total in the saddle) rides to an awesome 127-meter/418-foot waterfall and swimming hole, where they have dressing rooms and restrooms.

Quepos has a **Tití Canopy Tour** *(☎ 506/777-1020)* on the 73-hectare/180-acre grounds of Hotel Rancho Casa Grande *(☎ 506/777-0330)*, five minutes outside town across the street from the airport.

■ Nightlife

This place is renowned for its peace and solitude.
In fact, crowds from all over the world flock here to enjoy its solitude.
~ from an Italian hotel brochure

Nightlife along the Manuel Antonio road is confined to the hotel bars and restaurants and the lower-rent places near the park entrance. **Mar y Sombra**, a restaurant-bar that by all rights shouldn't be there because it's so close to the beach, is quite full at night. It turns into a dance hall on weekends during the dry season. The poolside bar at **Si Como No** attracts a more mature group, and the **Barba Roja** restaurant and bar is a very popular evening hangout. The casino and sports bar at **El Byblos** is in style – it even has billiard tables.

Costa Verde's two impossible-to-miss roadside eateries get big crowds, especially **La Cantina**, because it offers live music. **El Avion** is cool, too (see page 305).

Hard-core party animals head to Quepos' two discos and a bunch of happening bars for dance and romance possibilities. Unlike Manuel Antonio, which is spread out over a long distance, Quepos is perfect for bar-hopping. **Mar y Blues**, next to L'Aventura Multiboutique, is the place where you can get a real hot dog and a beer. **Wacky Wanda's** claims the cheapest food and drinks and the coolest air in town. The four-block area around here is home to most of Quepos' watering holes. The next block over, check out **Banco Bar**, a favorite of local ex-pats. **Dos Locos restaurant** gets a crowd for their live acoustic music every Wednesday and Friday night. There's a bit of open mike for musicians in waiting. The gambling casino is the Best Western Kamuk.

But the real nightlife in town is at Disco Iris. **Disco Arco Iris** is set in a grounded barge, approached across a gangplank, located just before the entrance to downtown. A small dance floor overflows with steamy dancers, and the music is a mix of Latin pop, salsa, reggae and rock.

> **WARNING:** The free-wheeling attitudes of Manuel Antonio and Quepos have attracted a few undesirables whose drug use and associated crime plague the area. Don't hesitate to go out anywhere in the evenings, but be aware of your surroundings and belongings and don't get so wasted as to be an easy target.

■ Odd & Ends

TICKETS: Lynch Travel Service *(☎ 506/777-1170, www.lynchtravel.com)* books tours, rents cars and arranges airline tickets.

LAUNDRY: There are two laundry services of note. One is alongside the bus depot and the other is on the road to Manuel Antonio, just past the soccer field. Drop-off and pay, and coin-op. Look for the *Lavandería* signs.

KEY TO HAPPINESS: If you lock your keys in your car as we once did (we're so embarrassed), phone **Vidrios y Cerrajeria Santa Martha** *(☎ 506/777-0776, cell 393-6124)* and they'll come and rescue you.

GIFTS: Get fragrances and more at **La Botanica** in Manuel Antonio. Also patronize the **Amazing Arts Gallery** at the Mono Azul Hotel, the **Mot Mot Boutique** in downtown Quepos, **La Buena Nota** near the beach, and Si Como No's **Regáleme Gifts**. In Quepos town, be sure to stop in **L'Adventura Boutique** *(☎ 506/777-1019)*. They have a big selection of products and gifts.

INFORMATION: ***Quepolandia*** is a locally published monthly magazine distributed free. Their office is next to two bars and an Internet café in Queops; you can't miss it unless you stop too long at the Epicentro and Tio Fernando's bars first. And Tio's is worth a stop.

DINNER AND A MOVIE: Jim Damalas, the owner of **Si Como No**, was once a big shot in Hollywood; he left the rat race for peace and quiet and found it here. Downstairs in the main building Jim now shows a flick in a super cool and comfortable 46-seat movie theater with Dolby stereo and popcorn. If you're not staying at Si Como No, eat dinner there and ask for tickets (which should be free to diners) as soon as you arrive. You can also buy tickets, but they often sell out. Check with the hotel *(☎ 506/777-0408)* to find what's playing.

DRY SEASON: Hotel Mono Azul in Manuel Antonio hosts daily **AA meetings**. Check with them for times *(☎ 506/777-1954)*.

EMERGENCY: Phone the **Rural Guard** *(☎ 506/777-0196)* or **Quepos Hospital** *(☎ 506/777-0922)*.

TALK IS CHEAP: Learn Spanish at **La Escuela de Idiomas D'Amore** *(☎ 506/777-1143, www.escueladamore.com)* between Quepos and Manuel Antonio.

WEB: One of the more helpful local community websites is www.maqbeach.com.

■ Places to Stay & Eat

Quepos Town

The first place we recommend is not really in Quepos at all, but well before it, 15 km/9.3 miles up a bumpy gravel road from the village of Damas. The new **Quepos Hot Springs** *(☎ 770/831-9189, www.queposhotsprings.com, $$)* is a wilderness retreat that offers comfortable riverside lodge rooms with satellite TV, a bar and an excellent restaurant. The river and waterfall offer good fishing and trails into the woods promise wildlife sightings. The very hot (48°C, 120°F) springs are an hour away on horseback. Soaking in the mineral waters is alleged to be a health curative. Turn inland at the second street south of the soccer field in Damas (which is 10 km north of Quepos) and proceed through San Rafael de Cerros, then go over three creeks.

Hotel Prices	
Prices are per night for two people	
[No $]	less than US $20
$	US $20-$40
$$	US $41-$80
$$$	US $81-$125
$$$$	US $126-$200
$$$$$	over US $200
Restaurant Prices	
Prices are for an average entrée	
$	less than US $5
$$	US $5-$10
$$$	US $11-$20
$$$$$	over US $20

Back in Quepos, the traditional hotel is the **Best Western Hotel Kamuk** *(US ☎ 800/528-1234, CR 506/777-0811, www.kamuk.co.cr, pool, restaurant, $$-$$$)*, with large comfortable rooms facing the sea in the center of town. It is attractive inside, with a small pool, restaurant and casino. The rooms are reasonably priced.

More accommodations can be had at German-owned **Hotel Villa Romántica** *(☎ 506/777-0037, www.villaromantica.com, pool, fan or air, breakfast included, $$)*, at the bottom of the hill before the hairpin turn for Manuel Antonio. Appealing, quiet hotel near Iguana Tours and Quepos Net Café.

One of the best values in town is the colonial-style, **Cabinas El Cisne** *(☎ 506/777-0719, $)*, a typical Tico hotel on the corner near the road to the airport, one block north of the church. They offer clean simple rooms, with air or fans, shared kitchen, cable TV and private parking.

Digs in the heart of downtown are basic and clean at **Hotel Malinche** *(☎ 506/777-0093, $)*. Painted bright pink, the **Hotel Quepos** *(☎ 506/777-0274, $)* is better than it looks from the outside. In fact, this was the first place we stayed in town, many years ago. Clean small rooms right across from the soccer field.

There are a growing number of new hostels around; our favorite is the **Wide Mouth Frog** *(☎ 506/777-2798, www.www.widemouthfrog.org, pool, air, $)*, an Aussie/English-owned place near the soccer field and post office downtown. They offer an open-air kitchen and communal area, Internet access, plus a private room and dorm sleeping. They are very nice folks and a pleasure to stay with. Say hi to Crispy and Maree. Up on a hilltop toward Manuel Antonio is **Vista Serena** *(☎ 506/777-5162, www.vistaserena.com, restaurant, $)*, a new hostel with a serene view of the Pacific. It has dorm-style sleeping in 12 rooms, plus a kitchen and laundry facilities.

For out-of-the-ordinary budget visitors there's **Finca Amanecer** *(Londres, ☎ 506/779-1123, www.fincaamanecer.com, $)*, located on an organic spice farm, 10 km/six miles inland on the Naranjo River. By car, go past the airport and straight at the gas station. From the Quepos station, take the Londres Abajo bus, specify "Finca Amancer." Taxis there are about US $10.

Quepos has numerous restaurants – many which double as bars. Eat local food cheap at **Restaurante El Pueblo**, a half-block north of Dos Locos, across from the Sansa office. It's our Tico food choice. **Dos Locos** is a gringo-happy institution that serves up savory Tex-Mex food at moderate prices. It's one block east of the Hotel Kamuk. It has live music twice a week and is a pleasant place to grab a window seat and relax. Across from the bus station.

The warm and friendly **El Banco Bar** also serves Tex-Mex food and cold beer in a former bank, but the most popular place in town is **El Gran Escape Restaurant and Fish Head Bar**. This place packs them in. If it's company you want, get it here, although food is less expensive almost everywhere else. Next door, the owner opened **Tropical Sushi**. Fresh fish in a sea port – what a concept. Our favorite bar food is a big hamburger at **Wacky Wanda's**, 50 meters/165 feet east of the bank.

Thankfully, you don't have to drive to the hotel zone to enjoy **Café Milagro**, the trendy coffee house that features excellent varieties of local coffee and grub. It's on the waterfront in the north end of town next to a hardware supply.

Manuel Antonio

Love thy neighbor as thyself, but choose your neighborhood.
~ Louise Beal

Because it's a long road, hotels are in two locations: close to the beach at the end of the road, or up the hill, sometimes quite far from the park. Each has its advantage. The beach and park side hotels are generally smaller, cheaper and hotter. What you gain by staying on the mountain are ocean breezes, more greenery, a view of the sea and much more appealing accommodations – but at a higher cost. If you're thinking of staying during high season, make a reservation. It's best to book a couple of months in advance, especially over holidays such as Christmas and for long weekends.

Gay-Friendly Options

It is Forbidden to Enter a Woman
Even a Foreigner if Dressed as a Man.
~ sign in a Bangkok Temple

Manuel Antonio has attracted a fair number of gays and lesbians – both Ticos and foreigners – who live here full time, and the word is out that it's a good, tolerant destination for gay and lesbian tourists. Most hotels, bars and restaurants are gay-friendly with mixed crowds. Four hotels that are exclusively marketed to gays, lesbians and their families are the posh **La Plantacíon** (☎ *506/777-1332, www.bigrubys.com, pool, $$$$)*, older **Casa Blanca** (☎ *506/777-0253, www.hotelcasablanca.com, pool, $$$-$$$$)*, **Kekoldi Beach Hotel** (☎ *506/248-0804, www.purpleroofs.com/kekoldi-crca.html, $$$)* and the intimate **Hotel Villa Roca** (☎ *506/777-2335, www.villaroca.com, $$$)*. Gay backpackers with little money and no sense of décor can find a bed at the Swiss-owned **Costa Linda**, near the beach.

Near the Park

Very few people can afford to be poor.
~ George Bernard Shaw

Closest to the park entrance is **Hotel Manuel Antonio** (☎ *506/777-1351, $$)*, a modern, two-story structure across from Espadilla Beach, at the very end of the road. Just before the turn-around, several shops, bars, and restaurants have opened at the corner. This area gets crowded with tourists who use the beach across the street instead of the park.

If you turn inland, up the road at this corner, you'll come to a series of quiet hotels in the woods. On the right is **Cabinas Espadilla** *(☎ 506/777-0416, www.espadilla.com, pool, air or fan, $$-$$$)* with clean rooms and gorgeous gardens.

Eat creative cooking in natural surroundings at the funky **Vela Bar Restaurant**, where the chef prepares daily dinner specials a notch above many pricier hotel restaurants. Few people realize the **Vela Bar Hotel** *(☎ 506/777-0413, www.velabar.com, $-$$)* cabins in back can be an excellent budget choice. Several sizes and qualities of rooms are priced accordingly. Best is the private one-bedroom house.

On the corner where the road turns to the park is **Hotel Villa Bosque** *(☎ 506/777-0463, air, $$)*, a colonial-style two-story hotel with a big bar and restaurant, as well as an upstairs pool. Very pleasing public areas and rooms. If you want comfort and park proximity, this is a good choice. **Hotel Playa Espadilla** *(☎ 506/777-0903, www.espadilla.com, air, tennis, pool, restaurant, $$-$$$)* sits across the street. This hotel is at the base of nine hectares/22.2 acres of rainforest of a private protected biological reserve, with hiking trails to overlooks. It's quiet and attractive, with kitchenettes and ice-cold air conditioning. Free-form pool, rancho restaurant. Lots of natural eye candy; check it out.

Turn toward the park where the Argentine steakhouse restaurant of **Los Almendros** *(☎ 506/777-5137, $$)* faces the road. The **steak house** is a meat-lover's paradise, featuring chorizo and BBQ. Behind it is a fairly large motel with a pool and sprawling grounds. Basic small rooms with fans or air and clean tiled baths. A good value; sleeps three. **Tico Lodge**, a small motel run by a local family, is just before the back entrance of the park. Built in 1999, accommodations are adequate, clean and very inexpensive for the area.

Worth the Splurge

Walking isn't a lost art – one must, by some means, get to the garage.
~ Evan Esar, American humorist, 1899-1965

Dollar for dollar, Manuel Antonio has some of the most stunning boutique and small luxury hotels in all the Americas – and that's really saying something with all the wonderful places in Costa Rica's Caribbean and Pacific coast destinations. Unfortunately, the price of the area's free natural beauty can easily slide into triple-digit nightly rates. However, when you compare location and amenities with prices for regular boring rooms in the US, Manuel Antonio's properties are bona fide bargains. We've stayed at many of them over the years and found them well worth the splurge. But not all are so pricey as to make you miss out if you're on a budget, especially in the green season. Get ready for a stay in paradise.

Costa Verde *(☎ 506/777-0584, fax 777-0560, www.costaverde.com, pool, restaurants, air, Internet, $$$-$$$$).* The signature phrase of this renowned hotel is "Still more monkeys than people"– and it's accurate. Owner Allan Templeton, an ex-Peace Corps volunteer, is an environmental activist who built his property in a considered, responsible manner – and what he did build is impressive. Tall trees and forest on the hillside hotel property, which includes 74 hectares/30 acres left as a biological corridor, separate two, three-story buildings, each with an inviting cliff-side pool. Inside the ambiance is Spanish colonial with bright fabrics and teak wood grown on their own plantation. The best rooms are the top-floor suites, with kitchenettes, but even the lowest-priced standard rooms are worthwhile. Hillside bungalows with views have been added to the property. The views of the park and ocean are breathtaking, and on our first day we encountered toucans, a sloth, plus tití and howler monkeys up close. Please don't feed the monkeys that come at dusk. A 10-minute walk from the beach. You can't miss Costa Verde: across the road from the entrance is their tile-roof, **La Cantina BBQ** restaurant (with live music) and Internet café in an old railroad car. Up the road apiece is **El Avion**, one of the two C-123 cargo planes used to smuggle arms to the Contras in Nicaragua, now converted into a light meals eatery.

Hotel Verde Mar *(☎ 506/777-1805, fax 777-1805, www.verdemar.com, pool, kitchenette, $$).* With the Verde Mar, you don't have to sacrifice beach location for an appealing, affordable and comfortable hotel. This one, with its small pool, is only 50 meters/164 feet from Playa Espadilla beach. Built down from the road, the hotel's large rooms, with a pleasing Mexican décor, are in two-story, bright yellow buildings. Nice owner.

Karahé *(☎ 506/777-0170, fax 777-0175, www.karahe.com, rooms or villas, restaurant, pool, breakfast included, $$$-$$$$).* Choose from older villas on the mountainside or big rooms in a modern building close to the beach. The inviting lower gardens have a jacuzzi and small pool. The older rooms are less expensive and have a view of the sea, about a quarter-mile from the park. Nearby is **La Buena Nota**, a good gift shop.

Villas de la Selva *(☎ 506/777-1137, villasdelaselva@latinmail.com, 5 rooms, pool, air or fan parking, $$-$$$).* You won't see a hotel along the sharp bend in the road below the Costa Verde, but you'll see a tiny office and sign for these quiet cliff-side accommodations. Steps lead to five picture-perfect rooms with balconies and dramatic views and a house that sleeps six (also for rent). Lovely pool. Steep walk to the beach.

Hotel Casitas Eclipse *(☎ 506/777-0408, fax 777-1738, www.casitas-eclipse.org, 30 rooms and suites, cable, pools, restaurant, air, $$$).* These magnificent, pure white Mediterranean-style villas are reminiscent of those along the Cote d'Azur. They are set along a steep garden hillside with views of the forest. These boldly designed buildings with drop-dead gorgeous rooms and three pools make the Eclipse one of Manuel Antonio's best looking hotels. Their El Gato Negro restaurant is Italian gourmet. Bar Cockatoo above the restaurant is gay-friendly, as it the hotel.

Si Como No *(☎ 506/777-0777, fax 777-1093, www.sicomono.com, restaurants, pools, air, jacuzzi, movie theater, breakfast included, $$$$).* Si Como No, which means, "Yes, Why Not?" is a deservedly much admired, luxury hotel that emphasizes a balance with the ecology. When they wanted to expand, instead of clearing any more land they bought the property next door. That home was originally built for Jim Croce, a singer from the 1970s, who died before its completion. That building has been replaced with a new luxury villa that houses stunning individual suites and three fabulous honeymoon penthouses. Impeccably clean with an engaging décor, the older hotel features pleasing standard rooms, but it is the duplex *casitas* scattered along the hillside with second-floor views of the jungle and sea that are the most desirable. The pool has a slide and swim-up bar. Dine at **Rico Tico Grill**, or the excellent **Claro Que Sí** restaurant. There's also a fabulous private movie theater with free shows for guests every night. Across the street, a **Butterfly Garden** and a private reserve are open to the public.

La Colina B&B *(☎ 506/777-0231, fax 777-1553, www.lacolina.com, restaurant, air, pool, breakfast included, $$).* La Colina is a small hotel with the feel of an enjoyable bed and breakfast. They rent giant one-room flats, standard rooms with garden views, or six top-floor, ocean-view suites with balconies. Big breakfasts go along with the owners' concept of high-quality, affordable lodging. The stylish but homey décor is a welcome change, and the dual-level swimming pool is a real treat! Long-term stays are available, and even short-term stays are a good value. eat at their romantic, candle-lit restaurant in the evening.

Hotel Las Tres Banderas *(☎ 506/777-1871, fax 777-1478, www.hoteltresbanderas.com, restaurant, bar, jacuzzi, pool, air, $$).* Neat-as-a-pin rooms and suites are found in this colonial-style hotel up on the hill. All offer guests a very good value. The pool is cool and blue, the water pressure is a pleasure, and the lovely rooms are spacious. Owned by a congenial American and a Pole (try an iced vodka at the bar), one of which is always available on premises. Meal plan.

Makanda-by-the-Sea *(☎ 506/777-0442, fax 777-1032, www.makanda.com, restaurant, pool, jacuzzi, in-room breakfast included, $$$$).* If you imagine the perfect luxury honeymoon hotel with private contemporary studios and villas, enticing infinity pool, tropical flower gardens, and a romantic candlelit, poolside tent restaurant with billowing sheer drapes, then you've imagined Makanda. Located on 4.85 hectares/12 acres of forest off the main road, down a side road toward Punta Quepos, Makanda is as close to paradise as you can get. Polynesia transplanted to Costa Rica. Plus, you can walk off the sumptuous Sunspot Grill restaurant lunches and dinners on the steep steps between villas. Reservations suggested for both hotel and restaurant (open to the public).

El Byblos Hotel & Casino *(☎ 506/777-0411, fax 777-0009, restaurant, pool, billiards, casino, air, cable TV, $$$-$$$$).* This unique local mainstay, one of the first in the area, offers big, exotic, recently remodeled bungalow suites set either side of a stream, or huge rooms in the main building. They are well appointed and beautifully decorated; you get a lot for your money here. Porches and balconies. Private and quiet, the property features a Tiki poolside restaurant that serves mouth-watering brick-oven pizza. The main building houses the international restaurant, casino and billiard hall, plus **Billfish Bar**. The charming French owner, Irene, who as a young woman was once denied entrance to the famous Byblos Hotel in St. Tropez, created a new Byblos of her own, where all people are made to feel welcome.

Hotel Plinio *(☎ 506/777-0055, fax 777-0558, www.hotelplinio.com, restaurant, pool, recreation room, $$).* Hotel Plinio is one of Manuel Antonio's most popular hotels with mid-range travelers. It's American owned and operated, with a 3.6-hectare/nine-acre private reserve above it and an observation tower on the mountaintop. The hotel itself is built up the steep hillside, so if you can't climb stairs, this one is not for you. Once you're up, it's like staying in a tree house in the forest canopy – only this tree house has suites. The Italian international restaurant is a well-known spot for dinner.

Mango Moon B&B (☎ *506/777-5323, www.mangomoon.net, pool, breakfast, air, cable, $$$+*) is an intimate and relaxed bed and breakfast perched aside a winding road, with a view of the sea. Gorgeous rooms are available, including one that's US standard handicap, but the best digs are in their one- and two-bedroom suites. Luxury and tranquility combine to make Mango a fruitful vacation hotel. Ask about their honeymoon package. Located past the Mariposa but before the Makanda.

Nature's Beachfront *(☎ 506/777-1473, fax: 777-1475, www.maqbeach.com/natures.html, studios and apartment, $$).* This small building built alone along a mangrove field and river estuary 50 meters/164 feet from the beach doesn't look like much from the outside. But the two studios downstairs and the two-bedroom apartment suite upstairs are much better than their rustic surroundings. Private and relatively inexpensive, these rooms down a bumpy road are not for everyone, but they are worth considering. Take the road for Hotel Parador, turn left at the fork for one km/.6 miles to the hotel.

Mimo's Aparthotel *(☎ 506/777-0054, fax 777-2217, www.hotelmimos.com, cable, coffeemakers, air, pool, jacuzzi, $$).* Italian-owned Mimo's is a lovely roadside inn that features three stories built into a garden hillside. Another good value hotel. In bright white stucco with brown tile roof, the junior suites all boast hand-painted drawings on the walls, kitchenettes, pretty tiled bathrooms and artistic touches all around. There's a deep rectangular pool and ranch-style restaurant. Best of all, they offer covered parking so your car isn't 160° when you get in it again. Motor scooter and bike rentals.

Mono Azul *(☎ 506/777-2572, www.monoazul.com, pool, restaurant, breakfast included, $$).* We first stopped at the Mono Azul because of its name, "Blue Monkey," the name of a pub in England once owned by June's father. The vivacious American owners, Jennifer and Chip, offer a variety of standard and deluxe rooms at reasonable prices. But our favorite digs here are across the street from their main hotel where they feature studios apartments up on the hill above the road behind the crafts shop. Environmentally active, Mono Azul is home to "Kids Saving the Rainforest" and their gift shop profits contribute to the cause. We give its roadside restaurant kudos for the mixed international menu.

Most of Manuel Antonio's dining choices are in the various hotels, but **Café Milagro** *($)*, along the road, is a wonderful exception that should not be missed. It has excellent coffee blends, tasty *pan dulce* and cinnamon buns, sandwiches and breakfast. This coffee house is a great addition to the mix. Visit their second location downtown.

The best restaurants for quality and ambiance are the **Sunspot Grill** *($$$$)* at Makanda-by-the-Sea and the informal **Rico Tico Bar and Grill** *($$)* at Si Como No. You can sit under the wing of Colonel Oliver North's cargo plane at **El Avion** (by Si Como No), or try **La Cantina's Railroad Car**, at Hotel Costa Verde.

The **Barba Roja** *($$)* happy hour is famous for its happiness, as is their restaurant. **Hotel Plinio's** *($$$)* place is a very popular spot, so is **Karaola's** *($$$)* for Mexican and seafood specialties. The poolside Tiki restaurant at **El Byblos** *($$)* is a sentimental and gastronomical favorite for pizza lunch, or the steaks in the main restaurant for dinner *($$$)*. Breezy **La Terraza Restaurant** *($$$)* offers views of the sea along with food of the sea at **Villas El Parque**. The **Mono Azul** *($$)* continues to get rave reviews from diners like us who stop for lunch or dinner. They do take out too. One of Manuel Antonio's more popular eateries is the funky **Marluna**. Their seafood specialties garner praise and the reasonable prices are an extra attraction; we go often.

■ Leaving Town

To leave Quepos and head south, take the first left after the soccer field, then make right and follow the road up the hill and out of town. You'll soon come to the **airport** and the graded but not fully paved Costanera Sur roadway. Oil palm plantations and sugar cane fields will accompany you along the coast, following the old railroad route.

Out of Town

Matapalo is the next beach, after about 25 km/15.5 miles of dust. This is a quiet, beach without end. Its strong surf attracts surfer dude pilgrims on an endless summer. There are two praiseworthy Swiss-owned places to stay along here. **El Coquito del Pacifico** *(☎ 506/228-2228, www.elcoquito.com, pool, $$)* is a German-owned hotel that offers luxury cabin rooms and a gourmet restaurant. Farther north by a kilometer or so is **La Piedra Buena** *(no phone, $)*, five cabins with a gourmet Swiss restaurant. Excellent. Great views from up on the hill are at the five **El Castillo Guest House** *(☎ 506/392-3460, www.elcastillo.net, pool, $$-$$$)*.

Dominical

Forty km/25 miles south of is the surfer village of Dominical, the beginning of the southern section of Costa Rica. It is the gateway for drivers heading to the Osa Peninsula; the road south is good from here. Or you can reach it via a four-hour drive from San José over the Talamanca mountains through San Isidro de General. If you're driving that way, try and time your mountain crossing for the mornings or take the more popular way along coastal highway through Quepos. The town's only gas station is two km/1.25 miles north of town near the Hacienda Barú (which happens to rent cabins; see below.)

DID YOU KNOW? A yellow fever outbreak in 1947 wiped out the howler and spider money population in the Dominical area, leaving white-faced capuchin monkeys as the only survivors.

Dominical gets strong winds off the Pacific; the coconut palms lining its "blue flag" beach are bent by its diligence. The pueblo and beach are near the mouth of the Río Barú, which, combined with the winds, makes a very strong and potentially dangerous surf. It's not suitable for swimming, and that's what attracts the surfers bent on riding its 10-foot waves. Learn to surf like the pros at the **Green Iguana Surf Camp** *(☎ 506/787-0192, www.greeniguanasurfcamp.com)*.

WARNING: When we say the surf is rough here, we mean rough. In 2001, 19 people drowned in the currents of Dominical beach. Finally, after a man drowned the day after his wedding, a lifeguard system was established. These heroic, but poorly paid, lifeguards saved 17 people during the first half of the high season of 2002. Never surf alone, but do contribute to the lifesaving service.

■ Places to Stay & Eat

The place that put Dominical on the eco-tourism map is **Hacienda Barú** (see below), owned by dedicated conservationist, Jack Ewing, a fixture in the area since the 1970s. You'll find the hacienda about three km/1.8 miles north of Rio Barú. Dominical itself begins just south of the bridge. Turn right there to the new "Pueblo" shopping center and on your right is the very appealing **Hotel Rio Lindo** (☎ *506-787-0028, www.dominicalbiz.com, air or fan, lively restaurant, pool, $$*) with newly remodeled rooms. In the Pueblo center is an Internet café, pharmacy, soda and a **Visitor Center** (☎ *506/787-0096*). Follow that dirt road around to the beach for the venerable **DiuWak Hotel** (☎ *506/787-0087, www.diuwak.com, pool, $$*), the best-known accommodations in town. The rest of Domincal's services are beachside, among the trees. For good food and rental rooms, check out **San Clemente Bar & Grill**, a family Mexican restaurant, or nearby **Thrusters**, the meeting, eating and drinking place, *de rigueur* to all visitors.

Hotel Prices	
Prices are per night for two people	
[No $]	less than US $20
$	US $20-$40
$$	US $41-$80
$$$	US $81-$125
$$$$	US $126-$200
$$$$$	over US $200

Restaurant Prices	
Prices are for an average entrée	
$	less than US $5
$$	US $5-$10
$$$	US $11-$20
$$$$$	over US $20

East of town, toward San Isidro, is the classy hotel of the area, **Villas Río Mar** *(☎ 506/787-0052, fax 787-0054, www.villasriomar.com, $$$)*, a "jungle and beach resort" set on the banks of the Barú River about 800 meters/2,631 feet from the beach. Luxury accommodations in thatched roof bungalows, huge pool, tennis courts and palenque restaurant.

About a kilometer south of town the place to dance, drink or hang out on the weekend is **Roca Verde** *(☎ 506/787-0036, www.rocaverde.net, 10 rooms, $$$)*, although service is spotty. For accommodations, we prefer **Costa Paraíso Lodge** *(☎ 506/787-0025, $$+)*, another kilometer south along the beach. The rocky shore has a safe swimming beach just down from the rooms in the beautiful villas, with separate bungalows above and a honeymoon cottage available. Peaceful.

For ecologically minded travelers, **Finca Brian and Emilia** *(cell ☎ 506/396-6206, or book through CIPROTUR, $)* is a working fruit and nut farm owned by long-time resident Brian Trentham, up in the mountains above Dominicalito. A simple cottage holds up to four, and there is a wood-fired, 1,000-gallon hot tub. Brian is devoted to the land and his rustic *finca* sets a standard that all eco-tourism lodges should follow, with attentive staff, gourmet veggie meals and first-rate horseback and hiking tours. To get there, four km/2.5 miles south of Dominical, follow signs up the steep rugged road toward **Bella Vista Lodge** *(US ☎ 877/268-2916, CR 506/388-0155, www.bellavistalodge.com),* another pastoral mountaintop lodge with *bella vistas*, beautiful views, on the same mountainside. Woody, the lodge owner, emphasizes a combination of comfort and adventure. Say hi.

Albergue El Silencio *(☎ 506/286-4203, restaurant, www.turismoruralcr.com)*, is a community cooperative lodge next to a palm plantation in El Silencio, a town six km/3.7 miles inland and 10 km/seven miles north of Dominical. Ten rustic rooms with private bath.

One of the most pleasing eco-retreat lodges is **Pacific Edge** *(☎ 506/381-4369, $$)*, owned by a Brit and an America. It's located on a wooded mountaintop plateau above the whitewater surf of Dominicalito. Four bungalows feature rustic but comfortable rooms and a big front porch, with breathtaking views of the ocean and a wide hammock. Tours and meal plan available. Make a left after the small bridge in Dominicalito (four km/2.5 miles south of Dominical) and head for one km/1.25 miles up a four-wheel-drive road.

ADVANCE OR TAKE A CHANCE

For all of the more remote properties, we recommend a reservation, or at least calling in an advance, to ask about availability. This is especially important in the dry season. The best way to do that is through either **Selva Mar** *(☎ 506/771-4582, selvamar@racsa.co.cr)*, a booking agency with agreements with most of the hotels, or through **CIPROTUR** *(☎ 506/771-6096, ciprotur@sol.racsa.co.cr)*, the regional official tourist information center. Both are in San Isidro. Hotels can also be booked at travel agents/visitor centers in Dominical.

■ Hacienda Baru

Lo, the poor Indian! Whose untutor'd mind
Sees God in clouds, or hears him in the wind;
His soul proud Science never taught to stray
Far as the solar walk or milky way;
Yet simple Nature to his hope has giv'n,
Behind the cloud-topp'd hill, a humbler heav'n.
~ *Essay on Man*, Alexander Pope, 1688-1744

This private national wildlife refuge has cabins and organized tours. **Hacienda Baru** *(☎ 506/787-0003, restaurant, www.haciendabaru.com)* comprises 324 hectares/800 acres of varied micro-eco-systems that includes over 81 hectares/200 acres of untouched old-growth forest plus some 300 acres of secondary forest. The balance is in mangroves, swamp and plantation grounds. The mix attract plenty of wildlife. The refuge's owner, naturalist Jack Ewing, offers a long species list of animals, birds and plants recorded here. It's available in the information lodge. Cute cabins with red roofs house overnighters in rustic accommodations near the beach. The refuge's staff offers a myriad trips, including a fabulous "Night in the Jungle" tour on horseback and foot. You'll find petroglyphs of the pre-Columbian Brunca/Boruca peoples that once lived here, and the hacienda's site yielded a rare, un-plundered Indian grave. The overnight trip allows you to look for nocturnal rainforest wanderers. Located about two km/1.25 miles north of the Dominical bridge. The clean gas station there makes a good restroom stop.

■ Nauyaca Waterfalls

The grass isn't greener on the other side of the fence.
It's greener where you water it more.
~ inspirational poster slogan

Don Lulo's jungle waterfalls are well off the road between Dominical and San Isidro, two km/1.25 miles before the town of Platanillo. Adventurers leave their cars at the roadside office and ride horseback with guides through the 20 hectares/50 acres of rainforest and /18 hectares/45 acres of replanted secondary forest to Lulo's house, where they serve a country breakfast. More winding trails lead to the two falls (one 50 meters/164 feet and the other 22 meters/72 feet) that cascade into a deep crystalline pool, perfect to dive in and swim. Dressing rooms and a small campsite are available. After your refreshing morning, ride back to Lulo's for a typical Tico lunch. Make reservations direct *(☎ 506/771-3187, www.nauyaca.com)* or contact a local hotel or tour agency. Cost is approximately US $40.

Uvita & Playa Ojocha (Playa Tortuga)

This little-visited area of the Pacific coast is only now opening up to tourism because of the newly improved road. Eventually it will be over-developed when the road is discovered and made into a shortcut for Osa and the South, so see it now before it's gone. Beaches are long and deserted. There are a fair number of Quebecois living and vacationing here so the ambiance takes on a bit of French Polynesia. And the cuisine has a decidedly French flavor. This area is the home of one of Costa Rica's newest national treasures, **Ballena National Marine Park**, of which only 50 meters/164 feet of land is part of the park; the rest is underwater surrounding the largest Pacific Coast coral reef in the country, and Isla Ballena ("*ballena*" is the Spanish word for "whale"). Past the small islands offshore, hundreds of humpback and pilot whales migrate in the dry season. Year-round water residents include spotted dolphins that frolic in the bay.

Just north of Uvita is **Rancho La Merced Biological Reserve** (no phone, $), for eco-tourists or travelers looking for a real Costa Rican forest and sea experience. **Oro Verde** (no phone, $) is another local biological reserve that offers interesting and informative day tours.

To check availability at any of the above attractions, check with **Selva Mar** *(☎ 506/771-4582, selvamar@racsa.co.cr)* or **CIPROTUR** *(☎ 506/771-6096, ciprotur@sol.racsa.co.cr)*, or your local hotel.

■ Places to Stay & Eat

The pride of Uvita – and perhaps of all Costa Rica – is **La Cusinga Lodge** (☎ *506/770-2549, www.lacusingalodge.com, $$+*), located about 22 km/13.5 miles south of Dominical.

> **DID YOU KNOW?** Cusingas are medium-size birds that resemble toucans, with black feathers and large colored beaks. They are also known as toucanets and are common visitors in the surrounding trees.

The rustic lodge, with comfortable and pleasant wooden cabins, perches on a ledge of forest overlooking the Ballena National Marine Park. Sunsets from the cabin porches are impressive. La Cusinga is an authentic eco-lodge and is frequently used by students and naturalists. It offers Modified American Plan service. Its edge-of-the-rainforest location doesn't make it difficult to get to, just hard to leave.

If you're tired and hungry, try the **Balcón de Uvita** (*one km north of gasolinera, ☎ 506/743-8034, www.balcondeuvita.com, pool, $$*), a lovely mountaintop restaurant and bungalow hotel with commanding views of the countryside. The cabins are warm and welcoming and the grounds are stunning. But El Balcón is best known for its restaurant serving Thai and Indonesian cuisine, Thursday through Sunday. People will take along detour to eat here and we know why! But if being close to the sea is paramount, then Steve Fisk's little hotel, **Hotel Toucan** (*☎ 506/743-8140, www.toucanhotel.com, restaurant, $*) near the village of Uvita itself, is a good choice. Its tropical gardens are beautiful and the rooms are cozy, if basic. Gringo owned.

Up a dirt road from the highway, 16 km/10 miles south of Dominical, is **Coco Tico** (*☎ 506/743-8032*) a rustic but pleasant motel-style property, 500 meters/1,644 feet inland from the road. It's been owned since 1995 by English-speaking José "Chepe" Díaz and his friendly family. They offer an inexpensive meal package. Camping allowed. About nine km/5.5 miles is south of La Cusinga is a notable luxury casita hotel, **Villas Gaia** (*Playa Tortuga, ☎ 506/244-0316, www.villasgaia.com, pool, air, restaurant, $$*). Its 12 lovely, comfortable, colorful wooden casitas pepper the hillside gardens and jungle. An all-inclusive dining option is available. One *casa* can sleep seven. Attractive accommodations and a great place to stay.

Hotel Prices	
Prices are per night for two people	
[No $]	less than US $20
$	US $20-$40
$$	US $41-$80
$$$	US $81-$125
$$$$	US $126-$200
$$$$$	over US $200
Restaurant Prices	
Prices are for an average entrée	
$	less than US $5
$$	US $5-$10
$$$	US $11-$20
$$$$$	over US $20

Last, but not at all least, the inviting four-star eco-hotel, **Whales & Dolphins Ecolodge** (*Playa Hermosa, ☎ 506/743-8150, www.whalesanddolphins.net, two pools, restaurant, air, $$$*) is 17 km/10.5 miles south of Dominical. This lodge, perched on the hillside overlooking the bay, is devoted to ecotourism while pampering guests in a hotel-style comfort. Eco-activities include whale-watching, horseback riding, birding and mangrove kayak tours. Check out the fascinating whale tail-shaped beach.

Talamanca Mountains

Up the airy mountain,
Down the rushy glen,
We daren't go a-hunting,
For fear of little men.
~ *The Fairies*, William Allingham, 1828-1889

The backbone of Costa Rica is the rugged majestic mountain range called Talamanca (**Cordilla de Talamanca**), which begins near Cartago and stretches down to the border with Panama. At the base of its western slopes is the **Valle de El General**, a long fertile valley that divides the mountains from the shore. The name, El General, refers to the valley's major river.

The Talamancas are not volcanic, but they are tall. One peak, **Mount Chirripó**, is the highest point in Costa Rica (3,820 meters/12,530 feet). The capital and largest city in the region is **San Isidro de General**, an agricultural center, 45 km/28 miles directly northeast (on a good road) of Dominical. To get to San Isidro and points south, the Inter-American Highway snakes along mountains, valleys and ridges from Cartago, eventually leading down into the valley. Between Cartago and San Isidro lies the dramatic section known as the "Ridge of Death," Cerro de la Muerte. This is the high land of resplendent quetzals and the Tapantí-Macizo de la Muerte Cloud Forest, a national park. Ethereal. From San Isidro the Inter-American Highway continues to Palmar Sur, Ciudad Neily, and finally to Canoas, the border town with Panama.

■ Along the Cerro de la Muerte

The bird of paradise alights only upon the hand that does not grasp.
~ *Flight of the White Crows*, John Berry

The intimidating name, Ridge of Death, refers to a section of the Inter-American Highway that climbs across the Continental Divide near Mount Chirripó, Costa Rica's highest point. It's not called that because of its modern driving record – although its sharp curves, sudden fog and sheer drops on the side of the road have taken their toll – but because among the early settlers who made the difficult mountain crossing, several died from the cold en route. The nickname solidified during the building of the Inter-American Highway when a number of workmen met their maker.

But we enjoy this drive every time we can fit it into our itinerary. It offers breathtaking views of the surrounding green mountains and deep valleys. As the altitude rises, the winds pick up and the temperature drops –

it is not unusual to fall to freezing. Above the tree line, the vegetation becomes windswept with heather, grasses, gorse and gnarled brush. Along the route are the Tapantí-Macizo de la Muerte Cloud Forest and several comfortable cloud forest lodges that specialize in finding quetzals (the bird of the gods) in the wild. The best drive begins in Dominical or San Isidro and heads north to Cartago. Dramatic!

SYMBIOTIC RELATIONS

The quetzal, which lives at altitudes between 5,000 and 10,000 feet, depends on the laurel tree, and the tree depends upon it for propagation.

The Talamanca area offers pleasant stays in lodges or B&Bs, either for a single night or as a productive eco-alternative to farther destinations such as Monteverde. The area also offers a ton of chances to cast your line into private stocked trout ponds. For a small fee, the owners loan you tackle and you can catch your supper.

So get on a bus or rent a car and head south from Cartago – or better yet, north from Sn Isidro. However, the caveat that applies to all major mountain crossings in Costa Rica is to do them in the mornings before clouds, fog and rain obscure the view and make for white-knuckle driving.

Despite its name, a drive along the Cerro de la Muerte offers beautiful views.

On the way, at Km 56 of the Inter-American Highway, is **Hamacas Y Souvenirs Café**, which has a wide selection of handicrafts and loads of hammocks for sale.

At about Km 64 the **Tapantí-Macizo de la Muerte National Park** *(macizo@sol.racsa.co.cr)*, a combination of two parks, back to back. This high-mountain section is obviously the cloud forest, located on the east side of the highway. Down a short gravel road, the park entrance features dormitory accommodations and a big welcoming fireplace. A walk in the woods here is a real treat. Cold clouds sweep up and obscure the tall soaring trees, dripping with epiphytes and thick with moss. Several well-marked trails provide fascinating walks into this unique ecosystem. Worth a stop on your way – or even as an extended exploration. Look for the small sign along the road.

■ Places to Stay & Eat

These places are listed in order heading south.

Down a steep side road around Km 58 of the highway is a turn for the town of Copey, at 7,000 feet, and the **El Toucanet Lodge** *(☎ 506/541-1435, www.eltoucanet.com, including breakfast, $$)*. It's set in an area of apple, peach, plum and avocados orchards, above an oak forest and stream. There's a restaurant with fireplace and an open-air jacuzzi.

At Km 62 is the **Albergue de Montaña Tapantí** *(☎ 506/232-0436, breakfast included, $$)*, a Swiss chalet-style lodge at 10,000 feet above sea level. The air outside is refreshingly brisk. This wonderful alpine hotel offers large, suite-like rooms with parquet wood floors and a big warm wood-burning stove in the lounge. The charming restaurant serves trout as its specialty. This is a popular birding lodge.

Hotel Prices	
Prices are per night for two people	
[No $]	less than US $20
$	US $20-$40
$$	US $41-$80
$$$	US $81-$125
$$$$	US $126-$200
$$$$$	over US $200

Restaurant Prices	
Prices are for an average entrée	
$	less than US $5
$$	US $5-$10
$$$	US $11-$20
$$$$$	over US $20

At Km 70, the **Albergue Mirador de Queztales** *(☎ 506/381-8456, recajhi@sol.racsa.co.cr, restaurant, $$)* offers the best deal for rustic wood cabin accommodations on the side of a steep cloud forest hillside. Its low price includes breakfast, candlelight dinner (with home-grown trout to die for), and a 90-minute quetzal walk. **Finca Eddie Serrano**, next door, has 43 hectares/106 acres of forest with hiking trails and a waterfall.

See those big lumbering trucks you seem to follow going uphill and run ahead of going down? They all stop at **Chespiritos** (Km 78), a truck stop restaurant that serves "blue plate" *casados*, snacks, water and other travel items. Save your appetite for sweets until Km 93 and **La Georgina**, a bakery famous for its tasty pies and muffins.

At about Km 80 there's a turn downhill for San Gerardo de Dota, an unspoiled valley at 7,000 feet above sea level. It's a rough nine-km/5.6-mile four-wheel-drive road, but you're heading to year-round quetzal country. **Trogon Lodge** *(☎ 506/223-2421, www.grupomawamba.com, $$)*, with hardwood cabins next to a cloud forest stream and hiking trails, also offers day or overnight trips to here from San José, including transportation. Guided hikes and horseback rides too. Along the river, stay at the Chacón family's **Albergue de Montaña Río Savegre** *(☎ 506/390-5096, restaurant, $$)*, which has an excellent reputation for frequent quetzal sightings. Basic rooms, but with a warm and welcoming buffet dining area. Waterfalls, fishing, horseback riding – all in the midst of an orchard farm growing apples, pears, plums and peaches and surrounded by 500 acres of primary forest.

The view from **Mirador Vista del Valle** *(☎ 506/771-9686, $$)*, Km 119, is incredible. The little wooden restaurant and local artisan gift shop hang over the mountainside along the highway with breathtaking vistas of the valley below. They have hiking paths down into the forest and a couple of small rustic rooms for rent with balconies. It was in one of these we met Marco Saborio, the famous Costa Rican nature photographer, fresh from capturing quetzals with his Nikon (see his work at www.agpix.com). We talked photography and ecology over delicious coffee. You can make reservations for this and the other lodges through **CIPROTUR** *(☎ 506/771-6096, cipro tur@sol.racsa.co.cr)* or **Selva Mar** *(☎ 506/771-4582, www.exploringcostarica.com)* in San Isidro.

San Isidro de General

Beautiful must be the mountains whence ye come,
And bright in the fruitful valleys the streams,
Wherefrom ye learn your song.
~ *Nightingales*, Robert Bridges, 1844-1930

The first cars to reach San Isidro, which was founded in 1897 at an altitude of 709 meters/2,332 feet above sea level, didn't come until 1945. Since then it has prospered with the surrounding agricultural boom in the Valle de General and, although still small, is an important hub and crossroads for the south. If we're making a circuit up from Dominical we like to spend a night here just to enjoy the non-touristy atmosphere and so we can leave early for the mountain crossing. The town sits down a bit west of the highway. It's clean and safe, with a pretty park in the center. The bustling new *mercado* between Calle Central and Calle 2 is very attractive (the old market is now the **Southern Regional Museum**) and although there is very little of interest for tourists, we like to stop and look around. Near Hotel del Sur on the highway are two bamboo furniture factories, open to tour. San Isidro can be a central base for exploring the area, rafting the challenging Río El General and Coto Brus rivers, nipping down to the Pacific beaches, or hiking Mt. Chirripó. Non-city folk have other accommodation choices outside town in the mountains.

For some very good eco-lodges and rural accommodations outside town, see Mt. Chirripó section, below.

■ Events

San Isidro has its fair share of unique events. The first week of February hosts the **San Isidro de General Fair**, with livestock shows, industrial expositions, bull teasing and an agricultural and flower exhibition. Around the same time is **Fiesta de los Diablos**, a Boruca Indian ceremony, in nearby Rey Curré. May 15th is **San Isidro Labrador's Day**, the Patron Saint of farmers and farm animals. The celebration brings parades and fairs into town. Also, a priest performs a blessing on all crops and animals.

Long-horned bull with decorative leaves.

■ Adventure Tour Operators

San Isidro is the headquarters for the area's adventure tour operators in the area. The official regional tourist information and booking center is **CIPROTUR** *(☎ 506/771-6096, www.ecotourism.co.cr)* on Calle 1, Av 1 and 3. They offer lots of information, in English, about the Brunca area, and can help make booking arrangements for guides and tours. In the same place is **SEPA** language school *(☎ 506/770-1457, www.sabalolodge.com),* and in Southern Costa Rica, it's totally a Spanish immersion. The school also offers courses combined with field programs in environmental education.

Selva Mar *(☎ 506/771-4582, www.exploringcostarica.com)*, Av 4 near the Parque Central, books rooms for all the area's hotels and adventure tours. David Mora's excellent **Brunca Tours** *(☎ 506/771/3100, www.ecotourism.co.cr/rafting)*, headquartered in the Hotel del Sur, offers tremendous white water rafting trips and guided tours to the major southern attractions including Osa and Mt Chirripó.

One of the ecological gemstones of the area is **FUDEBIOL** (Foundation for Development of Las Quebradas Biological Center) a few kilometers from San Isidro. The 750-hectare/1,853-acre preserve was established to protect San Isidro's water supply on the banks of the Río Quebradas. Community-run, it offers wonderful hiking trails and a butterfly garden, picnic area and camping (US $6). Ask at CIPROTUR.

■ Places to Stay & Eat

The **Hotel del Sur** *(☎ 506/771-3033, fax 771-0527, www.ecotourism.co.cr, pool, restaurant, air, $$)* is a spacious resort-style hotel that in the past catered primarily to businessmen. But because it's the only high-quality hotel in town, it increasingly attracts tourists, with large rooms, an inviting figure-eight pool and very good restaurant. South of town, about six km/3.7 miles on the left.

Downtown sister hotels, **Hotel Iguazu** and **Nuevo Iguazu** *(☎ 506/771-2571 & 771-0076, www.ecotourism.co.cr, $)* are pleasant, plain hotels with clean and comfortable rooms. The Iguazu is in the city center, diagonally across from the McDonald's, while the New Iguazu is about eight blocks away in the Boston *barrio*, 200 m/660 feet southeast of the Unesco highway intersection.

For very cheap in-town digs, we preferred the basic **Hotel El Jardín** *(☎ 506/771-0349, shared bath, $)* on Av Central, a half-block west of the park. The hotel and their little restaurant out front are very clean. Around the corner on Av 2 and Calle 2 is the upstairs **La Cascada**, an extremely popular modern bar and restaurant with balcony views of the street below and fairy castle-styled **Cine Valle**, the only movie theater for *mucho* miles, now closed. American rock and roll and two cable TVs attract a young crowd for *bocas* or a *plata fuerte* (full meal). Next to the cathedral facing the plaza is **Delji**, a winged design fast-food chicken shack – eat on picnic tables or take out. Colonel Sanders, eat your heart out.

For typical Tico food, the best place in town is **Restaurant Chirripó**, with an accompanying hotel behind. The old-fashioned eatery can be found on Av 2 facing the park. Next door is an Internet café, and on the corner is a good, inexpensive Soda. But what a pleasure to find authentic Mexican food at **Mexico Lindo**, a *taquería* (taco joint) 100 meters/329 feet south of the CIPROTUR and ICE (tourist) offices. Homemade guacamole and a range of Mexican specialties make this a required stop even if you're just driving through. Closed Wednesdays. *Se habla* English.

Chirripó National Park

And He's allowed me to go up to the mountain,
and I have looked over and I have seen the promised land.
I may not get there with you, but I want you to know here tonight
that we, as a people, will get to that promised land.
~ Martin Luther King, on the night before his assassination

Contiguous with Costa Rica's largest national park, Parque Internacional La Amistad, **Chirripó National Park** contains Mount Chirripó, Central America's tallest peak at 3,820 meters/12,530 feet, as well as two other peaks over 3,800 meters/12,500 feet. It became a

Pacifica

park in 1975 – yet only 25,000 years before, glaciers covered this mighty mountain range, leaving carved, U-shaped valleys and clear-water lakes. Atmospheric conditions can vary greatly by altitude and change unpredictably. It can be very windy and cold – the lowest recorded temperature is -9°C (16°F). At mid-level the humid and cold cloud forest is characterized by tree-size ferns, moss, bromeliads, orchids, and towering (50-meter/164-foot) oak trees. At about 3,400 meters/11,152 feet, a tundra-like plateau, known as a *páramo*, begins. The highest parts of the park contain the most-visited important geological and ecological zones, including Sabana de los Leones (Lion's Savannah), where puma have been spotted; Cerro Chirripó; and Valle de los Lagos (Valley of the Lakes). Chirripó means, "Land of the Eternal Lakes" in the indigenous language.

The best way to appreciate the park, other than turn the heater on in the car as you drive Cerro de la Muerte, is a two-day or longer hike to the summit. There is a new 60-person capacity, park-operated shelter, **Crestones Lodge**, six km/3.7 miles below the peak. Make reservations through the park office, below. Bring warm clothes, poncho and rain gear, food, plenty of water, dry clothes and a very good sleeping bag (you can rent blankets, sleeping bags and cook stoves at the lodge). Some tours operators will help outfit you. Reaching the top does not require any climbing skills, and the hike in general is not super difficult, but the first day is especially strenuous (nine-12 hours). Some hikers choose to camp en route and take two days to reach base camp. Dedicated naturalist hikers can stay at the lodge and do day hikes to the various fascinating zones within the park.

Reservations to Climb

You must make reservations to climb the mountain, and pay your entrance fees and overnight charges (US $6 per night). Phone the **Park Office** three days in advance *(☎ 506/233-4160)*, or call the **Ranger Station** *(☎ 506/771-3155)* to make reservations. **CIPROTUR** can help with arrangements, and **La Amistad Area Conservation Office** in San Isidro *(☎ 506/771-3155)* can make your reservations and accept payment. The most crowded time of year is mid-February, when the *Carrera a Chirripó*, a foot race to the summit, takes place. Other busy times are the school holidays in December and *Semana Santa* (Easter Week). Make your start very early in the morning, as early as 5 am.

If you're not hiking the mountain but want to do something ecological anyway, visit the **Cloudbridge Project** *(www.cloudbridge.org)*, a 450-acre effort by the Giddy family to reforest a section of the cloud forest bridging two mountainsides. When they bought the land, it that had been previously cleared for cattle. Visitor and volunteers are welcome. Located just outside park entrance – follow signs for Casa Amanzimoti.

Access Routes, the Night Before

The most common entrance is from **San Gerardo de Rivas**, an engaging alpine village under an hour from San Isidro by car, an hour and a half by bus (which leaves San Isidro at 5 am and 2 pm). Turn and go over the Río Jilguero bridge on the highway north of San Isidro. San Gerardo is home to basic accommodations and an overnight here is recommended for an early start – try **Albergue de la Montaña Pelícano** *(☎ 506/382-3000, all-wood rooms, restaurant, shared bath, eccentric "root" art gallery, $)*, **Albergue Vista del Cerro** *(☎ 506/373-3365, 4 rustic rooms including a "bed on a rock," camping, $)* or **Cabinas El Descano** *(☎ 506/771-7962, rooms and cabins, breakfast included, restaurant, hiking trails, $)*. One of the more comfortable hotels is **Río Chirripó Retreat** *(US ☎ 707/937-3775, www.riochirripo.com, $$)*, located between San Gerado and Canaan. It's a veggie, yoga, meditation retreat at the foot of the trail that leads up the mountain. Ommm. About 10 km/6.2 miles before San Gerardo, in the town of Rivas, is the **Albergue de la Montaña Talari** *(☎ 506/771-0341, breakfast included, $$)*, a very pleasant riverside hotel. They'll pick you up for free in San Isidro with advance notice.

The closest lodge to the principal entrance to Chirripó is also the newest. The **Urán Hotel** *(US ☎ 609/695-8879, CR 506/338-2333, www.hoteluran.com, restaurant)* has 21 rooms, eight of which have private bath and hot water showers.

GUIDING LIGHT

Another advantage of staying in San Gerardo or Rivas is the ability to arrange for **local porters** to haul your gear up to the lodge (US $ 25), either on horseback or their backs, while you schlep yourself up on winged feet. Although the trails are well marked and easy to follow, consider hiring a bilingual guide. Not only do they offer encouragement, but a good guide with knowledge of the ecology can add immensely to your experience. Ask CIPROTUR, or check with your hotel for guide recommendations.

■ Setting Out

If you go it alone from San Gerardo, follow the road over two bridges for the Chirripó sign at the edge of a coffee field. Markers every half-km give you encouragement or discouragement, however you read the rapidly increasing altitude and slowly shrinking kilometers left to the summit. The climb through the thick cloud forest is made even more captivating by the sweet clear song of the *jilguero* birds, high in the treetops. The first camping area with potable water is Llano Bonito, a flat grassy area about four or more hours up. Things only go uphill from there. If it's clear, the view from the top is indescribable. A new trail to Chirripó, now open from the town of Herradura, north of San Gerardo, requires hiring a guide and is a two-day ascent. CIPROTUR and any of the local tour operators should be able to recommend guides available for hire.

Costa Rica Trekking *(☎ 506/771-4582, www.chirripo.com)* has professional guides that run three organized treks in the Talamance Range, including one in Chirripó. This is an excellent hiking adventure operator.

Soak your cold bones (even if you haven't been to the mountaintop) in a refreshing **hot spring**, one km/.6 miles down the road between San Gerardo and Herradura. From the roadside sign, it's a 10-minute hike to the misty warm spring pool.

Caribbean Coast

The nuisance of the tropics is
The sheer necessity of fizz.
~ *The Modern Traveler,* Hilaire Belloc, 1898

IN THIS CHAPTER

Among the many natural blessings bestowed upon Costa Rica, one of the most precious is the lush Caribbean coast, an area of great natural riches. A blanket of green covers the warm and welcoming land year-round. Costa Rica's east coast , known equally as Atlantic or Caribbean, is as diverse in itself as it is from the rest of Costa Rica. Beside a hot and humid tropical climate, flat and swampy terrain, the Caribbean coast is home to an Afro-Caribbean culture that speaks English. Black workers from Jamaica and outlying islands were imported to work building the Atlantic Railroad in the late 1800s, with many more coming to work the banana plantations in the early 1900s. Today, their blended culture is what gives this coastline the laid-back, easygoing tropical island atmosphere.

There are two major overland routes to reach the area. One is the old route from San José through Cartago and Turrialba, down the mountain slopes to Siquirres, essentially paralleling the course of the mighty Reventazón River. It was replaced in convenience and importance in 1978 by the Guápiles Highway, a more direct and less twisty road across a mountain pass in the Braulio Carrillo National Park that leads down to Guápiles.

The coastline also has two major sections: Tortuguero and Sarapiquí in the north (which has a Puerto Viejo de Sarapiquí) and the more populated southern shoreline from Limón, through Cahuita, down to Puerto Viejo de Talamanca. Confusing the two Puerto Viejos (de Sarapiquí with de Talamanca) would result in a very long drive. Tortuguero, at the northern end, is reached by air or by boat only. The more visited destinations in the south are Cahuita and Puerto Viejo, towns so quintessentially Caribbean that they look and feel more like island villages.

Before we get to the Caribbean side, we need to cross the Continental Divide – the high mountain spine of Costa Rica. Two main routes overland, through Turrialba (see page 175) and the new road route (Guápiles Highway) through the Braulio Carrillo National Forest. *Vamos*!

Braulio Carrillo National Park

This we know: the earth does not belong to man;
man belongs to the earth. All things are connected.
Whatever befalls the earth, befalls the sons of earth.
Man did not weave the web of life: he is merely a strand in it.
Whatever he does to the web he does to himself.
~ attributed to the Chief of the Suquamish and Allied Tribes

Close to San José (only 20 minutes north), but a world away from city life, Braulio Carrillo National Park offers – if nothing else – spectacular views on a route to the Caribbean across the Continental Divide. When they first planned a highway to replace the twisty Turrialba route to the Atlantic side, forward-thinking environmentalists argued that a road through the mountainous area's primary rainforest would bring indiscriminate development (Costa Rica has an issue with squatters) and deforestation. In response, the government created a national park in 1978 on both sides of the road built to bridge the Continental Divide.

DID YOU KNOW? Braulio Carrillo park, named after the country's most famous president, is the only national park bisected by a major highway. It's shown as Route 32 on maps, but known by all as the **Guápiles Highway**.

Hiking trails are accessible from the road, but never hike here alone. Pay the entrance fee at the visitor's center, two km/1.25 miles after the toll coming from San José, or 500 meters/1,644 feet after the tunnel coming from Guápiles. Park at the ranger station and let them know where you're going. The short Zurqui Tunnel goes through the mountains. In rainy season, take extra care as mudslides have plagued this road.

As dramatic as the stunning scenery is along the highway, the park's 47,582 hectares/117,528 acres go far beyond what you can see from the mountain passes. The park now encompasses the **Barva Volcano** and the **Cacho Negro Volcano** (2,150 meters/7,072 feet), which descend into the Caribbean lowlands. The park stretches all the way up to **La Selva** near **Puerto Viejo de Sarapiquí**. Its diverse ecosystem boasts over 6,000 species of plants, half the total of the entire country – our favorite is the giant "Poor Man's Umbrella," with leaves that *campesinos* use for cover when caught in the rain. Its avifauna includes more than 500 species of birds, both migratory and resident, plus hundreds of mammal and other wildlife species. Eighty-four percent of the forest is still primeval, making Braulio Carrillo National Park the largest contiguous tract of virgin forest left in the Central Valley.

Braulio Carrillo National Park

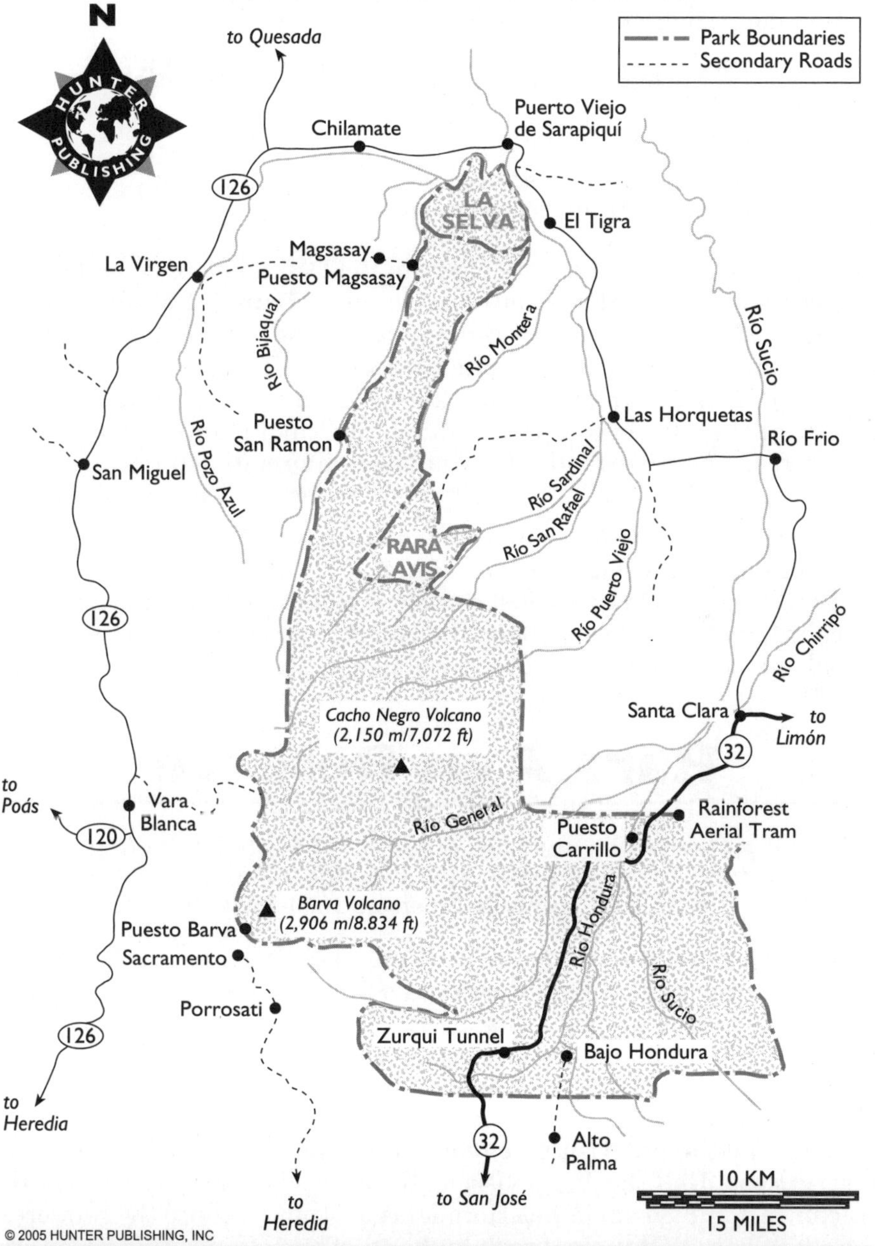

© 2005 HUNTER PUBLISHING, INC

Caribbean Coast

■ Outside the Park

Watch for the park's trademark, the rusty-colored **Río Sucio** (SUE-see-oh), which means "dirty" in Spanish. Under a bridge near the Caribbean side of the park, two rivers (the Hondura and Sucio) merge in a rocky riverbed – the Sucio is a dirty orange color from sulfuric deposits leeching from underground near the Irazú Volcano.

DRIVE RIGHT

Guápiles Highway is best traversed in the morning (true of all Costa Rica's mountain crossings) as fog and rain can make for white-knuckled driving in late afternoon and evening.

An hour's drive from San José, opposite Río Danta before Guápiles, is a turn for a dirt road, three km/1.8 miles in to **Tropical Magic Forest** *(☎ 506/392-2088, gmtours@sol.racsa.co.cr)*, one of the closest canopy tours to the capital. Past the Santa Clara, Guápiles, petrol station make a right for **Las Cusingas** *(☎ 506/382-5805)*, a small, private botanical garden at the edge of a forest, growing medicinal plants, ornamentals, fruits trees and bromeliads. They offer an informational tour; reservations are necessary.A 30-minute hike leads to the chilly Río Santa Clara.

Another garden of delight can be found at **Costa Flores**, one of Central America's largest wholesaler of tropical flowers. It's located about 15 km/nine miles east of Guápiles, just outside of Guacimo. Visitors can wander a beautifully landscaped path on the 360-acre farm that grows 600 varieties of precious heliconias. Food available.

Rara Avis Reserve

What lies behind us and what lies before us are tiny matters compared to what lies within us.
~ Ralph Waldo Emerson, 1803-1882

Abutting the eastern edge of Braulio Carrillo is the private **Rara Avis Rainforest Lodge & Reserve** *(☎ 506/253-0844, www.rara-avis.com, $$)*, which pioneered sustainable eco-tourism in Costa Rica. It was founded by Amos Bien, the former director of La Selva Biological Station. Rara Avis, which means, rare bird, is a thriving venture begun in 1983 to prove eco-tourism, conservation and biological study could co-exist to the benefit of all three. A visit here requires an overnight stay, with advance reservations. Half the fun of visiting Rara Avis is getting to it. Meet the tractor cart at 8:30 am in Las Horquetas for your ride into the rainforest (no cars allowed). It takes three-plus hours of bumps and grinds in an eco-version of a sadistic hayride. This trip is not for those with bad backs. Horseback trips in can also be arranged with the preserve. The first sign of civilization on the way in is **El Plastico**, a base station for education and research, named after the former prison colony there whose inmates slept under plastic tarps al fresco.

Guided nature walks are offered free of charge and birders will be pleased to hear that 362 species have been recorded here. Another blessing of the surrounding rainforest is the cacophony of nighttime sounds. The tremendous Catarata Rara Avis, a twin-level waterfall with an enjoyable swimming hole underneath, is the highlight attraction.

Rara Avis offers a real-world learning experience in sustainability and rainforest conservation. They have a butterfly farm and rope climbing apparatus for scaling into the canopy. At 700 meters/2,303 feet the climate stays cool (25°C, 77°F), but it rains a lot. For some reason, mosquitoes are not a problem. This is a rough journey with agreeable rustic lodgings and is best suited to the physically fit.

■ Places to Stay

Rara Avis' three basic accommodations are scattered around the jungle near the waterfalls. The **River Edge Cabin** *($$$)* features private rooms with baths, as does the most comfortable option, the **Waterfall Lodge** *($$$)*. Both are designed for groups, but you get your own room. All meals and tours are included. Budget/student/backpack travelers can stay in **Casitas**, two-room, dorm-style wooden cabins. Because of the difficulty of access to Rara Avis, we recommend a minimum two-night stay. *(☎ 506/764-3131 or 764-4993, or book on-line at www.rara-avis.com.)*

If rustic is not for you, an appealing luxury resort outside of Horquetas is **Sueño Azul** *(☎ 506/761 4244, www.suenoazulresort.com, pool, restaurant, pool, $$+)*, that features 25 well-appointed rooms (with terraces) built on the site of a century-old hacienda. It has a spa and offers yoga. Tours and private reserve over a narrow metal bridge built for oxcarts.

La Selva & Sarapiquí

One thing you can't recycle is wasted time.
~ anonymous

La Selva Biological Station *(☎ 506/524-0607, fax 524-0608, www.ots.ac.cr, dining hall, gift shop, $$)*, three km/1.8 miles south of Puerto Viejo de Sarapiquí, is famous throughout educational, scientific and tourist circles. It is primarily a research facility run by the **Organization for Tropical Studies** (OTS), situated near the confluence of the Sarapiquí and Puerto Viejo rivers. OTS is a not-for-profit consortium of more than 55 universities in the United States, Latin America and Australia. They also operate Las Cruces and Palo Verde. Limited accommodations are available. Their rule is no sandals on nature walks.

■ Adventures

La Selva's **Sarapiquí Rainforest** shares a border with Braulio Carrillo National Park, effectively enlarging it by 1,516 hectares (3,746 acres), for a total UNESCO biosphere of 49,000 hectares (121,000 acres). Since the 1950s the researchers at OTS have identified key organisms in the complex rainforest biological process. You get to share in the wonder by exploring part of the 57 km (35 miles) of well-maintained trails looking for peccaries, agoutis, sloths, howler monkeys, spider monkeys, mandible or keel-billed toucans, white-crowned parrots, red and blue poison dart frogs, orchids, philodendron or any of the 100 species of trees in the old-growth forest. And, if you're staying overnight, chow down with scientists and students who talk shop over the communal dining table in the evening. No sandals allowed due to the risk of snake bites.

Oasis Nature Tours *(☎ 506/766-6108, www.oasisnaturetours.com)* is a Puerto Viejo nature tour agency running trips on the Sarapiquí and San Juan rivers, plus Tortuguero canals. Half- and full-day excursions, plus you get a Nicaraguan passport stamp on the San Juan trip.

A few kilometers south of town in El Tigre you'll find a cooperative of local women, known as **MUSA**, who run a medicinal herb farm. They sell health products made from their own harvests.

Just down the road from the Selva Verde Lodge (see below) is the **Sarapiquí Learning Center** *(www.learningcentercostarica.com)*, where local community members, activists and visitors sit down for an afternoon coffee break, called a *charla*, or "chat." A plate of goodies, some good Costa Rican coffee and a bilingual sharing of culture. They also teach Latin dance plus arts and crafts.

Two nature films were recently produced by the BBC on the "Hummingbird Hotel" and "Weird Nature" of **Heliconia Island** *(☎ 506/397-3948, www.heliconiaisland.com),* located between Las Horquetas and Puerto Viejo de Sarapiquí. Former graphic artist and New Yorker, Tim Ryan, created a marvelous living botanical garden of flowers and indigenous plants that attracts flocks of birds, butterflies, and hummingbirds to his two-hectare/five-acre island. He offers day and night tours. Entrance fee is US $10, tasty meals are offered in a riverside Japanese restaurant. Located outside Puerto Viejo.

The longest suspension bridge in Costa Rica (262 meters/862 feet) spans the great Río Sarapiquí from Tirimbina to the **Sarapiquís Centro-Neotropic** *(☎ 506/761-1004, www.sarapiquis.org, $$-$$$),* an archeological/ecological project managed by a Belgian foundation. **Sarapiquís** is an eco-model of a 15th-century pre-Columbian village designed using ecologically sustainable technologies. It's an innovative concept in ecotourism for Costa Rica, a country not known for its pre-Colombian archeology. The architecture, inspired by the indigenous building style of the early peoples, features four huge palenques, built with cone-shaped roofs made of thatched palm. Three units contain the attractive guest rooms, while a fourth holds the restaurant, bar, offices and gift shop. A cultural museum boasts a film theater. Visitors can explore the gardens and ongoing archeological dig, where stone formations and four large tombs, dating from 800-1550 A.D., are being excavated. It makes for a unique departure from the common tourist experience.

In 1995, the Milwaukee Public Museum and Wisconsin's Nature Center opened the 304-hectare/750-acre **Tirimbina Rainforest Center** *(☎ 506/761-1418, www.tirimbina.org),* to preserve endangered tropical forests and ecosystems, two km/1.25 miles north of La Virgen. The area was selectively logged in the 1960s and is an example of rainforest regeneration now serving as an active research and study destination for school groups, teacher training and scientific research projects. The center also offers overnight accommodations.

Find another uncommon tourist experience at **Rancho Leona** *(☎ 506/761-0048, www.rancholeona.com, restaurant),* an exotic art retreat in La Virgen, on the edge of the Braulio Carrillo National Park. The artist-owners create incredible stained glass art in their studio and sponsor varied art retreats (bring your own medium supplies). It's a creative atmosphere and participants can take long tours, kayak trips or study Spanish in a supportive environment. Check their website for schedules.

■ Places to Stay

In Chilamate, **Selva Verde Lodge** *(US & Canada ☎ 800/451-7111; CR ☎ 506/766-6800, www.holbrooktravel.com, $$)* is a popular forest lodge close to, but not affiliated with, La Selva biological station. Giovanna and Juan Holbrook, of Holbrook Travel in Miami, built

Hotel Prices

Prices are per night for two people

[No $]	less than US $20
$	US $20-$40
$$	US $41-$80
$$$	US $81-$125
$$$$	US $126-$200
$$$$$	over US $200

this large lodge in 1985 as a way to save 202 hectares/500 acres of secondary rainforest. Located on the banks of the Río Sarapiquí, Selva Verde offers above-ground rustic bungalow rooms connected by covered walkways to the main lodge, restaurant and bar. One evening a peccary ambled nonchalantly across our path. It's a friendly and comfortable place to use as a local base. They offer eco-tours and a super friendly atmosphere. Holbrook Travel is an excellent agency that will make all your Costa Rica travel arrangements for you.

Good food and comfortable accommodations can be found at the popular, family-run **La Quinta de Sarapiquí** *(☎ 506/761-1052, www.laquinta-sarapiqui.com, restaurant, pool, $$)* along the Río Sardinal. This charming country inn hosts a local natural history display within a large art gallery that features local artists and handicrafts. And for living specimens, it's a good place to find the tiny poison dart frog (*Dendrobates pumilio*) in their "frog garden." You'll get personal attention from Beatriz Gámez and her husband Leo. Good people. Great place to stay.

If you're taking the river boat trips or just floating around, a quality inn in the center of Puerto Viejo de Sarapiquí is **Hotel El Bambú** *(☎ 506/766-6005, www.elbambu.com, pool, restaurant, $$)*, a large and agreeable hotel with a big kidney-shaped pool and first-rate restaurant. West of Puerto Viejo, **Ara Ambigua Lodge** *(☎ 506/766-6281, www.tourism.co.cr/hotels/ara_ambigua, $-$$)* features ranch-style rooms (we preferred the "rustic" rooms to the standard ones) on a rise above their gardens, outdoor restaurant and private lake's tranquil waters.

Tortuguero National Park

You are today where your thoughts have brought you;
you will be tomorrow where your thoughts take you.
~ James Lane Allen, 1849-1925

The fertile Atlantic slope drains into the Caribbean Sea at Costa Rica's remote northeastern border with Nicaragua, along the San Juan River. Here, the **Barra del Colorado Wildlife Refuge** and the 20,000-hectare/49,400-acre Tortuguero National Park, do their bit to protect their famous network of canals, tributaries and the vast green, alluvial plain.

Dark brown waters in the 160 km/99 miles of inland canals crisscross untamed, low-lying forests and mangrove swamps of the northeast region, allowing traditional means (motor boats) of transportation between tiny villages and plantations. The Amazon-like canals provide a window on the environment and are a wonderful way to see birds, monkeys and other wildlife. Of the 16 endangered mammals in Costa Rica, 13 of them are found in the parks or nearby. The 405 bird species recorded here, dwarf the number found in the entire European continent.

But it is **sea turtles** that everyone comes to see. No matter how far they roam, turtles, like children, invariably come back to nest on the same beach where they born. Four of the world's eight turtle species nest at Tortuguero (tor-two-GERO) at various times of the year. The largest, leatherbacks, nest in March and April and hatch May through June. Far more common are green turtles, who nest July through September – the best time to visit. Turtles lay their eggs at night and are easily disrupted, so the 100 people maximum allowed on the protected beach must be with a guide (usually a trained local) who uses a dimmed flashlight and stealth to get his or her group close to the mother turtle without disturbing her. Visitors can spend overnights in one of the handful of quality jungle lodges or the few *cabinas* available in the little village.

You can fly in and out of Tortuguero but, unless you're in hurry to have life pass you by, take the slow boat tour and stay one or two nights. Fly back if you want. A navigable canal parallel to the shore connects with the Río San Juan to the north or Moín, near Limón. Our boat once ran aground on a shallows and we had to get out and pull, à la Humphrey Bogart in the movie, *The African Queen* (without the leeches). A travel experience that can't be bought – only lived. Seeing turtles at Tortuguero is a total eco-experience and one you'll talk of for many years.

THE TALE OF THE TURTLE

In the early 1950s, Dr. Archie Carr, a zoology professor from the University of Florida, searched the Caribbean for details of sea turtle biology. In Costa Rica, he found that the 40-km/25 miles of black sand south of Barra del Colorado was a globally important nesting sight for green turtles. He established a research station near Tortuguero (derived from the Spanish for turtle, *tortuga*) – even as locals continued to harvest the turtles and their eggs. Dr. Carr published a fascinating chronicle of his expeditions in 1956, entitled, *The Windward Road*. The book so touched Joshua Powers, a NY publisher's representative, that he and 20 others, lead by John Phipps, formed the **Caribbean Conservation Corporation** *(www.cccturtle.org, ☎ 800/678-7853)* in 1959. Their mission is to ensure the survival of sea turtles within the wide Caribbean basin. At the edge of the village CCC operates H. Clay Frick Natural History Visitor Center to educate and inform the 40,000 people who visit annually.

Since its inception, CCC's efforts very likely saved the green turtle (*Chelonia mydas*) from immediate extinction and were instrumental in the creation of Tortuguero Park in 1970. It may have been just in time; the beach here had been fertile turtle hunting grounds since pre-Columbian times. By the 1700s the Spanish organized harvests of turtle meat, eggs, oil and shells. Wholesale hunting reached a crescendo in 1912, when an 18-ton ship transported turtles captured by *veladors*, men who turned turtles over on their backs, tied them to a log and floated them out to the ship. By the time the national park was created, both logging and turtle harvesting had fallen to an unprofitable level, and the village of Tortuguero diminished to less than 100 people. By that stage, turtles were endangered. Rusting machinery scattered around town are reminders of the old days. The park's turtle population now, at least, is on the rise and the community supports itself from the eco-tourists who come to see the nesting cycle, fish or cruise the canals.

Despite the success of conservation and ecosystem protection, reality takes it toll on naturalists' work. "Miss Junie," a green turtle tagged with a sophisticated tracking device when she nested at Tortuguero in September 2000, was caught and killed in August 2001 by a Nicaraguan fisherman near Miskito Keys, 420 km/261 miles away. Teachers, students and conservationists had tracked Miss Junie via satellite links with the Internet. Her US $ 3,200 transmitter was bought back for US $10. You can adopt a turtle at the CCC website listed above.

■ The Way Here

Tortuguero should be an overnight adventure. Book a tour with a travel agency in San José or directly with one of the good lodges mentioned below. You can fly in and out, but that sort of defeats the purpose. Flying is cheaper and much faster, but for us it's just not the same. Tour vans take you from San José, through the banana plantations around Batán (no marching required) or Moín, near Limón, to the docks. Canopied launches then carry you 3½ hours up river to the park. Infrequently, tours take the old long cut from Muelle, north of Puerto Viejo de Sarapiquí. This route, via the San Juan River, gets your passport stamped in Nicaragua. If you're interested in this route, check with **Restaurant Muelle** *(☎ 506/766-6019)*, if you have a good command of Spanish, or better yet, **Oasis Tours** in Puerto Viejo *(☎ 506/766-6108)* if you're in the north. Or ask the tour companies in San José if they're taking this longer six-hour route.

If you're a real adventurer and want to stay in one of the low-rent village *cabinas* and get to Tortuguero without a group, you'll have to make your way to Moín or Cariari. A bus from the Radio Casino building, one block north of the Mercado in Limón, takes you to nearby Moín. So will a taxi. Morning departures (8-9 am) are the norm; shop around for prices. A fast boat, the **Riverboat *Francesca***, owned by Fran and Modesto Watson *(☎ 506/226-0986, www.tortuguerocanals.com)*, offers lodge overnights and carries passengers daily from the Moín dock. A one-day, one-night package with meals and round-trip transportation on their riverboat runs US $184.

A cheaper boat goes by way of Cariari, a small village infamous in the land struggle. Ignore the road sign for Tortuguero near Guápiles – the road leads to Cariari. You almost could drive all the way to Tortuguero at one time, though. In 1996, a couple of corrupt politicians and local people in a nearby small village bulldozed a road through the rainforest as close as six km/3.7 miles to Tortuguero village – part of it in the park! They were caught and the road was closed, but it will take years to heal the scar. The incident reinforces what has been strongly recommended since Tortuguero's inception. The park boundaries need to expand – at the very least to connect it with the Barra Colorado to the north and west. Tortuguero is the only Costa Rica Park that has never been increased in size. Write a letter to the Costa Rica government, but don't hold your breath awaiting a response.

To reach Cariari, take the 9 am bus from the Terminal del Caribe (Calle 1 between Av 11 & 13) in San José. Take another bus from Cariari toward Geest banana plantation until it ends at the dock (ask the bus driver for *el barco a Tortuguero*). The daily boat meets the bus for a 1:30 pm departure, and you're stuck until the next day if you miss it or it's not running. You can check, in Spanish, via radiophone with the park's info center *(☎ 506/392-3201)*.

THE LEGEND OF TORTUGUERO

Der Aberglaube ist die Poesie des Lebens.
Superstition is the poetry of life.
~ Goethe, 1749-1832

An indigenous legend explains why turtles always come back to mate and nest at the same beach where they were born. The native peoples believed a carved image of a turtle, located in a cave in Tortuguero Mountain, a 117-meter/385-foot hill rising above the tree line at the north end of the park, directed them. The turtle statue swivelled toward the land when it was time to come in and nest, and turned 180° to the sea when it was time to head away. Not surprisingly, there is a cave in the hill, but an earthquake collapsed its entrance; only bats know what is inside. This hill makes a fun morning hike for its views of the forests and seashore.

■ Places to Stay

Hotel Prices	
Prices are per night for two people	
[No $]	less than US $20
$	US $20-$40
$$	US $41-$80
$$$	US $81-$125
$$$$	US $126-$200
$$$$$	over US $200

Tortuguero village itself is located on a narrow sandy peninsula of land hard against the ocean's rough surf and the placid main canal. On either side of the canal are jungle lodges that specialize in multi-day package tours. Guests in lodges on the sandy side have the advantage of being able to walk to the village and beach without crossing the canal by launch. The inland hotels are cooler and feature forest hiking trails. The jungle lodges offer all-inclusive packages that include round-trip transportation, all meals, a tour and one or two night's accommodations. (Unless you're a true nature lover, there's not much to do for two nights if it's not green turtle season.) Prices listed are for double occupancy, and are for two nights unless otherwise noted.

The new **Pachira Lodge** *(☎ 506/256-7080, www.pachiralodge.com)* has an elegant, Caribbean-casual buffet dining room and attractive bedrooms in bungalows built on pilings above the jungle floor. The rooms connect by covered walkways and the beautiful pool is heart-shaped. Canal side. They offer a boat and plane combo, two nights, for around US $350 per person. Via boat, at US $ 250, is a good deal. One night and boat transfers runs US $ 180. They also feature a one-day trip (12 hours) from San José for approximately US $100.

Tortuga Lodge *(☎ 506/257-0766, www.costaricaexpeditions.com)* is a canal-side hotel owned by the experienced Costa Rica Expeditions, and it sets the standard for service and luxury in Tortuguero. There's excellent dining, family style, in a new canal-side restaurant and a new pool uses a non-chemical purification system. Desirable rooms in two-story buildings with verandas upstairs (preferred). Around US $400 for two nights, plane and water transfer combo (less if by boat both ways).

On the ocean side is the smaller **Laguna Lodge** *(☎ 506/225-3740, fax 283-8031, www.lagunatortuguero.com),* which offers rooms built of old almond wood. From here you can walk to town. Their free-form pool is inviting, as is an edge-of-the-canal, thatched-roof dining room. Booking through Riverboat Francesca can be a bit cheaper. Packages run US $250, with student discounts offered.

Mawamba Lodge *(☎/fax 506/710-7282, www.grupomawamba.com).* For US $ 350 you get three days, two nights, transportation, tours and all

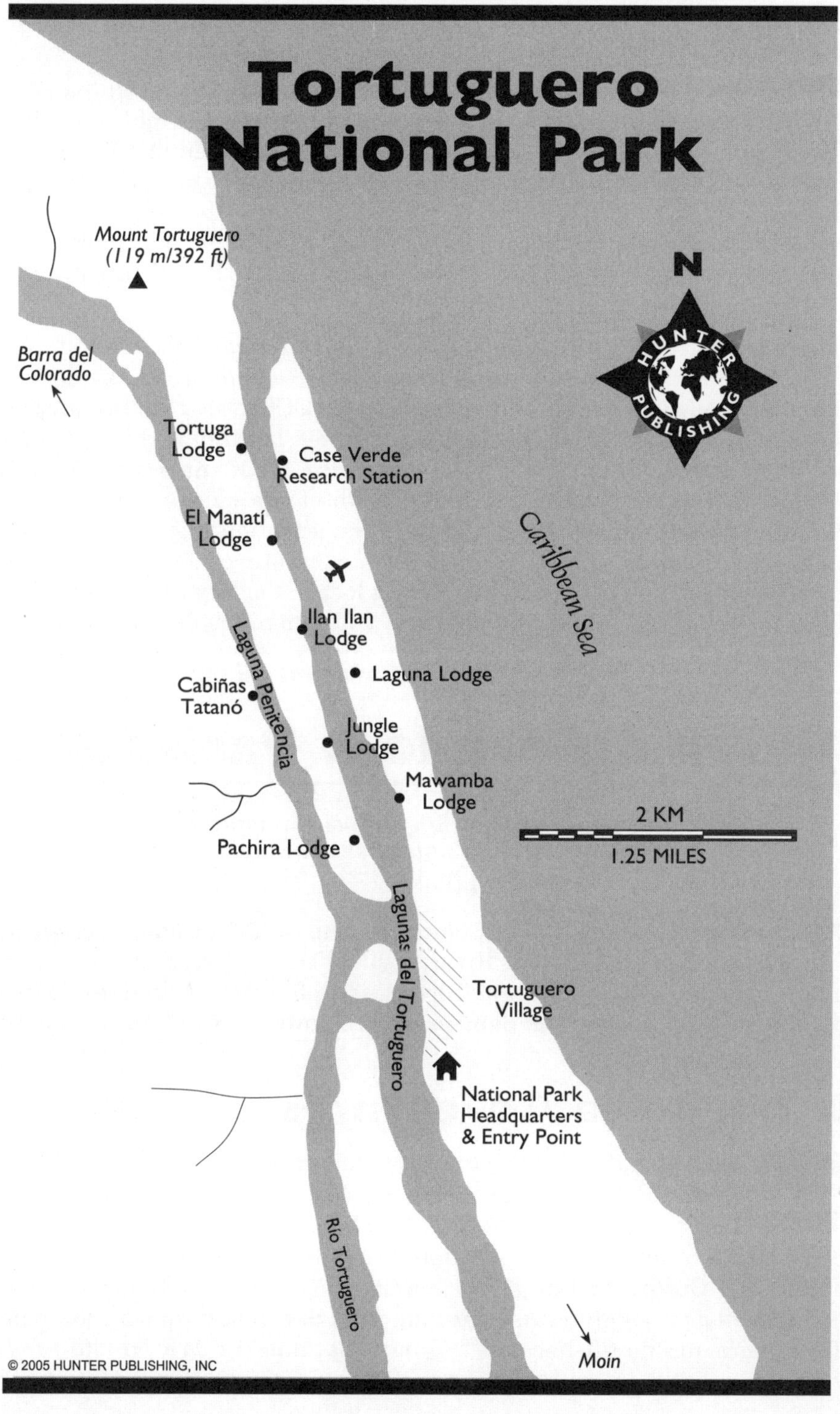

© 2005 HUNTER PUBLISHING, INC

meals at this oceanside lodge. They also have a free-form pool, jacuzzi, and game room. Verandas have rocking chairs and hammocks.

Another canal-side lodge is **Hotel Ilan-Ilan** *(☎ 506/255-2031, fax 255-1946, www.ilan-ilanlodge.com),* a series of very rustic Caribbean *cabina* rooms. This was our first hotel stay, back when it opened in 1989. Packages run US $250 for two nights. Our most recent stay was at the **Jungle Lodge** *(☎ 506/233-0133, cotour@sol.racsa.co.cr)*, the area's largest lodge, with a large pool featuring "waterfall" circulation, bar and dance floor, and a good restaurant. It has basic rooms. Rates run around US$ 200 for a two-night package.

In town, ask locals for directions to the **Cabinas Aracari** *(no phone, $)*, near the soccer field, or **Cabinas Miss Junie** *(no phone, $)*, the best of the *cabinas*, owned by the former cook for the CCC research station. The famous "Miss Junie" turtle was named after her. Located at the north end of the village. Ask her to feed you. Otherwise, **El Manatí** *(☎ 506/383-0330)*, canal side, two km/1.25 miles north of the village, offers rustic rooms in the headquarters of the local conservation effort to save the manatees. Clean but basic. What it lacks in creature comforts it makes up with the illusion of staying at a jungle lodge at a very appealing price. Good for families, but has no pool. US $45 per night for two; includes breakfast.

Barra Colorado & Parismina

The weather for catching fish is that weather,
and no other, in which fish are caught.
~ William Blake, 1757-1827

These two locations, Barra Colorado, near the Nicaraguan border at the mouth of the Colorado River, north of Tortuguero, and Parismina, a small fishing village south of Tortuguero, offer high-priced, exclusive deep-sea fishing lodges and depressed little towns.

■ Deep-Sea Adventures

Tarpon and snook are the big fish, but deep-sea catches include barracuda, jack, tuna, dorado, sailfish and marlin. **Silver King Lodge** *(US ☎ 800/847-3474, CR 506/381-1403, www.silverkinglodge.com)* has a good reputation and is the most luxurious. The venerable **Río Colorado Lodge** *(☎ 506/232-8610, www.riocoloradolodge.com)* offers a one-night cruise package from San José through Moín and Tortuguero for non-fishermen. **Samay Lagoon Lodge** *(☎ 506/384-7047, www.samay.com)* has three-day, two-night tours for fishermen or nature lovers that include a stop in Tortuguero. Bring a bunch of buddies and rent Dr. Alfredo Lopez's 65-foot ***Rain Goddess*** *(☎ 506/231-4299, www.bluwing.com)* a houseboat with crew for tours, fishing or relaxing.

■ Parismina

The well-known **Río Parismina Lodge** *(US ☎ 800/338-5688, CR 506/229-7597, www.riop.com, $$$$)* caters to fishermen and has all the amenities, including beautiful garden grounds. **Jungle Tarpon Lodge** *(US ☎ 800/544-2261, www.jungletarpon.com)* offers fishing and adventure packages, as does **Caribbean Expedition Lodge** *(US ☎ 361/992-9614, www.costaricafishing.com)* at the mouth of the Río Parsimina.

■ Nicaragua

In Nicaragua, everybody is considered a poet until he proves to the contrary.
~ Daniel Ortega, former Sandinista President.

The meandering San Juan River, the border between Nicaragua and Costa Rica, has played more than a minor role in history. During the California Gold Rush, Cornelius Vanderbilt used the river, in combination with a overland railroad, to transport thousands of men and tons of gold on the 42-day journey back and forth between New York and San Francisco. It was this economic interest that put him squarely up against the 19th-century carpetbagger, William Walker (see page 76). Vanderbilt used huge dredges in the bay to gather sand for railroad beds in low-lying areas. One of these gigantic rusting pieces of machinery is still visible in the shallows of the river mouth from a new $2.1 million dollar sportfishing and eco-tourism resort, **Río Indio Lodge** *(www.crica.com/hotels/rioindio.htm or www.bluwing.com, US ☎ 800/788-7857, CR 506/256-3222, $$$$$)*. It's a premier eco-lodge and a major effort on the part of Nicaragua to imitate Costa Rica's success in attracting eco-tourists. Developed by Dr. Alfredo Lopez, a Costa Rican physician and eco-entrepreneur, the lodge is a deep-sea fishing destination as well as a base for eco-tours of Nicaragua's virtually unexplored **Río Indio-Maiz Biological Reserve**.The swampy wilderness encompasses 640,000 acres of the largest lowland tropical rainforest in Central America. Vanderbilt's abandoned railroad beds now function as trails for diverse wildlife and intrepid hikers. If eco-tourism catches on, Nicaragua may be the next Costa Rica.

Limón

And finds with keen discriminating sight,
Black's not so black; – nor white so very white.
~ George Cannon, 1770-1827

Sadly, Puerto Limón (lee-MON) city has little to offer tourists, especially eco-tourists. Except for Carnaval, a mid-October, week-long celebration that attracts party animals, cultural aficionados, Ticos, tourists, backpackers, musicians, dancers, students, locals and hippies among the over 100,000 people that pack the city streets. Its official title is **Dia de las Culturas**, named to celebrate the mixed culture of Costa Rica, but it's a wild, totally Caribbean Mardi Gras – set to the rhythm of an Afro-Caribbean beat. The rest of the year Limón is a working port.

Established where Columbus first anchored offshore in 1502, Limón has proved over the years to be the premium deep-sea harbor for Costa Rica exports. The government built the Atlantic Railroad across the mountains to bring highland coffee and bananas here to ship, and the city became a major port.

Fortunately, new revenue from cruise ship passengers is beginning to spur renovations to make the town more interesting to travelers.

■ Sights & Adventures

The reintroduction of the **Jungle Train** *(☎ 506/758-5852)*, or at least the Limón to Siquirres section of the old run, may be worth a ride through the banana plantations. Transportation from San José for a day-trip is offered. Contact a local travel agent for arrangements. A full-day tour, including lunch, is about US $80.

To explore the historic old seaside city can be a satisfying experience. We especially enjoy people-watching from under royal palm trees in **Parque Vargas**, or strolling the long **seawall**. We admired the slowly disintegrating but charming century-old wooden buildings downtown, and then wandered around in the refurbished **Mercado Central**, filled with fruits and vegetables, fish and hard goods. Another spot that piqued our curiosity was the **graveyard** on the way into town. What had caught our eye was the large Chinese burial section, complete with oriental designed graves, on a hillside under tall pine trees.

Horseback riding in the tropical forests is as close as La Bomba, a town on one of the roads south from Limón to Cahuita. **Ranch Cedar Valley** *(☎ 506/798-2637, www.bluelimbo.com)* offers very interesting individually guided forest rides.

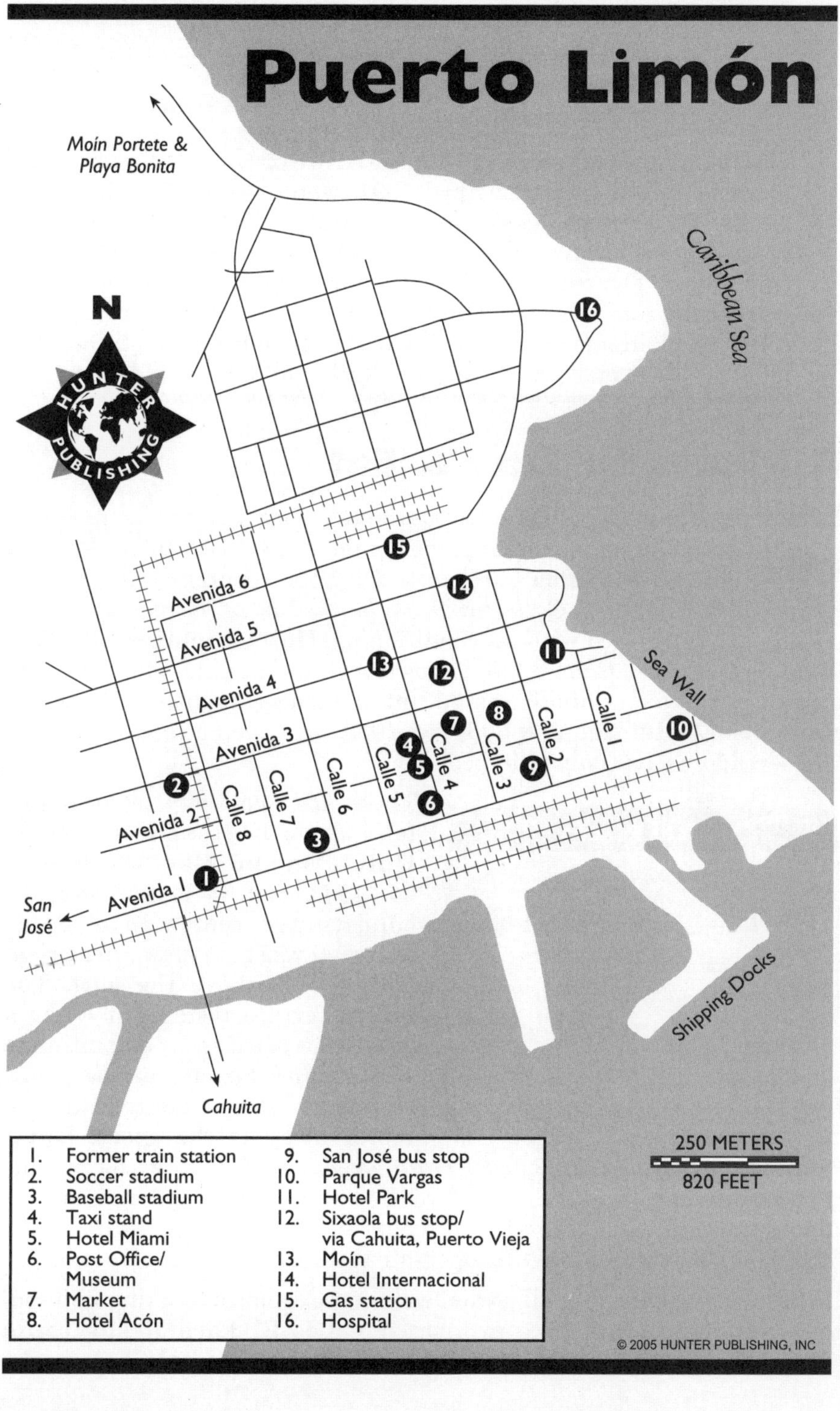
Puerto Limón
Moín Portete & Playa Bonita
N
HUNTER PUBLISHING
Caribbean Sea
Sea Wall
Avenida 6
Avenida 5
Avenida 4
Avenida 3
Avenida 2
Avenida 1
Calle 8
Calle 7
Calle 6
Calle 5
Calle 4
Calle 3
Calle 2
Calle 1
San José
Cahuita
Shipping Docks
250 METERS
820 FEET
1. Former train station
2. Soccer stadium
3. Baseball stadium
4. Taxi stand
5. Hotel Miami
6. Post Office/ Museum
7. Market
8. Hotel Acón
9. San José bus stop
10. Parque Vargas
11. Hotel Park
12. Sixaola bus stop/ via Cahuita, Puerto Vieja
13. Moín
14. Hotel Internacional
15. Gas station
16. Hospital
© 2005 HUNTER PUBLISHING, INC

TEREMOTO

All things have second birth.
The earthquake is not satisfied at once.
~ *The Prelude*, William Wordsworth, 1770-1850

Limón has always been poor, but the entire province got quite a financial jolt at 4 pm, 22 April 1991, when a powerful earthquake, or *teremoto*, destroyed the Atlantic Railroad, many buildings and cut roadway and communications with the Central Valley. The epicenter that destroyed nearly 3,000 homes was inland and west of Cahuita. North of Limón, the ground rose or sank in some places as much as two meters/6.6 feet. One fishing village on a river near the beach had to be abandoned.

■ Places to Stay & Eat

If you want a room for Carnaval, book months in advance. Better accommodations are found north of town in Moín or Portere (see below). In Limón City, the venerable **Hotel Park** *(☎ 506/798-0555, Av 3, Calle 1 & 2, 2½ blocks east of the mercado, $)* is an old-world standard with air conditioning. **Hotel Miami** *(☎ 506/758-0490, Av 2, between Calle 4 & 5, $)* has few vices, but it does have air conditioning. There's a Chinese restaurant and a noisy nightclub in the air-conditioned **Hotel Acon** *(☎ 506/758-1010, near the northeast corner of the* mercado, *$)*, but only twin beds.

Hotel Prices	
Prices are per night for two people	
[No $]	less than US $20
$	US $20-$40
$$	US $41-$80
$$$	US $81-$125
$$$$	US $126-$200
$$$$$	over US $200

There are some very pleasant hotels just north of the city near Moín. **Hotel Maribú Caribe** *(☎ 506/758-4543, pool, $$+)* offers nice, air-conditioned rooms in circular thatched-roof bungalows on a promontory cliff above the surf. The breezy terrace restaurant offers a great view of the rocks and intimate white sand beach between outcroppings below. Less than a kilometer away is the green **Hotel Matama** *(☎ 506/758-1123, www.matama.com, pool, $$+)*, with gorgeous garden grounds, air-conditioned rooms, a pool, bar and restaurant. Across the street is the beach and **Apartotel Cocorí** *(☎ 506/798-1670, fans, $$)*, which has a relaxing open-air restaurant and small rooms.

Downtown has many small sodas featuring Caribbean cooking and seafood, but the coolest place in town to eat is **El Faro Restaurante** *(☎ 506/758-2269, www.blueliimbo.com, $-$$)*, high on a hill above the city. It serves an international menu on a terrace with a view of Isla Uvita. Located above the cemetery; turn uphill by the railroad tracks west of the cemetery. Near the college of Limón and technical school. **Restaurante Placeres** *($-$$)* is a fine choice, 25 meters/82 feet east of the entrance to Emaus, on the road to Portete, just north of the city.

Cahuita

The fair breeze blew, the white foam flew,
The furrow followed free;
We were the first that ever burst
Into that silent sea.
~ *The Æolian Harp*, Samuel Taylor Coleridge, 1772-1834

Cahuita (pronounced cow-EAT-ah) town is an increasingly desirable destination to all kinds of tourists. It's best known for the beautiful oatmeal sand beaches in **Cahuita National Park** – set aside in 1978 to protect the shoreline. The town offers laid-back luxury hotels and backpacker hostels, near the beaches or deep in the inland rainforests – all at generally at lower prices than Pacific coast destinations. Surfers, bohemians and adventurers share the streets here with eco-tourists. Come soon and stay long.

The park's ecological significance can not be overstated as it protects one of the most important coral reefs in the country and the most developed reef of the Caribbean coast. Its more than 1,000 hectares/2,470 acres stretch east of the road from the town south to the mouth of the Río Carbon. The 22,400-hectare/55,328-acre offshore park limits include 600 hectares/1,482 acres of coral reef and a myriad of marine life. If you snorkel or dive the reef, you'll find more than 300 species of mollusks, lobsters, three species of turtles and thousands of colorful fish such as French angelfish, Isabelita and queen angelfish. Unfortunately, pesticide and silt runoff from plantations have killed most of the coral.

DID YOU KNOW? The park's wetland flora, dominated by yolillo palms and sangrillo trees, gave the town and park its name, derived from the indigenous language. "Kawa" means sangrillo tree, and "ta" means point, which became known as *Cahuita*.

About 45 km/28 miles south of Limón, along a coastal road that passes forests and banana plantations, Cahuita's beach-bum small town is set off the road to the left. The sandy streets have many more bicycles and pedestrians in shorts than cars. It's a town of loping dogs and blond surfers, black-skinned fishermen kicking a soccer ball and visitors bent under big backpacks, small shops and sodas, B&Bs and restaurants. Hotels can be found in town or north of the village along a dirt road that follows the shoreline. The park is at the southern town limit, across a wooden pedestrian bridge. If you're day-tripping, surfing or swimming, enter here. If you're coming to camp or snorkel, enter at the south end of the park, Puerto Vargas, the official entrance about seven km/4.4 miles south of town.

■ Adventures

The best swimming is on Playa Negra, a black, volcanic sand beach beginning past the north end of town, with the softest sand farther on. You can get all kinds of services at locally owned, friendly **Mister Big J's** *(☎ 506/755-0328)*. Laundry, surfboards and snorkel rentals, maps, books, trips to Tortuguero, horseback riding, hiking, fishing and iguana farm tours – all are offered at Big J's. Other tour agencies are **Cahuita Tours and Adventures** *(☎ 506/755-0232)*, **Roberto Tours** *(☎ 506/755-0117)*, and **Turística Cahuita** *(☎ 506/755-0071)*.

Aviarios del Caribe *(☎ 506/382-1335, www.ogphoto.comaviarios, breakfast included, $$$)*, an hour south of Limón, is a renowned sloth sanctuary, wildlife rehabilitation center and nature reserve, about 10 km/.6 miles north of Cahuita. As you might guess from the name, this riverside B&B is a birder's delight, and their canal canoe tours (US $30) into the surrounding Estrella River Delta Wildlife Refuge (6 am or 2 pm) are recommended even if you're not staying here. Excellent private rooms or dorm-style accommodations at a reasonable cost. There's a US $5 entrance fee to the sloth sanctuary.

■ Places to Stay & Eat

South of town on the way into Cahuita are some note- worthy lodgings. **Selva Bananito** *(☎ 506/253-8118, www.selvabananito.com, meals included, $$$$)* is run by a brother-sister team, Jurgen and Sophia Stein, that converted their family farm bordering La Amistad International Park, into a nature lodge. In 1994, they set aside the untouched two-thirds of their land (810 hectares/2,000 acres) as a biological reserve. It's an adventure to reach; we surprised them – and ourselves – by driving for an hour up 15 km/9.3 miles of rough logging roads, across two creeks and two rivers, far into the forest. You don't have to though; they'll pick you up in their huge four-wheel-drive vehicle. School children in the area commute to class on horseback.

Hotel Prices	
Prices are per night for two people	
[No $]	less than US $20
$	US $20-$40
$$	US $41-$80
$$$	US $81-$125
$$$$	US $126-$200
$$$$$	over US $200

Restaurant Prices	
Prices are for an average entrée	
$	less than US $5
$$	US $5-$10
$$$	US $11-$20
$$$$$	over US $20

The Steins are very ecologically minded. There's no electricity and the rustic but comfy cabins are built with "seconds," wood that would have been wasted by local loggers. A delicious dinner is served family-style by candlelight. Horseback ride, hike upstream to a waterfall and swimming hole, climb a huge Ceiba tree, camp overnight in the forest, hike or birdwatch, or mountain bike. Ten percent of the lodge's income goes to a non-profit foundation for watershed conservation. Two- or three-night visits are the norm, great for combining eco-tourism with a beach vacation. We met honeymooners here. From March to August Bananito offers an enticing five-night package including a two-night stay at the Reserva Field Station, where you work with a research team on the main nesting beach on the Caribbean coast for leatherback turtles.

Behind the church in tiny Penshurt (Km 35), Judith and Pierre Dubois run the fine little open-air **Piedmont Restaurant** *($)*. It offers tasty French cuisine, served in the open-air in family-style, next to Pierre Dubois' tropical orchid garden. They offer riverside camping spots, with shower and toilet, and a night tour to see and hear frogs.

BANANA FIELDS OR OIL FIELDS?

Laid-back local Caribbean communities found their environmental concerns ignored in 1998 when the Costa Rican government gave oil exploration rights to several North American companies, including Harken Energy Corporation of Texas, in which American President George W. Bush had been a principal stockholder. Finally, in February 2002, after a long court battle, a federal judge halted the test drilling and called into question the legality of the licensing process, but he did not rule on their environmental objections. A coalition of groups is trying to fight oil exploration and exploitation in the fragile eco-systems of Costa Rica. They include **ADELA** (Anti-Petroleum Struggle), www.grupoadela.org.

In Cahuita itself, we liked the Caribbean-style four-room mansion, **Kelly Creek Hotel** *(☎ 506/755-0007, kellycr@sol.racsa.co.cr, $$)*. Owned by a couple from Spain, the exotic all-wood building and small Iberian restaurant (an unforgettable *paella*) are located right at the entrance to the park, downtown. Entrance to the beach opposite, just outside the park, is by donation. More secluded, right behind Kelly's, is **Alby Lodge** *(☎ 506/755-0031, $+)*. Austrian hostess, Yvonne, welcomes guests to her thatched-roof bungalows set around a lovely garden. Communal kitchen and hang out area; an excellent value. Out of town at Playa Negra is **Cabinas Iguana** *(☎ 506/755-0005, $)* on beautiful grounds. Waterfall pool. The enchanting American-owned **El Encanto Bed & Breakfast** (*☎ 506/755-0113, www.elencantobedandbreakfats.com, rooms, house, cable, breakfast included, $$*) offers comfortably serene accommodations, set in a beautiful tropical garden near the beach. Peace and quiet can be enhanced in the Japanese-style meditation room.

North of town again is the **Atlántida Lodge** *(☎ 506/755-0115, www.atlantida.co.cr, jacuzzi, pool, $$)*, next to the soccer field, where pleasing bamboo bungalow rooms, opposite a good swimming part of Playa Negra, can be found on the lodges' lush garden grounds. Cool pool with a bar and an aromatic BBQ.

Another good deal is the luxury hotel in Cahuita, the **Magellan Inn** *(☎ 506755-0035, www.toposrealestate.org, $$)*, an intimate inn on a rolling lawn, with nice pool and all the comforts of home – in paradise. Their **La Casa Creole Restaurant** offers seafood, Creole, and international cuisine in candlelit ambiance. Two km/1.2 miles north of downtown, the Magellan is an oasis in paradise and is the class of Cahuita.

RUNDOWN STEW

You'll see signs in Cahuita and Puerto Viejo for "rondon" soup or stew. It is a dish ubiquitous in Caribbean cooking. This coconut-based concoction is made with whatever the cook could "run down" or, as locals say, "rondon."

Eat at Swiss-owned, Mexican-sounding **El Cactus** for Italian-style pizza. Go figure. The **Sol y Mar Restaurant** fills up if only for its location opposite the park entrance. We loved **Miss Edith's** and her coconut flavored, homemade Creole cuisine. It's a *de rigueur* stop for dinner or lunch at the north end of town. Tables are shared, so mingle and relax while you wait. It's worth it if you like new tastes. A Québécois owns **Cha-Cha-Cha**, a good place for international cuisine when you've had your fill of coconut rice and beans. Although the food didn't impress us at **Sobre Las Olas**, north of town, it does overlook the beach.

■ Hitoy-Cerere Biological Reserve

Surrounded by the indigenous reserves of the Talamanca, Tayní, and Teliere peoples, this 9, 950 hectare/24,577-acre reserve protects biological diversity in a relatively unexplored wilderness region of rainforest on the eastern slopes of the Talamanca mountains. There are no camping facilities; only day-trips are permitted. Although there are trails, hiking is best done along the streambeds. Do not hike to the waterfall when it's raining as the narrow canyon is prone to flash floods. To get there is half the fun, if you can call it that. Follow the road from Limón south toward Cahuita; turn at the tiny village of Penshurt into a labyrinth of banana plantations (about nine km/5.6 miles along) in the Estrella River Valley. Enter the shady maze of bananas and travel for another 10 km/6.2 miles through La Guaria, Pandora (don't open any boxes here) and Finca 12, turn south at Finca 16, then to the west, where the dirt road rises above the valley. A four-wheel-drive will get you up the last bumpy haul to the reserve, five km/three miles past Finca Cartegena. Maybe not.

Puerto Viejo

We are the music-makers, And we are the dreamers of dreams,
Wandering by lone sea-breakers, And sitting by desolate streams;
World-losers and world-forsakers, On whom the pale moon gleams;
We are the movers and shakers Of the world, for ever, it seems.
~ *Ode*, Arthur O'Shaughnessy, 1844-1881

Half an hour south of Cahuita is Puerto Viejo, "Old Harbor," the most laid-back Caribbean beach town in Costa Rica. This lazy village, settled as a fishing center in the earliest part of the last century, didn't get electric until 1987 and only had three phones by October 1997. While surfers come December through March to challenge the six-meter/20-foot surf at Salsa Brava, "Angry Sauce," others come for the *laissez-faire* lifestyle and boutique accommodations along the ribbon of road that hugs the coast. They come for reggae nights at Stanford's, a run-down dance bar on the beach where candles illuminate tree stump tables on the sand and the smell of ganja hangs in the breeze.

The clock ticks slowly here, and things are never hurried. It's an easy blend of Afro-Caribbean culture, escapist North Americans, indigenous families, intense eco-tourists heading on to explore the inland forest reserves or Gandoca-Manzanillo Preserve, beach-bums, pot heads, surfer dudes and travelers like us, here to absorb some sun and slow the pace of our lives down to simplicity.

SUPPORT MEETINGS

AA meetings are here held on Tuesdays and Saturdays at 4 pm at the PV Fellowship Church. An **NA meeting** is also held here on Thursday.

As you enter town, look for the grounded barge (now growing weeds) that was once a commercial venture to ship the beach's black volcanic sand. Be aware that the town's popularity has grown immensely in recent years and many hotels recommend reservations on weekends, holidays, and at times during the dry season. Others are first-come, first-serve.

■ Adventures

At the center of town is the headquarters for the **Asociación Talamanqueña de Ecoturismo y Conservacióon**, ATEC *(☎ 506/750-0398, www.greencoast.com/atec.htm)*, across the street from Restaurant Tamara. Outside this alternative tourism's tiny office there's a bulletin board and local map. They have an Internet connection and information about various tours, projects, and sights. Their focus is on eco and cultural tourism. It's here you can arrange for local guides for the nearby Gandoca-Manzanillo Wildlife Refuge, rainforest hikes, or visits to the indigenous communities, such as the Bribri, banished to inland reserves, as well as short nature walks (recommended) or serious rainforest hikes, camping trips, plus they serve as facilitators for taking fishing, diving, and snorkeling trips. **Geo-Expediciones** *(☎ 506/272-2024, almonds@sol.racsa.co.cr)*, at Almonds and Coral Lodge, offers kayaking and adventures, including an excursion to Bocas del Toro in Panama. In Manzanillo, divers and kayakers should head to **Aquamor Talamanca Adventures** *(☎ 506/759-0612, aquamor@sol.racsa.co.cr)*. In addition to those watersports, they also offer trips into the wildlife preserve.

Friday through Monday, the **Finca La Isla Botanical Gardens** back in Penhurst are open for wandering the tropical, fruit and medicinal gardens (US $4). Or you can opt for a guided tour full of interesting observations (US $10). Nervous? Get the butterflies out at the new **Butterfly Garden**, high on the hill above Punta Uva. Or do yoga at **Samasati Nature Retreat** *(☎ 506/224-1870, www.samasati.com)*, about two km/1.25 miles north of town. The retreat has daily classes and meditations, or you can stay there. Ommmmmm.

During February and March the **Southern Caribbean Music Festival** *(☎ 506/750-0062, wolfbliss@sol.racsa.co.cr)* attracts musicians and island music lovers from all over to the **Playa Chiquita Lodge** *(☎ 506/750-0408, www.playachiquitalodge.com, $-$$)*, a popular budget *cabinas* hotel on the beach. Good sand and volleyball.

The "in" place to watch the sunset (which way is west?) is the **Red Stripe** bar near the bus station. If you don't drink, try an ice cream from the place next door.

■ Places to Eat

In-town dining highlights include **Restaurant Tamara**, the biggest place downtown, serving Tico food and owned by a local family. Here, the fish is fresh and the *casados* are inexpensive. Upstairs is a breezy evening bar. You'll find a number of local women have set up dining ar-

Restaurant Prices	
Prices are for an average entrée	
$	less than US $5
$$	US $5-$10
$$$	US $11-$20
$$$$$	over US $20

eas on their porch or house and offer Caribbean cuisine such as the ubiquitous Rondon stew (see page 334). They stick up a little sign and name their establishments with their first names, a folksy tradition that guarantees informal, home-style dining. **Miss Sam** serves a traditional Caribbean lunch, plus hot jonney cakes (called "Johnny" in the Caribbean) every day at 6 pm, except Sunday. Equally tasty is the food at a little restaurant across the street called **Lidia's Place**, which also features local home-cooked Caribbean cuisine in an open-air dining area.

Hotel Pizote *(☎ 506/750-0088, $$)*, just before the bridge into town, has a good restaurant and is the most traditional cabin hotel in town. And for seafood we recommend the Mediterranean-style **Marisqueria Oro** *($$)*, near the soccer field. Vegetarians can try **The Place** *($-$$)*, around the corner from the bus station. It's closed Thursdays. Just for the hot weather they serve a hot curry. We liked **Café Rico** for breakfast.

Our favorite PV eateries are side by side, opposite the ATEC office. **Chile Rojo**, or The Red Chile, features a very tasty large Thai/veggie menu in its tiny storefront restaurant. Since it only has a few tables, this place fills up and stays filled for dinner. Try and get a table under a fan. Their much larger next-door neighbor is **Café Viejo**, a delightful Italian restaurant run by three brothers from the Rimimi in the old country. Chef, Maurico, and his brothers, Marco and Mimi, have created a wonderful candle-lit ambiance in their semi-enclosed dining room, with romantic music, soft lighting and sheer curtains that flow in the breeze. Meals are expensive, but excellent.

Pizza and more can be had at the popular **Pizzeria Rusticone**, around the corner, one block. And check out **Max's** if you get down to the great beach at Manzanillo. Cool off when it gets hot with a "Super 12," a cold soda made in Limón, at **El Buen Precio**, supermarket, near the sunken barge.

■ Places to Stay

In Town

Find hotels in the town or one of the many down the various beaches to Manzanillo, at the end of the road.

> **AUTHOR TIP:** On the Web, check out options at www.puertoviejo.net, www.greencoast.com or www.puertoviejaweb.com.

In town, our all time favorite is **Cabinas Casa Verde** (☎ *506/750-0015, www.cabinascasaverde.com, pool, breakfast included, $$*), a block off the main drag. Rene and Carolina have created a magic tropical garden in which they built 16 clean and welcoming rooms, many with private bathrooms, in Caribbean wooden cabinas. In 2005 they finished their long-anticipated swimming pool, a deep, cool pool accented by faux rocks with a mini-waterfall. Indulge yourself in simple comforts, but book ahead; their high season is December through May, and also July and August. Spotlessly clean.

Hotel Prices	
Prices are per night for two people	
[No $]	less than US $20
$	US $20-$40
$$	US $41-$80
$$$	US $81-$125
$$$$	US $126-$200
$$$$$	over US $200

We also like **Cabinas Tropical** (☎ *506/750- 0283, www.cabinas-tropical.com, 5 double, 2 two triple rooms, $*) owned by tropical botanist, Rolf Blanke. This small, cute and very clean hotel has tiled bathrooms and individual garden patios. Rolf offers personal custom tours with his naturalist's knowledge. Uphill from the soccer field is **Cashew Hill** *(☎ 506/750-0256, $)*, with cooler climes and welcoming rooms featuring ocean views. Private or shared baths and kitchens. Two acres of woods to explore; mud in the rainy season.

Before town are the **Chimuri Beach Cottages** (☎ *506/750-0119, www.greencoast.com/chimuri.htm, $-$$*) on the palm-fringed black-sand beach. Take the left turn near the Pulperia Violeta, one km before you reach Puerto Viejo, for these three Victorian Caribbean beachside rustic cottages. Mauricio Salazar, one of the area's more experienced nature guides, will take you on hikes into the Keköldi Indigenous Reserve (US $25 per person) or on birding tours in a private nature reserve. He's associated with the Chimuri Nature Lodge.

Playas Cocles, Chiquita, Uva & Manzanillo

We have lumped together the white-sand beaches south from Puerto Viejo – Playas Cocles, Chiquita, Punta Uva and Manzanillo – as they are only 10-20 minutes down the road. A shuttle bus plods the route, but don't hold your breath.

Farthest from town, but one of the most fascinating options, is **Almonds and Corals**, Almendros y Corales, **Tent Lodge Camp** *(Playa Mazanillo, ☎ 506/272-2024, www.geoexpeditions.com, $$-$$$)*, which offers a unique jungle hotel experience. Private bungalows are actually huge tent houses with screened walls, all set on wooden platforms above the forest floor. They are connected to other buildings and the beach by raised wooden walkways. Inside each one, a second tent contains beds, a fan, hammocks and, behind curtain number two, the bathroom and shower stall. Not exactly roughing it, but more in tune with its surroundings than an air-conditioned room. An immensely popular lodge, Almonds and

Corals run a high-class operation with lots of tours, a good restaurant and a Caribbean bar.

Aguas Claras *(Playa Chiquita, ☎ 506/750-0131, www.aguasclaras.net, $$)* is a secret hideaway; we really don't want too many people to know about it. Imagine 19th-century mini-Victorian seaside villas – complete with gingerbread trim and painted in Caribbean colors – with open-air, screened kitchenettes and sitting areas. Now put them behind a bit of primary forest from a fabulous little bit of beach, right in the middle of a gorgeous garden divided by hibiscus bushes. Welcome to "Clear Waters." There's a three-night minimum stay, but we'd love to live here.

Designed like a primitive indigenous village, **Shawandha Lodge** *(Playa Chiquita, ☎ 506/750-0018, www.shawandhalodge.com, breakfast included, $$$)* features gorgeous thatched-roof wood bungalows nestled among huge Ceiba trees and tropical flowers. Refinement and comfort in a neo-primitive style. Tile bathrooms and ceramics designed by French artist Filou Pascal. Lots of primitive artwork; the Bribri never had it so good as Shawandha's creature comforts and romantic elegance. Private; 200 meters/658 feet from the road. Open-air French gourmet restaurant.

Cariblue Bungalows *(Playa Cocles, ☎ 506/750-0035, www.cariblue.com, breakfast included, $$$)* are owned by a congenial Italian couple, Leonardo Preseglio and Sandra Zerneri. These spacious hardwood bungalows are nestled in the tropical jungle across from a blue flag beach. Wooden porches have Yucatecan hammocks for lazy afternoons in this quiet, serene environment. Have dinner in their candlelit *rancho* restaurant or hang out and watch big-screen satellite TV in the open-air lounge and bar. A two-bedroom house is also available for families or two couples. Reservations recommended.

Miraflores Lodge (*Playa Chiquita, ☎ 506/750-0038, café restaurant, $-$$*) is an inviting forest lodge, set back about 500 meters/1600 feet from the beautiful Chiquita beach. Bilingual owner Pamela Carpenter built her lodge inspired by architecture of the local indigenous peoples, and she indulged in her love of flowers to landscape the surrounding property. The accommodations at Miraflores are known as Garden rooms on the first level and Lodge rooms on the upper. Pam says she wants to attract guests who are committed to sustainable development, nature adventure tourism and cultural diversity. That's us!

The owner of the small **La Costa de Papito** bungalow hotel *(Playa Cocles, ☎ 506/750-0704, www.greencoast.com/papito.htm, $$)* escaped from New York and brought his 13 years experience at the Carlton Arms Hotel to Costa Rica. The African-looking bungalows are beautifully designed with polished wood and tiled showers. The garden grounds have the same attention to detail. Breakfast ordered the night before (US $5), will be delivered to your room. This is a very interesting and pleasing place to stay. The cabins of **Yaré Hotel and Restaurant** *(☎ 506/750-*

0106, www.hotelyare.com, $$+) are bright, colorful and topped by sharply angled roofs with long overhangs that form a shade porch. Scattered in the woods, raised walkways connect the cabins with the restaurant/bar.

■ Nightlife

Nightspots are somewhat limited to **Stanford's**, a local bar/restaurant/dance hall on the beach, the bar upstairs above **Tamara's**, **Johnny's Place**, a tavern facing the beach in the central part of town, or **Elena's Bar & Restaurant**, down the road at Playa Chiquita. All the way in Playa Manzanillo, you can hang around upstairs at **Max's**, a local hangout on the beach at the end of the road, which also has a good reputation as a restaurant.

■ Gandoca-Manzanillo Wildlife Refuge

The sleepy village of Manzanillo, named after the towering Manzanillo tree that used to dominate its center, lies inside the Gandoca-Manzanillo Wildlife Refuge, a 9,500-hectare park between the mouths of the Cocles and Sixaola rivers. The refuge protects swamps, lagoons, flooded forests and the only remaining red mangrove, or Cativo (*Prioria copaifera*), forests in Costa Rica. Its marine element is meant to provide a refuge for endangered manatees (*Trichechus manatus*), plus green turtles (*Chelonia mydas*), hawksbill (*Eretmochelys imbricata*), loggerhead (*Caretta caretta*) and *baula*, or leatherback, turtles (*Dermochelys coriacea*).

The Gandoca section is accessed from the road to Sixaola. There are guided tours during leatherback turtle nesting season between February and May. Along the edge of the reserve are truly one-of-a-kind accommodations: the **Costa Rica Treehouse and Beach House** (*Punta Uva, ☎ 506/750-0706, www.costaricatreehouse.com, $$*). Yes, we're talking about a real treehouse, built around a tall and sturdy Sangrillo tree, located on 10 acres of oceanfront property in front of the **Iguana Verde Foundation** (www.iguanaverde.com). The foundation is trying to save the endangered green iguana by working with and educating local children. Adopt an iguana while you're there and sleep well in the heights of arboreal comfort. The treehouse, built on two levels connected by swaying steel cable bridges, boasts two large rooms and can sleep up to five people. The Beach House, overlooking the Caribbean, features open living space and sleeps up to four. Both unusual accommodations are built entirely of hardwoods from fallen trees. Reserve well in advance.

The Frontier - Panama

In a little café just the other side of the border,
She was sitting there giving me looks to make my mouth water.
~ *Come a Little Bit Closer,* Jay and the Americans

You're only a few kilometers from Panama. If you'd like to visit, follow the road through **Bribri**, a town in the midst of banana plantations – not an indigenous town of the Bribri Indians, although some live there. There is little reason to stop here, except for perhaps the bank or post office. On the frontier with Guabito, Panama, is the poor Costa Rican town of **Sixaola**, a minor border crossing.

For a two- or three-day trip, the **Bocas del Toro**, an island village in an archipelago of islands, is a good choice. Bring dollars, not *colones*, to Panama. To get to Bocas, cross the border at Sixaola and continue to Changuinola (15 minutes, US $4 by taxi). Take a taxi or bus (US $2) to the nearby banana port of Almirante and catch a water taxi (US $5) to Bocas. When you get there you'll find a quaint Caribbean village on the order of Puerto Viejo, only not as well-to-do. If your visa is expiring, hang out here for three days and return for a new stamp at the border. **Basitimientos National Marine Park** is on another of the archipelago's nearby islands.

For a cultural eco-adventure in the interior, look to **Wekso Ecolodge** *(☎ 507/620-0192, www.ecotour.org/destinations/wekso.htm, $$)*, set deep in the rainforest among the Naso/Teribe indigenous peoples. What remains of the Naso community consists of 11 villages stretched along the Teribe River, into **La Amistad Park**. The wooden communal lodge is up on stilts above a cleared meadow in the tropical forest. A true eco-cultural adventure in rustic but comfortable lodgings. Follow the directions to Changuinola and take a taxi (US $5) or bus (50¢) to El Silencio. From there, it's a hour upstream to Wekso, an ODESEN (Naso Organization for Sustainable Ecotourism Development) sponsored lodge. Arrange for a boat with them in advance, hire one (US $30) or wait for the commuter boat (US $5).

Zona Sur, the South Pacific

'There lies Peru and its riches;
here, Panama and its poverty.
Choose, each man, what best becomes a brave Castilian.
For my part, I go to the South.'
So saying, he stepped across the line.
~ *The Conquest of Peru*, William Prescott, relating Pizarro's famous "line in the sand" challenge to his men.

IN THIS CHAPTER

The southwestern corner of Costa Rica is the least populated, and by far the least visited, area of the country. It borders a largely unpopulated part of Panama; consequently, large sections of primary rainforest have been spared from development. Sizable areas of the coastline are also undeveloped, especially north and south of Golfito. Areas that are developed are mostly banana plantations. In a cooperative effort with neighboring Panama, a huge chunk of Costa Rica and a part of Panama was set aside as La Amistad International Park, a mountainous wooded area that straddles the Continental Divide. Contiguous with Chirripó, Hitoy Cerere, Tapantí-Macizo de la Muerte, and other parks, the total area under an umbrella of protection is 340,000 hectares/839,800 acres.

The pride of Tico conservationists is Corcovado National Park, a huge primary and secondary park on the Osa Peninsula. It's been a hotbed of contention between gold miners and loggers, who wish to exploit its natural resources, and ecologists trying to protect the incredible bio-diversity. Into the mix are the eco-tourism folks who wish to forge a balance.

On the other hand, the southern mountain slopes from San Isidro in the Valle de General, down through Buenos Aires and south to San Vito in the Valle de Coto Brus, have been extensively cultivated for agriculture, especially pineapples and coffee. Banana and oil palm plantations line the hot lowlands from Golfito south into the Valle de Colorado.

It's a wild country in parts – with deluges of warm rain. Dress appropriately. The South Pacific region boasts wonderful scenery, fabulous beaches, dramatic mountains, endless rainforests, big tube waves, white

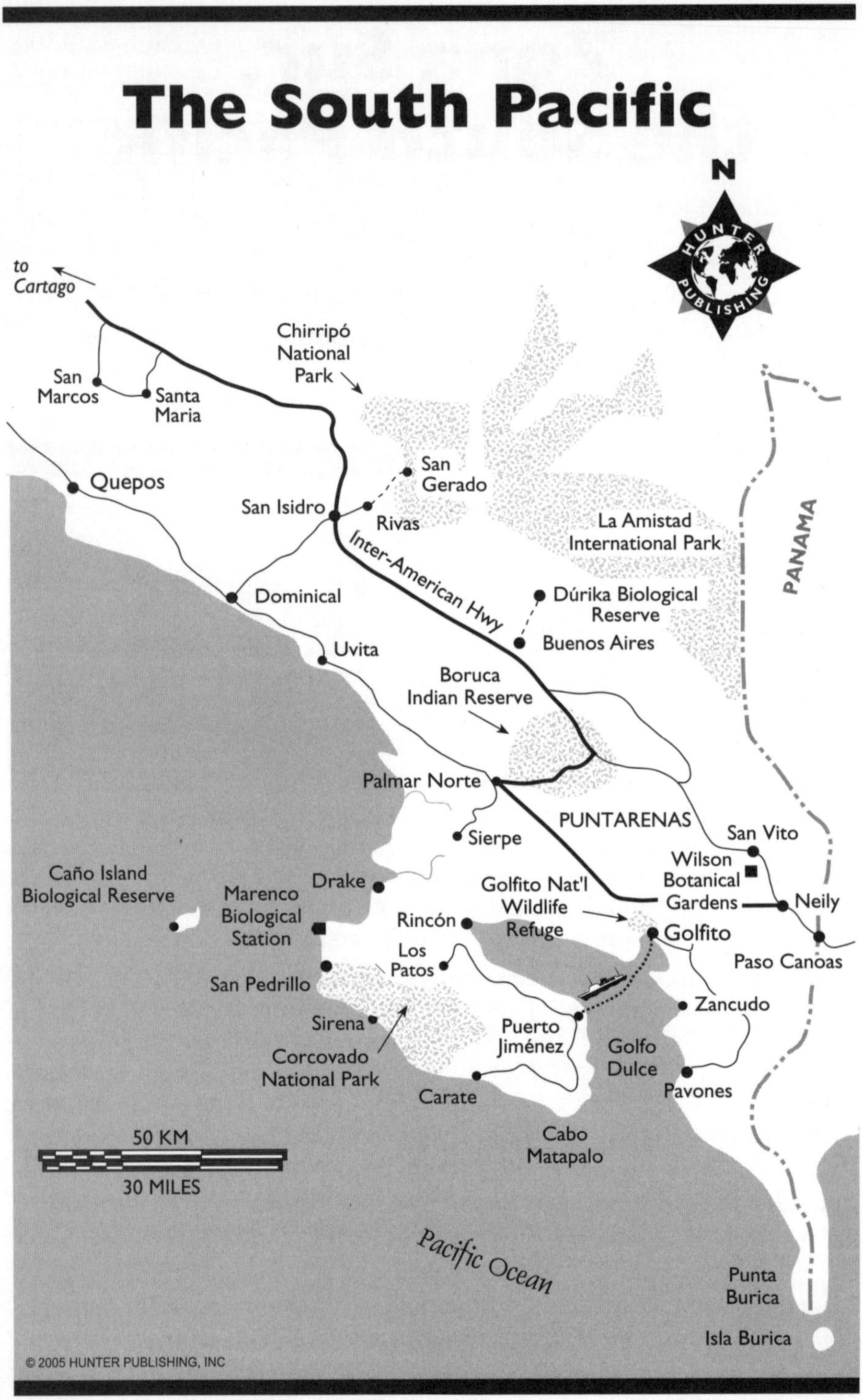
The South Pacific
N
HUNTER PUBLISHING
to Cartago
Chirripó National Park
San Marcos
Santa Maria
San Gerado
Quepos
San Isidro
Rivas
La Amistad International Park
PANAMA
Inter-American Hwy
Dominical
Dúrika Biological Reserve
Buenos Aires
Uvita
Boruca Indian Reserve
Palmar Norte
PUNTARENAS
San Vito
Sierpe
Wilson Botanical Gardens
Caño Island Biological Reserve
Drake
Marenco Biological Station
Golfito Nat'l Wildlife Refuge
Neily
Rincón
Golfito
Los Patos
Paso Canoas
San Pedrillo
Zancudo
Sirena
Puerto Jiménez
Golfo Dulce
Corcovado National Park
Carate
Pavones
50 KM
30 MILES
Cabo Matapalo
Pacific Ocean
Punta Burica
Isla Burica
© 2005 HUNTER PUBLISHING, INC

sand beaches, dark green hillsides of coffee, plus a dolphin and whale playground – with tons of budget hotels as well as remote luxury eco-lodges to enjoy.

Whale watching season is best from August through November, when whales pass in migration, breed and frolic in Golfo Dulce. Many tours and hotels offer whale watch cruises. Off the coast from Drake Bay is the Isla del Caño Biological Reserve, a popular, year-round day-trip to explore and hike, visit the cemetery of a prehistoric peoples, or snorkel and dive in the clear waters. Difficult to reach and far from San José, the south of Costa Rica is a very special, out-of-the-ordinary vacation. You can go with a wide range of package stays that include transportation from San José, or just wander around on your own.

Osa Peninsula

The herd may graze where it pleases or stampede where it pleases,
but he who lives the adventurous life
will remain unafraid when he finds himself alone.
~ Raymond Fosdick, 1883-1972,
author and former undersecretary-general of the League of Nations

Once isolated from the mainstream of tourism – and Costa Rica in general – the beautiful Osa Peninsula is now drawing eco- and adventure tourists in increasing numbers. Stunning Drake Bay and its northern section of pristine Corcovado National Park can be reached by charter plane from Pavas, San José, or by an 1½-hour boat trip from the Sierpe River. The small airport at Puerto Jiménez, the peninsula's most important town, has made it a lot easier to access the beach towns and southern Corcovado Park. Bordered to the west by the warm Pacific Ocean and the east by a marine paradise, the Golfo Dulce (Sweet Gulf) is where playful dolphins follow your launch and huge humpback whales come to breed. Osa is a memorable eco-adventure, just waiting for you. However, it is not for everyone – the environment can be physically demanding and requires some strenuous exertion.

Cross the Gulf by launch from Golfito or fly into Puerto Jiménez, where there are many good choices for lodging in the southern half. To get to Drake Bay, take a boat from Sierpe, a lazy little town along the river of the same name that marks the beginning of 14,555 hectares/35,950 acres of mangrove forest. If you stay overnight in Sierpe, 30 minutes off the Inter-American Highway, you may find it has a lot more to offer than simply an overnight stop. The **Valle de Diquis** area around Sierpe contains many mysterious stone spheres (*esferas de piedra*) made by unknown pre-Columbian peoples. You can easily visit **Isla Violin**, an important archeological site at the mouth of the river, just north of Drake Bay. Years ago, large nuggets found on this island started a "Gold Rush" that is still ongoing on the Osa Peninsula. The little town is a convenient mainland base for those who don't want the expense of Drake's more remote resorts but still want to see the area's sights.

DID YOU KNOW? Osa hosts at least 50% of the species found in Costa Rica and 2.5% of all species worldwide. Help save them at www.osacampaign.org.

■ Places to Stay, Guides Available

Two km/1.25 miles before the Sierpe is the **Eco-Manglares Lodge** *(☎ 506/786-7414, www.ecotourism.co.cr/docs/ecomanglaresrestaurant, $$)* a bungalow hotel with rustic and appealing wood rooms. The lodge offers day tours of Drake, Corcovado Park and Caño Island, plus fishing trips and mangrove estuary boat rides. North Americans run **Estero Azul Lodge** *(☎ 506/786-7422, www.samplecostarica.com, restaurant, $$$)*, on the banks of the river, just past Eco-Manglares. It has individual bungalows on beautiful grounds. Tours are offered by experienced guides.

In Sierpe town, **Hotel Pargo Rojo** *(☎ 506/786-6092, $-$$)* is surprisingly modern and a best bet if you're here on the cheap or just for a night waiting to go to Drake Bay. There's an excellent Tico restaurant. Tours available in their own boats (they'll watch your car for the day). Across the bridge is the intimate **Veragua River House** *(☎ 506/786-7460, restaurant, breakfast, $$)*, recently remodeled by its Italian owners, Ilana and Benedetto. Delightful Ilang Ilang flowers fill spacious grounds. Bungalows are simple and charming. The restaurant serves Italian food.

Accessible only by boat is Mike Stiles' **Río Sierpe Lodge** *(☎ 506/225-8553, www.riosierpelodge.com, restaurant, $$)* catering to fishermen, divers and eco-adventurers. It's near Isla Violin at the mouth of the Sierpe River, just north of Drake. Day tours or package rates for multiple nights, including transportation from San José, are a good buy.

■ Drake Bay

There is no frigate like a book,
To take us lands away.
~ *XCIX,* Emily Dickinson, 1830-1886

Drake Bay is one of the most stunningly beautiful areas of Osa and a gateway way into Corcovado park, also a base to explore Caño Island. Costa Ricans pronounce it "DRA-kay," but it's named after the British pirate, Sir Francis Drake, who visited in 1579 and allegedly buried treasure somewhere on its beaches. Don't bother digging for it; the richness of the location is more than enough.

To get to the northern part of the peninsula, the most convenient way is by a daily, 45-minute charter flight from Pavas airport, near San José, to the tiny Drake Bay airport (contact Drake Bay Wilderness Resort, below). But many guests come by boat from Sierpe near Palmar Norte, through the trackless mangrove swamps along the river. This area contains a wonderful wealth of birds and swampland wildlife. Most of Drake's accommodations are all-inclusive lodges, spread out on several kilometers of beach so it would be a mistake to go without reservations, unless you're traveling lightly to the little village of Drake. Since you're reserving, the lodges will arrange your transportation, which for some lodges means wading ashore or fording a stream. Because the Osa Peninsula and inland coast get a tremendous amount of precipitation, some of more remote hotels close during the rainy season when there are insufficient tourists to make them profitable.

YO HO HO AND A BOTTLE OF RUM

The English pirate, Sir Francis Drake, plagued Spanish shipping during his time in the Pacific (1577-1580) until given amnesty by the Crown. In 1588, his game of bowls on the Hoe in Plymouth was interrupted by the news that Spain had sent an "invincible" Armada to avenge the execution of Marie Stuart and restore Catholic rule. The 48-year-old captain finished the game and then went out and defeated their superior forces in a great sea battle. He died in 1596.

Places to Stay, Guide Available

In the village of Drake itself (it's official name is Agujitas), **Jinetes de Osa** *(US ☎ 800/317-0333, CR ☎ 506/236-5637, www.costaricadiving.com, closed September and October, full board)* is a cozy, nine-room lodge that faces a black beach. It specializes in PADI dive packages, regular "adventure" packages, or just spacious rooms, with meals, by the night. Dive packages for five nights are about US $750, adventure packages $650; rooms run US $80 per night, per person, with meals. Tours available. There are also Tico-style accommodations around. Our favorite was **Cabinas Mirador Lodge** *(☎ 506/831-1488, www.mirador.co.*

Hotel Prices

Prices are per night for two people

[No $]	less than US $20
$	US $20-$40
$$	US $41-$80
$$$	US $81-$125

cr, all-inclusive). Wade the river and scramble up the steep hill to their rustic cabins with dramatic bay views. A room and three meals a day (the owners are vegetarian, as is most of the food) is only US $40 per person. They offer tours and hiking, plus covered camping spots available on the hill above (US $15 per person, with meals).

The **Aguila de Osa Inn** *(US ☎ 866/924-8452, www.aguiladeosa.com, closed October, full board)* sits on a bluff overlooking the bay and the Agujitas River. It's the most luxurious place on Osa, with rooms high up the hill. The seafood and pasta restaurant is noteworthy, and the cabin rooms all feature hardwood floors, ceiling fans and lovely tiled bathrooms. All major tours are offered. Rooms start at US $175 per person, with meals and round-trip air and transfers from San José. You can opt for one of their various package deals catering to horseback riders, divers, sport fishermen, nature lovers and more.

On its own little peninsula at the head of the Agujitas River, **Drake Bay Wilderness Resort** *(☎ 506/770-8012, www.drakebay.com, pool, free laundry, full board)* offers 20 simple oceanside bungalows, or tenting platforms with beds and electricity. Many customers take advantage of the daily, 45-minute charter flight from San José. There are a ton of things to do here – snorkel, swim at the beach (a 15-minute walk away), languish in a natural rock pool, kayak or canoe the river, visit their iguana farm and butterfly garden, dine at the renowned restaurant, take a tour or simply hang out at the bar. A four-day package with two tours costs US $650 per person, including transportation. Rooms with meals are US $80 per person.

Canadian-owned **Cocalito Lodge** *(Canada ☎ 519/782-3978, CR 506/284-6369, full board)* claims it's "living in harmony with nature." Well, there's plenty of nature around. The main lodge and rustic cabins face the best lazy surf swimming beach on Drake Bay, and they have a private 45-acre reserve with its own small river. Cost is US $465 for three nights and two tours. Camping sites are offered.

Delfin Amor Eco Lodge *(☎ cell 506/394-2632, www.delfinamor.com, full board)* is an extension of Divine Dolphin, a group that researches and protects dolphins in the Caribbean and the Pacific. The emphasis is on the interactive "connection that humans have with dolphins and whales. " The dolphins seem to know the groups' boats and come right up to them. Their quest to create a marine sanctuary was the subject of a Discovery Channel program. The accommodations are tents and the food is good and plentiful. An eight-night suggested itinerary is US $1,250, including a night in San José, tours and transportation. Simple room and board is $85 per person, per night.

The rustic accommodations of **Proyecto Campanario Biological Reserve and Field Station** *(☎ 506/158-5778, www.campanario.org, all meals)* are within walking distance of the national park and a stone's throw from a lovely beach. They attract tourists who have a genuine interest in learning about the rainforest and the diverse ecology of Corcovado. It's a no-frills research field station with four private tent cabins or lodge with shared bath, dorm-style living. Director and educator Nancy Aitken offers a variety of enjoyable ecological field trips such as kayak trips in the mangroves, horseback riding, hiking, tide pool studies, snorkeling, scuba diving and Caño Island day tours. She also offers special Conservation Camps and Tropical Ecology courses. An unforgettable experience for just US $100 per day. Say hello to Klaus.

La Paloma Lodge *(☎ 506/239-0954, www.lapalomalodge.com, restaurant, pool, $$$$)* stands guard over Drake Bay from its crow's nest location on the mountaintop. Gorgeous hardwood guest cabins face the cool and soothing westerly breeze and sunsets over Caño Island can be breathtaking. The best digs are the two-story cabins, but everything here is classy. Tours and much more are offered.

Casa Corcovado *(☎ 506/256-3181, www.casacorcovado.com, pool, closed September through November 15)* is the closest resort to the park itself. Caribbean designed, well-appointed private bungalows are scattered in a wooded area of their private reserve, defining the best of "soft" adventure travel. This place was built by a naturalist for serious nature lovers. They offer tours, plus hiking to the park and waterfall, fine dining and a spring-fed swimming pool. It is billed as a "Lost World" resort, but you might like to find it. Fishing and diving packages. A two-night package with air from San José and boat transfer is around US $700 per person in high season.

■ Isla Caño

Oh it's a snug little island!
A right little, tight little island.
~ *Snug Island*, Thomas John Dibdin, 1771-1841

Tiny Caño Island, whose entire 196 hectares/480 acres are a designated biological reserve, lies only 20 km (12 miles) off Drake Bay. It's covered by virgin and secondary tropical rainforests and has several hiking trails. It was once used by pre-historic Diquis people as a graveyard and repository for those huge stone balls. Although the graves have long been looted, the spherical stones remain and you can see them up close on your hike. Isla Caño also served as another pirate hide out and pigs left to forage by the buccaneers were finally exterminated in the 1970s.

Today, the island is a natural playground, with deliciously clear waters ideal for snorkeling and diving. All the lodges mentioned above run day tours to the island and, if you have the opportunity to go, it is a delightful day.

■ Corcovado National Park

When one tugs at a single thing in nature,
he finds it attached to the rest of the world.
~ Gerald Durrell, 1925-1955

Olof Wessberg, the impetus behind the Cabo Blanco Absolute Reserve and an early backer of the national park system, went to the Osa Peninsula in 1975 to investigate the potential for an immense new park, "Corcovado." He was murdered under mysterious circumstances while on a walk with his host's son. Local business interests, gold miners, loggers and squatters didn't want to see more land taken off the market. Osa was a poor territory and many locals believed the government cared more for its wildlife than its people. Shortsighted as that was, it took some time before the local *campesinos* saw any financial benefit to protecting the rainforest. Wessberg's senseless murder galvanized proponents to create the Corcovado National Park and it was declared as one shortly after.

An old campesino *rests on his walking stick.*

Without regard, gold miners used destructive water-wash methods to pan for gold, and illegal logging plagued the park enough to force it to close so police could evict the trespassers in 1994 and 1995. As you fly over the peninsula, look down and you'll see clouds of brown silt washed downstream from deforested areas muddying the pristine gulf.

On the bright side, eco-tourism is bringing more money into local communities and people are realizing the benefits of the park land in many ways. Former gold diggers, called *oreros*, know the forest very well and are now working as guides. See below to stay in a former gold miners' camp in the heart of the peninsula at a rural cooperative.

Corcovado is a very important park; The diversity of wildlife there is staggering. It has been called "the most biologically intense place on earth" by the National Geographic Society and boasts the densest population of jaguars, scarlet macaws and tapirs in Meso-America. There are

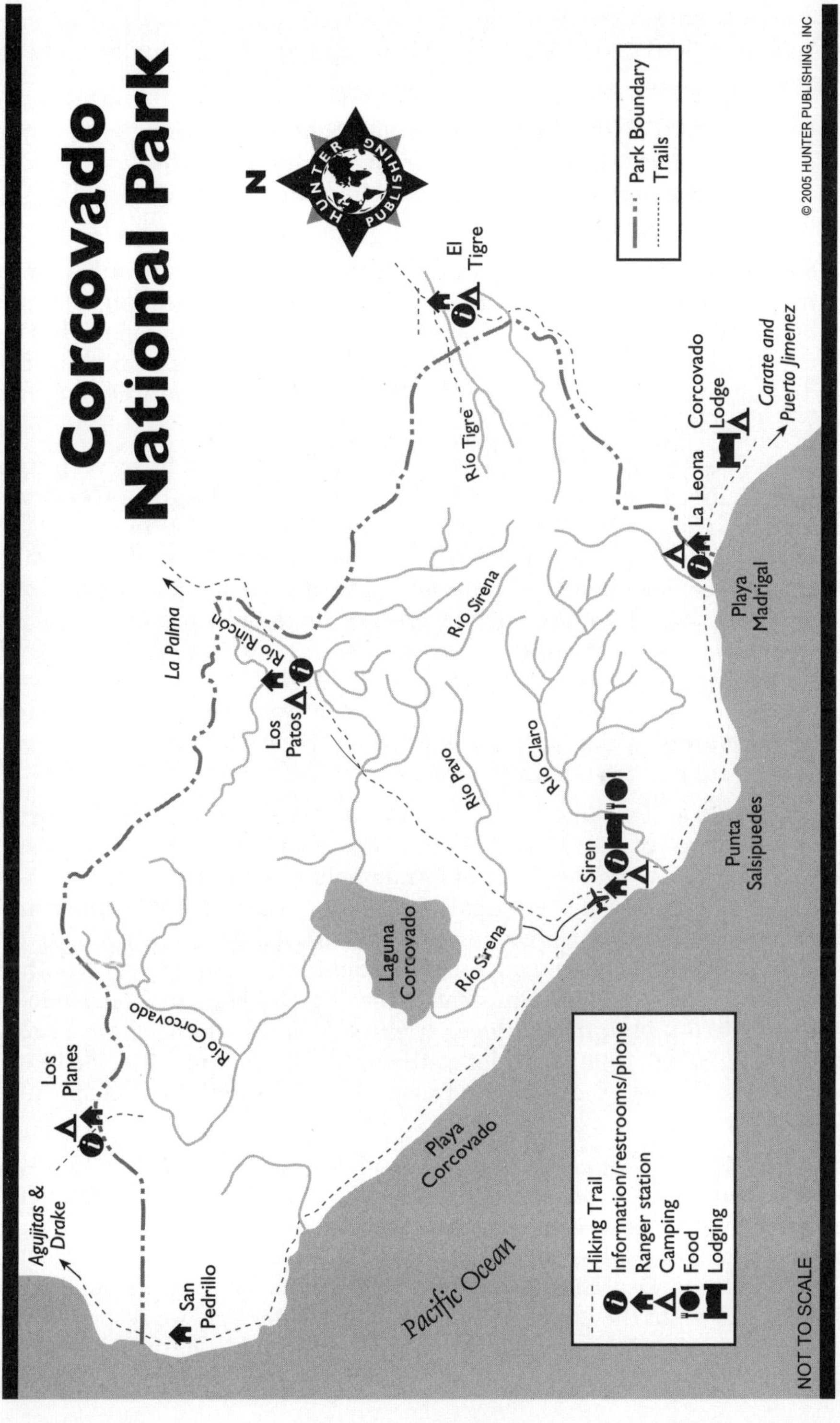
Corcovado National Park
N
HUNTER PUBLISHING
Park Boundary
Trails
© 2005 HUNTER PUBLISHING, INC
El Tigre
Río Tigre
Corcovado Lodge
Carate and Puerto Jimenez
La Leona
Playa Madrigal
Río Sirena
La Palma
Río Rincón
Los Patos
Río Pavo
Río Claro
Punta Salsipuedes
Siren
Laguna Corcovado
Río Sirena
Río Corcovado
Los Planes
Agujitas & Drake
San Pedrillo
Playa Corcovado
Pacific Ocean
Hiking Trail
Information/restrooms/phone
Ranger station
Camping
Food
Lodging
NOT TO SCALE

many trails within the park, but the most efficient way to get from one place to another is to walk the long beach. Entry is traditionally made at one of three points: La Leona, Los Patos or San Pedrillo.

> **AUTHOR TIP:** We recommend going into Corcovado with a guide for day or overnight trips, although lots of people go it alone and stay in the rustic shelters provided.

The Beachside La Leona research station, two km/1.25 miles from Carate, has the new Río Madrigal trail that starts at the station, ascends into the rainforest, and then follows the river back down again to the beach. To reach La Leona entrance, take the daily *colectivo* taxi that departs Puerto Jiménez in front of the Mini-Tigre every morning at 6 am, except Sunday.

Between Corcovado Park and the Golfo Dulce Forest Reserve is the Guaymí indigenous reserve, officially called the **Alto Laguna Reserve**. Located in the foothills of the highest point in Osa (750 meters/2,467 feet), its upper reaches are within a rich cloud forest. The Ngöbe (Guaymí) peoples have lived here for centuries. In an effort to improve their lives, the 22 families of the reserve have built a palenque-style **Ecoturistic Lodge Ngöbe**, where tourists are invited to observe and learn traditional handicrafts and lifestyle. A visit here offers a view of Costa Rica only a handful of tourists have ever seen. Book through **Galería Namu** in San José *(☎ 506/256-3412, www.galerianamu.com,* *see page 140)* and tell Aisling we recommended it.

Two km/1.25 miles outside the Los Patos park entrance is **Albergue Cerro de Oro** *(☎ 506/775-0033, www.turismoruralcr.com, $$)*, a rustic lodge run by a local community of former gold miners. The eight families involved will show you their medicinal plant garden and offer hikes and birdwatching tours of Corcovado. Also boat trips on the Rincón River, where you'll see the rusting abandoned gold-mining equipment. Another interesting rural eco-tourism destination on the edge of Corcovado is **Tesoro Verde** (*Agujitas village, 4 km/2.5 miles south of the beach, ☎ 506/770-8209*), a small eco-lodge that owns 36 acres of protected forest with trails. Visits to these kinds of coops directly benefit local people and the environment.

Bosque del Río Tigre Sanctuary and Lodge *(US ☎ 888/875-5045, www.osaadventures.com, tour and food included, $$-$$$)* is not a community project, but the work of two cheerful naturalists, Liz and Abraham Gallo. The Gallos built this attractive rustic lodge in the cooler foothills above Dos Brazos del Río Tigre, a half-hour outside Puerto Jiménez, close to the entrance to Corcovado. Their private reserve overlooks the river and the atmosphere is very comfortable, natural and tranquil.

■ Puerto Jiménez

Everybody talks about the weather, but nobody does anything about it.
~ attributed to Mark Twain, 1835-1910

Tiny, funky Puerto Jiménez, (PWERTO he-MEN-ez) is located on the southeast coast opposite Golfito and serves as the main town and gateway to the park and coastal villages of Cabo Matapalo and Carate. Most visitors come by way of the 50-minute scheduled flights from San José. The town has its own character and sense of humor (after you arrive in your puddle-jumper plane, notice the graveyard at the edge of the runway), and offers a variety of lodgings for explorations on the south coast. Beach lovers push on from here to Cabo Matapalo or Playa Platanares, and beach-park people wind up in Carate (but not Kung Fu – ha!). If you get lucky panning for gold at any of the places you visit, please remember to send us our cut.

> **PHOTO OP:** Search for the tiny bridge in town, with relatively low trees on either side of the road. We found it by chance. The bridge is a popular pathway for monkeys as they cross through the village. It's a great place to see them up close if you're there at the right time. Take your camera.

Tours & Adventures

For active tours and expeditions, look to **Escondido Trex** *(☎ 506/735-5210, www.escondidotrex.com)*, who been in operation since 1992. They offer multi-day adventure packages, kayak trips, mountain biking, hikes with park guide, plus half- and full-day tours. They claim their "back door entrance" into Corcovado encounters more wildlife. For marine tours you might as well **Let George Do It** *(☎ 506/735-5313)*. Jorge Espinoza has several boats and lots of experience. He also has an air hose so you can swim with the dolphins without loading up full scuba gear. Fancy a five-hour horseback and kayak day tour of the **Preciosa-Platanares Wildlife Refuge**, three km/1.9 miles outside Puerto Jiménez, with your own guide? Phone Lidiette or Magda *(☎ 506/735-5007)* or stop by the **Osa Natural** office. Osa Natural is a community-based agency that handles travel to Drake and can help with your arrangements almost anywhere.

Playa Platanares has good waves for boogie boards and body surfing (we're addicted) and May through November the local residents and businesses have a nest protection effort for four species of turtles. They released 15,000 baby turtles last year.

Places to Stay & Eat

Hotel Prices	
Prices are per night for two people	
[No $]	less than US $20
$	US $20-$40
$$	US $41-$80
$$$	US $81-$125
$$$$	US $126-$200
$$$$$	over US $200

If you get the munchies in town, we suggest you try the authentic Mexican food at **Juanita's Bar & Grill**, adjacent to CafeNet in the newly remodeled Epicenter. Born and raised in Monterrey, Mexico, Sanjuana cooks up fajitas, chimichangas, nachos and tequila chicken; and then serves it with a choice of seven Mexican beers or Osa's only frozen fruit margaritas.

Near the airport and the beach is a charming little motel on four acres of woods and gardens. **La Choza de Manglar** (*fax 506/735-5002, www.manglares.com, restaurant, Internet, $$*). It was completely remodeled in 2004 as a pair of artists created a natural art gallery in an all-around good place to stay.

Near Puerto Jiménez, the seaside cabinas at **Agua Luna** *(☎ 506/735-5431, air, cable, restaurant, $$-$$$)* are clean and modern. It has a rural setting, 30 minutes by foot, or 15 by car, from downtown. Their restaurant serves the best food around. **Parrot Bay Village** *(☎ 506/735-5180, www.parrotbayvillage.com, $$$+)* are located on a tranquil beach 10 minutes outside Jiménez. These beautiful accommodations with kitchens are set on expansive garden grounds, and the American owners also run a fresh seafood restaurant on the premises. Sportfishing is a big draw.

In Puerto Jiménez there are a fair number of low-cost housing options; stop in at Osa Natural and ask for a current recommendation. Ask about **Cabinas Marcelina**, **Hotel Bosquemar**, **Cabinas Iguana Iguana** *(pool, restaurant)*, or **Cabinas Puerto Jiménez** (on the beach). Backpackers, try **Pension Quintero** or **Cabinas Thompson** (near the bus station).

A star on the **Playa Platanares** shore (which is an excellent swimming area and one of the best beaches on Osa), is **Iguana Lodge** *(☎ 506/735-5205, www.iguanalodge.com, $$$)*. It has a big main lodge built in an indigenous style, and separate luxurious, African-style bungalows. Two stories, with a big covered porch, the bungalow rooms are romantic and delightful. Iguana's restaurant kitchen is well known for delicious dishes.

A short walk away is the same North American owners' lower-cost alternative, **The Pearl of the Osa** *(☎ 506/735-5205, www.thepearlofthe osa.com, $$$)*, a lime-green, Caribbean-looking hotel with large rooms that face either the surf or the jungle. Both are good deals and promise wonderful vacations on Osa's best beach.

If you can handle the road to get there, head for **Cabo Matapalo**, where the forest comes right to the deserted beach. It's been newly discovered by surfers who come to ride its right break. **Everyday Adventures** (☎ *506/353-8619, www.psychotours.com*) is an active tour agency in Matapalo run by professional adventurers. They also rent a rustic beach house. Inland, you'll find Brian Daily's fabulous little wilderness beach hotel, **Encanta La Vida** *(US ☎ 805/735-5678, CR ☎ 506/735-5678, www.encantalavida.com, $$$-$$$$)*, with choice of three different lodgings. Tours and meals are included, but there's a two-night minimum.

With a dramatic view of the sea and a vision to protect the rainforest, **Lapa Rios** *(☎ 506/735-5130, www.laparios.com, full board)* is the most luxurious place on Osa. It was created to pamper guests in paradise – a paradise preserved through the income from the luxury hotel. Lapa Rios proves projects that appreciate natural resources are viable alternatives to those that deplete them. Miles of private trails in the 405-hectare/ 1,000-acre private primary rainforest reserve, with cool waterfalls and river pools, beckon adventuresome travelers for the trip of a lifetime. Romantic, spacious, luxury polished-wood bungalows have fabulous views of the ocean, and the restaurant serves gourmet meals. About US $225 per person, per night, including meals. Honeymoon to die for.

A fascinating alternative is **El Remanso** (☎ *506/735-5569, www.elremanso.com, $$$*), set on 140 acres of mountaintop primary rainforest. Conservation activists, Joel Stewart and Belen Momene, took eight years to build their hotel and they've done an admirable job. They offer cabinas or a gorgeous house, Casa Osa. Each private cabin is surrounded by a running water moat, which keeps insects at bay. Then there's new "simple life lodge" on Playa Carbonera, near Cabo Matapalo. **Ojo del Mar** *(ojodelmar@yahoo.de, breakfast, $-$$)* is rustic, with airy, comfy cabins.

High atop Cabo Matapalo, overlooking the Gulfo Dulce and Pacific Ocean, where lush jungle spills down to the sea, sits **Bosque del Cabo Rainforest Lodge** (☎ *506/735-5206, www.bosquedelcabo.com, pool, restaurant, $$$*). Bosque features rental houses and private bungalows, all with porches that offer endlessly stunning views of the ocean. Solar-powered electric highlights the lodge's ecological efforts, which include more than 500 acres of private forest, full of birds and wildlife. This was the first lodge to open in this part of Osa and remains a privileged part of paradise, about halfway between Puerto Jimenéz and the park.

Carate is the last beach area before Corcovado, and the most popular entrance into the park. Its natural beauty is amazing. Jungle cascades down to the deserted black beach and scarlet macaws swoop from tree to tree. On the way here you'll pass **Terrapin Lodge** *(US & Canada ☎ 800/600-0751, CR 506/735-5211, www.terrapinlodge.com, $$$)*, a

Tico-owned lodge built in early 2000. In a rustic style, the six cabins are made with warm woods and bamboo and have metal roofs. The price includes three meals. A taxi from Puerto Jimenéz costs about US $65 so look to share. The *colectivo* shuttle bus costs $10. The local guide for tours has 25 years experience.

Nearer the entrance to the park, the **Lookout Inn** *(☎ 815/955-8783, CR 506/735-5431, www.lookout-inn.com, pool, all meals, $$$$)* is nestled on the side of a bluff overlooking the ocean and black sand beach. It offers the quaintness of a country B&B in the midst of a jungle environment. Very ecologically sensitive, this inn delivers both comfort and pleasurable ambiance. Try the impressive five-night personalized nature package for US $850 per person. When we were there the guide was a misplaced Irishman with the unlikely, but deserved, nickname of Crocodile Dundee. Beautiful **Luna Lodge** *(US ☎ 888/409-8448, CR 506/380-5036, www.lunalodge.com, meals included, $$$$)* sits on a mesa over the Carate River Valley in 24 hectares/60 acres of forest. Five secluded upscale bungalows with views are available, as are tent platforms ($$). Just outside the park entrance is the **Corcovado Lodge Tent Camp** *(☎ 506/257-0766, www.costaricaexpeditions.com, meals included, $$)*, a Costa Rica Expeditions beachfront hotel that features luxurious tent cabins up on platforms. This experienced tour company delivers a truly unique and rewarding experience. Recommended.

■ Golfito

Sittin' on the dock of the bay, watching the tide roll away.
Sittin' on the dock of the bay, wasting time.
~ *Sitting on the Dock of the Bay* (1967), Otis Redding, 1941-1967

In 1938, the United Fruit Company carved the town of Golfito (goal-FEET-o) out of the rainforest on the edge of the Golfo Dulce to serve as a harbor from which to ship bananas from their extensive Zona Sur plantations. Thousands came and settled the area for the work provided in the fields but, after a series of worker strikes among other problems, Chiquita Brands decided it wasn't worth it and pulled up stakes in 1984, leaving the local economy devastated. The Costa Rican government came up with an economic recovery scheme in 1989 that opened a "duty-free" zone at the north end of town in an air-conditioned mall called the **Depósito Libre**. Now, many *Josefinos* come down to spend the $500 per month they're allowed without paying duty. Appliances and luxury imported goods fill the majority of wish list items, but the opening of buyers clubs, such as Sams or Costco, in the Central Valley is eating into the duty-free trade. Still, in the holiday season and on weekends, many Ticos come down just to shop. Fifty minutes in the air, or five hours by TRACOPA bus (three runs daily from San José). Otherwise, Golfito has little to offer except the stunning jungle backdrop of the surrounding Golfito National Wildlife Reserve primary rainforest, plus easy access to the gulf. The waterfront is littered with rusting hulls and houses of the poor lean over the water on stilts like they do in Southeast Asia. Ecotourism money trickles down slowly here.

That said, we kind of liked Golfito's easy lifestyle and see hope for the future when more sport fishermen and eco-tourists discover the gulf's magnificent offerings.

Tours & Adventures

Buildings left by the banana company were given to the University of Costa Rica to establish a Tropical Studies Program, an offshoot of which is **University Tours** *(☎ 506/775-1249).* Young university faculty, biologists and naturalists run specialized tours into the surrounding Golfito Rainforest. They offer a Tropical Nature Tour, Town and Jungle Tour, "Banana Plantation Tour and a Mangrove Aquatic tour – all under two hours – plus a Jungle Hike and Butterfly Tour (three hours). A biologist will also escort a Wilson Botanical Garden Tour (4½ hours).

Besides natural wonders, this part of Costa Rica is where those mysterious round stone balls, made by the pre-Columbian Diquís people between AD 800 and 1500, are found. The smallest is about the size of a golf ball and the largest, called El Silencio, is over two meters/10 feet in diameter and weighs about 15 tons. These amazing spheres are almost perfectly round, quite a feat for "primitive" people. The community of **Palmar Norte**, about 31 miles north of Golfito, is hoping to open a Stonehenge-type park to display them. Let us know if it comes about.

Places to Stay & Eat

Try **Hotel Las Gaviotas** *(☎ 506/775-0062, lasgaviotas@hotmail.com, restaurant, cable, pool, $-$$)*, which has a waterfront location, a good restaurant, tropical garden grounds, a long dock and air-conditioned rooms. This is a good base for fishing and tours. Across the street is **Hotel El Gran Ceibo** *(☎ 506/775-0403, www.ecotourism.co.cr/elgranceibo, restaurant, cable, pool, air, $).* It's clean and neat and also offers tours. We sat for hours at the bars between these two hotels when the heavens opened up one evening, lending new meaning to whiskey and water.

Mar & Luna is an appealing little restaurant that sits on the waterfront near where fishing boats dock. The Hernández family serve up the freshest fish around at very good prices. They also have some rustic cabinas.

Golfo Dulce

That it will never come again
Is what makes life so sweet.
~ Emily Dickinson, 1830-1886

The area north of Golfito has clean beaches and several remote jungle lodges on the edge of the **Las Esquinas/Piedras Blancas National Park**. The closest beach to Golfito is Playa Cacao, out at the edge of the protected inner harbor. It's a US $5 boat ride; ask at the dock. If you would like to stay awhile, you could do boat trip to Zancudo Beach or Casa Orquídeas gardens from here.

Casa Orquídeas is a strikingly landscaped private tropical garden on the gulf shore. All the surrounding lodges offer trips here for guided tours of the lovely botanical gardens. Tropical Costa Rica's tropical fruit is both delicious and different from what you may get on your supermarket shelves, and the ornamental plants, thought of as house plants in the US, grow huge here. This is the beginning of Parque Las Esquinas/Piedras Blancas, Costa Rica's newest national park. The Austrian Government bought a large portion of the rainforest around here and donated it to the Costa Rican Park service. Another gift to visitors is the **Esquinas Rainforest Lodge** *(☎ 506/293-0780, www.esquinaslodge.com, pool, full board)*, a model in sustainable development where students from the University of Vienna do biological research. It features enchanting rooms with shady verandas, plus a clear-stream swimming pool nearby. Guided adventure tours available. About US $ 85 per person, meals included. Profits benefit local communities. In the US, ☎ 888/504-2190 for information (no reservations). From the Inter-American Highway, turn at Villa Briceño (Km 37) and head up the dirt road for four km/2.5 miles. Free airport pickup in Golfito.

Golfo Dulce Lodge *(☎ 506/232-0400, fax 232-0363, www.golfodulcelodge.com, pool, restaurant, full board)* is a complex of intimate accommodations informally grouped together as a tiny village on the edge of the rainforest, about 250 meters/822 feet from the rocky gulf beach of Playa Josecito. It's a 30-minute boat ride from either Golfito or Puerto Jiménez and is surrounded by the Piedras Blancas National Park. The property consists of more than 300 hectares/741 acres of virgin rainforest where the German biological program Profelis, a wildlife rehabilitation center, re-introduced margays (*Leopardus weidii*) and ocelots (*Leopardus pardalis*) into their natural habitat. They also work with Zoo Ave on a long-term project to develop a local self-sustaining scarlet macaw population. These beautiful birds were once plentiful before being decimated by pesticides and poaching. The lodge has five detached bungalows with covered porches in the deluxe category; plus, there are three standard furnished rooms. The on-site restaurant at this Swiss-owned establishment serves delicious European fare with a tropical twist. Lodging only runs between US $95 and US $105 per person, based on double occupancy. Packages with rainforest and botanical tours are US $325-355, per person. Three-night minimum.

South of Golfito, the Golfo Dulce opens to a Pacific coastline shared with Panama, where some very remote beach areas attract surfers from all over the world. If you're not a surfer, you can still enjoy the incredible beauty and laid-back sensibilities of these unspoiled beaches. If you're an eco-tourist, you can enjoy both worlds by heading for the award-winning eco-lodge, Tiskita.

■ Playa Zancudo

This black sand, "blue-flag" beach is long, flat and good for swimming. It lies 19 km/11.8 miles by *colectivo* boat, or 35 km/21 miles by bus on poor roads, from Golfito. Although still a secret to most tourists, it's full of young people from Europe and North America. At Zancudo, beach location and reasonable prices aren't dichotomies.

Places to Stay

Latitude 8 Lodge *(☎ 506/776-0168, www.latitude8lodge.com, 2 cabins, charter trips offered, $$)* has two newly remodeled cabins facing the Pacific Ocean and right on the beach. Fishing is what attracts folk here. Penn International fishing equipment or similar top-quality tackle is provided for all offshore charter fishing trips. Surfboards are available for rent and boogie boards are available on request.

The most expensive place to stay is the interesting **Roy's Zancudo Lodge** *(US ☎ 877/529-6980, CR ☎ 506/776-0008, www.royszancudolodge.com, pool, air, restaurant, all meals)*, which holds 25 world record catches. If you're a sport fisherman, or want to be one, ask them about their package tours. Multi-day packages run US $2,150 for three full days fishing and transfer from Golfito. Gringo-owned, **Oasis Beach Club** *(☎ 506/776-0087, www.oasisonthebeach.com, restaurant, $$)* offers clean and neat cabins or rooms on the beach. Completely remodeled and upgraded in 2004. Popular bar and restaurant too. The owners were pleased the Red Sox finally won!

Cabinas Los Cocos *(☎ 506/776-0012, www.loscocos.com, $$)* has four cool cabins with kitchenettes, two of which were once banana worker housing. The owners run **Zancudo Boat Tours**, so staying here is perfect for stay-and-play vacationers. Next door is **Cabinas Sol y Mar** *(☎ 506/776-0014, www.zancudo.com, $)*, with just four nice rooms that fill up early; guests stay longer here too. Breezy *rancho* restaurant draws lots of people.

■ Pavones

Despite a good break near Zancudo, all the surfers boogie down to Pavones for its famous long break, especially in the green season when it is at its zenith. The beach is a two-hour drive from Golfito by bus, less by boat. It's a destination that focuses on surfers and where their heads are. If you're not a surfer, it may not please you very much, although **Punta Banco Reserve**, six km/3.7 miles down the road, is a natural wonderland.

The bed and breakfast where it is always Sunday is **Casa Siempre Domingo** *(☎ 506/775-0131, www.casa-domingo.com, including breakfast and dinner, $$)*, an airy ranch-style home with sweeping views. Constructed in 1995, accommodations include four large rooms each with private, modern baths and air conditioning. Intimate and secluded. We also fancied **Cabinas La Ponderosa** (*US ☎ 954/771-9166, www.cabinaslaponderosa.com, air, meal option, $$-$$$*) with casual, comfortable wooden cabins. This is a pretty good bargain as the optional three meals a day are cooked to order for just $20 more per person. A house is also available for rent. For those who are really into Tai Chi, Pavones is the home to **Chen Taiji International Tropical Vacations** (US ☎ *305/931-0918, www.chentaijiinternational.com*). They arrange study vacations.

TISKITA JUNGLE LODGE

Like a kite, Cut from a string,
Lightly the soul of my youth Has taken flight.
~ Ishikawa Takuboku, 1885-1912

If you're looking for a jungle lodge experience in southern Costa Rica, you might choose the remote biological reserve and experimental fruit farm just south of Pavones, **Tiskita Jungle Lodge** *(☎ 506/296-8125, fax 296-8133, www.tiskita-lodge.co.cr, pool)*. Back in 1993 Peter Aspinall, an agronomist, and his family purchased 51 hectares/125 acres to create an experimental fruit farm. Their open orchard has attracted a plethora of bird life. Once the Aspinalls settled here, their focus expanded to help the local town of Punta Banco, where, with the help of Rotary Clubs, they helped fund a community health clinic. Philanthropic guests have contributed to support the village school. And the breezy hillside location, overlooking the Pacific and surrounded by rainforest, led the lodge to expand its land purchased for conservation to 222 hectares/550 acres.

A private airstrip makes access easy – it's an hour's flight from San José – and the lodge maintains a high degree of ecological practices. Upgraded rooms make this a comfortable, not luxurious, jungle experience. Traditional Costa Rican hospitality mixed with a beautiful environment, ecological focus and the best of rainforest/beach combinations, make Tiskita a real pleasure. Bring your kids to an educational family eco-camp stay during the summer. Package prices – including meals, transportation from Jiménez or Golfito, and guided tours – run US $375 or less per person, per diem.

Panama or Bust

I have seen other countries, in the same manner,
give themselves to you when you are about to leave.
~ *Out of Africa*, Isak Dinesen (Karen Blixen), 1885-1962

The Inter-American Highway bobs and weaves its way down the mountains on its way to Panama. These sections of high valleys contain several interesting attractions, although few tourists get down this way. To visit Native American reserves is an opportunity we appreciate, plus there's rich coffee in the Italian influenced San Vito town. Not to mention the largest park in Costa Rica – La Amistad – a park so vast it overflows into Panama.

■ Parque Internacional La Amistad

Equal and exact justice to all men,
of whatever state or persuasion, religious or political;
peace, commerce, and honest friendship with all nations...
~ First Inaugural Address (1801), Thomas Jefferson, 1743-1826

La Amistad means "friendship," and it is one of the rare instances of international environmental cooperation in the world. Panama's Park Service protects and supervises 40,000 hectares/98,800 acres of Parque Internacional La Amistad's land. Not only does the park boast a tremendous biodiversity of plant and animal life in its 193,000 hectares/476,710 acres, it also hosts seven indigenous reservations within its expansive borders. In 1982, UNESCO granted it World Heritage Status.

If you're searching for the exotic wild animals such as tapirs, jaguars, ocelots, giant anteaters, quetzals or any of the many rare animals in Costa Rica, La Amistad is the place to go. Go, but go only with an experienced guide, as this is a trackless and unforgiving jungle. There is little or no infrastructure for tourism. If you'd like to hike here, start at the main entrance, Altamira, about 30 km/18.6 miles northeast of San Vito, where guides are available and hotels are at hand.

For a more detailed view of this World Heritage Site, check out www.wcmc.org.uk/protected_areas/data/wh/talamanc.html.

■ Boruca Country

This country was a lot better off when the Indians were running it.
~ *NY Times* magazine, Vine Victor Deloria, Jr.

The Boruca are one of Costa Rica's last indigenous peoples. Their small community, on a government reservation, is surprisingly accessible from the Inter-American Highway at the agricultural town of Buenos Aires. The area around Buenos Aires was largely deforested for pineapple plantations, and it's a major supplier of the country's tasty fruit. The indigenous Boruca village is 18 km/11 miles up in the mountains and can be reached by a daily bus from Buenos Aires.

The Boruca are known for their handicrafts. Women weave a special cotton cloth dyed with natural pigments, including one extracted from a rare, cliff-dwelling mollusk that excretes tiny amounts of a purple dye when in danger. Boruca men create intricate ceremonial masks from native wood and carve scenes into gourds. The village offers many great photo ops, and your purchases will support a culture struggling to survive. We were lucky to see the men perform the re-enactment of their own people's defeat at the hands of the Spanish conquistadors using a bullfight as an allegory. The Fiesta de los Diablitos features two young men in a burlap and wood bull costume (representing the Spanish) and several young men wearing burlap bags and masks (representing the Boruca warriors). The combatants have a life-and-death struggle to the sound of flutes and drums. This is a remarkable New Year's tradition.

Please buy some of their beautiful handicrafts – perhaps from the coop in the village of Rey Curré, along the highway, or one of the many hotels in the district that sell Boruca crafts. In addition, stop at the indigenous museum in Térraba, south of Buenos Aires.

Also near Buenos Aires is the **Dúrika Biological Reserve** *(☎ 506/730-0657, www.durika.org)*, a private community-led effort to achieve agro-eco-cultural sustainability. The guides here can take you where no one else goes – through their biological reserve or to an even more remote indigenous village. Two other guided hikes lead to the Cerro Dúrika, a 3,280-meter peak, or take you on a fantastic cross-country hike from there across the mountains to the Atlantic coast. Talk about cross-country. Make advance reservations.

■ San Vito

You people married to Italian men, you know what I mean.
~ Geraldine Ferraro, former US Vice-Presidential candidate

Refreshing cool mountain air and great views announce your arrival in San Vito, an area in the Coto Brus Valley where Italian immigrant farmers settled in the early 1950s. The region, 287 km/178 miles from San José, is very fertile and produces some of Costa Rica's finest coffee. Don't leave without drinking up and stocking up on the *granos de oro*. Alas, there are no fine Italian wines to be had. There is little Old World flavor in the buildings of San Vito, but plenty in the best Italian restaurant in town, **Pizzeria and Restaurante Lilliana**.

Most visitors come to San Vito for the **Wilson Botanical Garden** and **Las Cruces Biological Station** *(☎ 506/240-6696, or 773-4004, www.ots.ac.cr)*. Overlooking a mid-elevation forest, the Wilson Botanical Garden and its peaceful, secluded and magnificently landscaped setting remains a near-secret paradise. In 1961, Robert and Catherine Wilson, owners of a tropical plant nursery in Florida, settled here. With the help of Roberto Burle-Marx, a renowned Brazilian landscape designer, they created gentle paths through their 10 hectares/25 acres of colorful gardens. Visitors now stay in a modern lodge with spacious guest rooms and private baths. They eat healthful international cuisine with a special emphasis on fruits and veggies. Wilson is owned and operated by the Organization for Tropical Studies, who also run La Selva and Palo Verde Biological Stations.

Fifty km/31 miles farther on is **Ciudad Neily**, an unremarkable town with good bus connections. What is remarkable, however, is the drive down from San Vito to Neily. On clear days the views are breathtaking. Make this run during the day (mornings are better) and you'll be impressed. Guaranteed.

If you're just passing through this last major town before Panama, don't miss **Eurotica Hotel & Restaurant** *(☎ 506/783-3756, $)*, where Belgian-born Liliana and her Tico husband, Mauro, offer pleasant accommodations in their basic hotel on 10 hectares/24.7 acres of land. Stop in for lunch or dinner; their kitchen has an excellent reputation for Italian and Costa Rican food. Plus, they offer Belgian beer! Say *bonjour* for us.

The last stop on the border is **Canoas**, a somewhat seedy border town that attracts bargain shoppers looking for imported goods. Say goodnight, Costa Rica.

AUTHORS' FAREWELL

No matter where you go or what you do in Costa Rica, we hope you return with a suitcase full of fabulous memories. We've tried to organize and evaluate the many attractions in a way you'll find useful – and hope we've provided both seasoned and first-time adventurers with unique experiences in a magical destination you'll want to visit again and again. *Pura Vida.*

There's a whole world yet to be discovered,
not of mysterious unknown places
but of relationships among places.
A century ago, great explorers were busy
filling the blanks on the globe.
What we haven't filled in yet are the blanks
of knowledge about those places,
their peoples and environments, and how they fit together
– an even greater challenge.
~ Gilbert M. Grosvenor, National Geographic Society

Appendix

Useful Websites

US Booking Agents

Any good US travel agent can make advance arrangements with a number of reputable tour wholesalers. Try one of these companies.

Costa Rica Experts . . . . www.costaricaexperts.com
Holbrook Travel . . . . www.holbrooktravel.com
Nature Tours Inc . . . . www.naturetoursinc.com
Costa Rica Connection . . . . www.crconnect.com
Tropical Travel . . . . www.tropicaltravel.com
Rico Tours . . . . www.ricotours.com
Costa Rica Discover . . . . www.costaricadiscover.com

General Information

There are many sites on Costa Rica to help your planning and research. Here are a handful we recommend.

www.costarica.tourism.co.cr – Costa Rica Chamber of Tourism.

http://costaricabureau.com – Costa Rica Tourism & Travel Bureau. Visit this site for all kinds of details about Costa Rica's offerings.

http://costarica.com – A good general site. Information offered includes real estate, weather, retirement, shopping, culture and news.

www.costaricahomepages.com – Find out what's happening all over the country.

www.amcostarica.com is a great news resource. Updated daily.

www.cocori.com. Offers "All the facts and fun of Costa Rica."

www.visitcostarica.com – The Costa Rica Tourism Board (ICT) provides general background information. Reach them by e-mail at tourism@tourism.co.cr.

www.nacion.co.cr – *La Nacion*, the country's leading newspaper, runs this site.

www.ticotimes.net – Run by the ***Tico Times***, Costa Rica's English-language newspaper.

www.insidecostarica.com – A daily newsletter.

■ For Volunteers

For volunteer information, check out **www.globalvolunteers.org/cstrmain.htm** or **www.amerispan.com**, which has everything from ecology to Habitat for Humanity programs.

■ Lots of Links

Below are some interesting websites that offer hundreds of links to Costa Rica-related websites.

www.web-span.com/tropinet/. This site offers a plethora of links for flights, accommodations, tours, language schools and more.

www.gksoft.com/govt/en/cr.html. A comprehensive list of Costa Rican government agencies on the Web.

http://directory.centramerica.com/costarica_asp/main.asp. Directories for professionals and businesses in Costa Rica.

Rainy Day Reading

Rain is not unheard of in Costa Rica; you'll need a good book for rainy days. Some books from large presses are available outside the country, but books from smaller presses can often be bought only in Costa Rica. The books below are all English-language and worth looking for in San José bookstores.

■ Nature

Ecotourism and Sustainable Development: Who Owns Paradise?, Martha Honey, Island Press. A good overview of the impact, rewards and problems associated with eco-tourism.

A Guide to the Birds of Costa Rica, Gary Stiles and Alexander Skutch, Cornell University Press. A birder's bible. Download a birding checklist, courtesy of Rara Avis, at www.interlog.com/~rainfrst/birds.html.

Travel & Site Guide to Birds of Costa Rica, Aaron Sekerak, Lone Pine Press. Who, what and where.

The Resplendent Quetzel, Michael & Patricia Fogden, 1996. Beautiful color photos of this sought-after mystical bird.

Green Republic: A Conservation History of Costa Rica, Sterling Evans, University of Texas, 1999. A insightful history about the conservation ethic.

Life Above the Jungle Floor, Don Perry, 1991. Groundbreaking insight into the biodiversity of the rainforest canopy. Out of print in the States.

Windward Road: Adventures of a Naturalist on Remote Caribbean Shores, Archie Carr, University of Florida Press. In print since its publication in 1956, the book is a fascinating chronicle of Dr. Carr's Caribbean expeditions following endangered sea turtles to Tortuguero.

Costa Rica Native Ornamental Plants, Barry Hammel, INBio Pocket Guides, 1999. It grows on you.

Costa Rica Mammals, Eduardo Carillo, INBio Pocket Guides, 1999. In living color.

■ Living Large in Costa Rica

The New Golden Door to Retirement and Living in Costa Rica, Christopher Howard, Globe Pequot, 2000. Opens doors.

Official Guide to Living and Making Money in Costa Rica, Christine Pratt (*Tico Times* editor). She should know.

The Legal Guide to Costa Rica, Roger Petersen. Look for this in San José bookstores.

Choose Costa Rica for Retirement, John Howells, Costa Rica Books, 2000. In its fifth edition.

Ticos: Culture and Social Change in Costa Rica, Mavis, Richard & Karen Biesanz, Lynne Rienner Publishers, 1999. A complete social history of Costa Ricans and an unvarnished look at today's society.

You Can Drive to Costa Rica in 8 Days, Dawna Rae Wessler, Harmony Gardens Publishing, 2000. The intrepid author did this trip in her ancient VW bus. Amusing tips on how (or how not to) make the journey.

■ Fiction

The Lonely Men's Island, José León Sánchez, translated by Michael Jensen, 1997. A novel portraying life in Costa Rica's notorious San Lucas island prison.

Tata Mundo, Fabián Dobles, translated by Joan Henry, 1998. Short stories.

Costa Rica: A Traveler's Literary Companion, edited/translated by Barbara Ras, Consortium Books, 1994. Twenty-six stories by 20 of Costa Rica's better-known authors reflect the culture by location.

Assault on Paradise: A Novel, Tatiana Lobo, 1998. This historical novel, winner of the 1995 Sor Juana Inés de la Cruz Prize, is set in the early 18th century. It uses its swashbuckling hero for an eloquent and moving indictment of the conquistadors and their bloodthirsty legacy.

Years Like Brief Days, Fabián Dobles, Dufour Editions, 1996. Not-so-nostalgic journey of an old Costa Rican man whose return to his village evokes childhood memories.

■ Culinary Delights

Sabor!, Carlina Avila & Marilyn Root, 1997. A wonderful guide to tropical fruits and vegetables. Includes lots of regional recipes. Black-and-white illustrations.

My Kitchen, Nelly Urbina,1999. Simple recipes of typical meals.

Costa Rican Typical Foods, Carmen de Musmanni and Lupita Weiler, 1994. Sixty traditional meal recipes.

A Bite of Costa Rica, or *How We Costa Ricans Eat*, Oscar Chavarría, 1992. Traditional recipes.

Showtime

A handful of movies were made on location in Costa Rica. They include:

Corner of Paradise (1997), starring Penelope Cruz.

New Adventures of The Jungle Book (1998), starring Sean Price McConnell as Mowgli.

1492, Conquest of Paradise (1992), starring Gerard Depardieu with Sigourney Weaver portraying Queen Isabella of Spain.

Spy Kids 2 – The Island of Lost Dreams (2002), starring Antonio Banderas and Ricardo Montalban. Filmed on and around Lake Arenal.

Blue Butterfly (2002), starring William Hurt, is based on the life of Georges Broussard, founder of the Montreal Insectarium.

Jurassic Park. We're cheating a little on this one. The dinosaurs in the *Jurassic Park* movies were supposed to live on Costa Rica's Coco's Island. However, the movies were actually filmed in Hawaii.

Recipes to Try at Home

When you get home, you'll awake one morning with an overwhelming craving for Gallo Pinto, the same breakfast fare you once thought you were sick of in Costa Rica. Don't despair. Put on a pot of that good coffee you brought home and rummage in the cupboard for the bottle of Salsa Lizano. Then follow this recipe, and relive the time you spent in paradise.

Gallo Pinto (Red Rooster)

- 1-2 cups black beans
- 1½-2 cups white rice, uncooked
- 1 large bell pepper
- 1 small onion
- fresh cilantro, to taste
- olive oil
- Salsa Lizano, to season (now available at US supermarkets)
- Tabasco sauce or chopped jalapeño pepper (optional)

In separate pots, fully cook one to two cups of black beans and one and one half cups of white rice. Set aside, but do not drain beans. Coarsely chop bell pepper, onion and some fresh cilantro, and fry together in cooking oil. When onion and pepper are cooked, add four to five serving spoons of the cooked beans, stir, and then add 2 cups of the bean broth. Add cooked rice; mix together. Mix in the remainder of cooked beans, season with Salsa Lizano and serve hot. Give Gallo Pinto an extra kick with a little Tabasco sauce or chopped jalapeño pepper. Serves six.

Mango Ginger Chicken

- ¼ cup olive oil
- 2 cloves of crushed garlic
- 2 pounds of skinless chicken breasts
- 1 cup peeled and diced mangoes
- ¼ cup dark brown sugar
- ¼ tsp. ground cloves
- 2 Tbsp. grated fresh ginger
- ¼ tsp ground nutmeg
- 1 tsp. soy sauce

Heat the oil in a frying pan. Add the garlic and sauté lightly. Add the chicken breast. Cook until the chicken is done. Mix other ingredients in a bowl and spoon over the chicken. Salt and pepper to taste. When all ingredients are heated thoroughly (about five minutes), serve with rice, slaw salad and black beans. Serves four.

Appendix

Spanish Vocabulary

What a great language I have,
It's a fine language we inherited from the fierce Conquistadors....
They carried everything off and left us everything....
They left us the words.
~Pablo Neruda, author, 1904-1973

■ DAYS OF THE WEEK

domingo . . . . . . Sunday
lunes . . . . . . Monday
martes . . . . . . Tuesday
miercoles . . . . . . Wednesday
jueves . . . . . . Thursday
viernes . . . . . . Friday
sabado . . . . . . Saturday

■ MONTHS OF THE YEAR

enero . . . . . . January
febrero . . . . . . February
marzo . . . . . . March
abril . . . . . . April
mayo . . . . . . May
junio . . . . . . June
julio . . . . . . July
agosto . . . . . . August
septiembre . . . . . . September
octubre . . . . . . October
noviembre . . . . . . November
diciembre . . . . . . December

■ NUMBERS

uno . . . . . . one
dos . . . . . . two
tres . . . . . . three
cuatro . . . . . . four
cinco . . . . . . five
seis . . . . . . six
siete . . . . . . seven
ocho . . . . . . eight
nueve . . . . . . nine
diez . . . . . . ten
once . . . . . . eleven
doce . . . . . . twelve
trece . . . . . . thirteen
catorce . . . . . . fourteen

quince	fifteen
dieciséis	sixteen
diecisiete	seventeen
dieciocho	eighteen
diecinueve	nineteen
veinte	twenty
veintiuno	twenty-one
veintidós	twenty-two
treinta	thirty
cuarenta	forty
cincuenta	fifty
sesenta	sixty
setenta	seventy
ochenta	eighty
noventa	ninety
cienone	hundred
ciento uno	one hundred one
doscientos	two hundred
quinientos	five hundred
mil	one thousand
mil uno	one thousand one
mil dos	two thousand
un millón	one million
mil millones	one billion
primero	first
segundo	second
tercero	third
cuarto	fourth
quinto	fifth
sexto	sixth
séptimo	seventh
octavo	eighth
noveno	ninth
décimo	tenth
undécimo	eleventh
duodécimo	twelfth
último	last

■ CONVERSATION

¿Como esta usted?	How are you?
¿Bien, gracias, y usted?	Well, thanks, and you?
Buenas dias.	Good morning.
Buenas tardes.	Good afternoon.
Buenas noches.	Good evening/night.
Hasta la vista.	See you again.
Hasta luego.	So long.

¡Buena suerte! . . . . Good luck!
Adios. . . . . Goodbye.
Mucho gusto de conocerle. Glad to meet you.
Felicidades. . . . . Congratulations.
Muchas felicidades. . . . . Happy birthday.
Feliz Navidad. . . . . Merry Christmas.
Feliz Año Nuevo. . . . . Happy New Year.
Gracias. . . . . Thank you.
Por favor. . . . . Please.
De nada/con mucho gusto. . . . . You're welcome.
Perdoneme. . . . . Pardon me.
¿Como se llama esto? . . . . What do you call this?
Lo siento. . . . . I'm sorry.
Quisiera... . . . . I would like...
Adelante. . . . . Come in.
Permitame presentarle... . . . . May I introduce...
¿Como se llamo usted?. What is your name?
Me llamo.... . . . . My name is...
No se. . . . . I don't know.
Tengo sed. . . . . I am thirsty.
Tengo hambre. . . . . I am hungry.
Soy norteamericano/a . . . . I am an American.
¿Donde puedo encontrar...? . . . . Where can I find...?
¿Que es esto? . . . . What is this?
¿Habla usted ingles? . . . . Do you speak English?
Hablo/entiendo un poco. . . . . . I speak/understand a little Spanish.
Español
Hay alguien aqui que hable ingles? . . . . Is there anyone here who speaks English?
Le entiendo. . . . . I understand you.
No entiendo. . . . . I don't understand.
Hable mas despacio por favor. . . . . Please speak more slowly.
Repita por favor. . . . . Please repeat.

■ TELLING TIME

¿Que hora es? . . . . What time is it?
Son las.... . . . . It is...
... cinco. . . . . five oíclock.
... ocho y diez. . . . . ten past eight.
... seis y cuarto. . . . . quarter past six.
... cinco y media. . . . . half past five.
... siete y menos cinco. . . . . five of seven.
antes de ayer. . . . . the day before yesterday.
anoche. . . . . yesterday evening.
esta mañana. . . . . this morning.
a mediodia. . . . . at noon.

en la noche. . . . . in the evening.
de noche. . . . . at night.
mañana en la mañana. tomorrow morning.
mañana en la noche. . . . . tomorrow evening.
pasado mañana. . . . . the day after tomorrow.

■ DIRECTIONS

¿En que direccion queda...? . . . . In which direction is...?
Lleveme a... por favor. Take me to... please.
Llevame alla ... por favor. . . . . Take me there please.
¿Que lugar es este? . . . . What place is this?
¿Donde queda el pueblo?. . . . . Where is the town?
¿Cual es el mejor camino para...? . . . Which is the best road to...?
Malécon . . . . Road by the sea.
De vuelta a la derecha. . . . . Turn to the right.
De vuelta a la isquierda. . . . . Turn to the left.
Siga derecho.. . . . . Go this way.
En esta dirección. . . . . In this direction.
¿A que distancia estamos de...? . . . . How far is it to...?
¿Es este el camino a...? . . . . Is this the road to...?
Es.... . . . . Is it...
¿... cerca? . . . . near?
¿... lejos? . . . . far?
¿... norte? . . . . north?
¿... sur? . . . . south?
¿... este? . . . . east?
¿... oeste? . . . . west?
Indiqueme por favor. . . . . Please point.
Hagame favor de decirme . . . . Please direct me to...
donde esta...
... el telefono. . . . . the telephone.
... el bano. . . . . the bathroom.
... el correo. . . . . the post office.
... el banco. . . . . the bank.
... la comisaria. . . . . the police station.

■ ACCOMMODATIONS

Estoy buscando un hotel... . . . . . . I am looking for a hotel that's...
... bueno. . . . . good.
... barato. . . . . cheap.
... cercano. . . . . nearby.
... limpio. . . . . clean.
¿Dónde queda un buen hotel? . . . . Where is a good hotel?
¿Hay habitaciones libres? . . . . Do you have available rooms?
¿Dónde están los baños/servicios? . . . . Where are the bathrooms?
Quisiera un... . . . . I would like a...
... cuarto sencillo. . . . . single room.

... cuarto con baño. .... room with a bath.
... cuarto doble. .... double room.
¿Puedo verlo? .... May I see it?
¿Cuanto cuesta? .... What's the cost?
¡Es demasiado caro!. .... It's too expensive!

Transportation Schedules & Fares

As mentioned earlier, you should steer clear of the area around the old Coca-Cola plant at night. If you'd rather avoid the area altogether, a new company called **A Safe Passage** can help you out. Offices are located in Alajuela, just a few minutes from the airport. Board their van here and you'll be delivered to the bus terminal. A Safe Passage will also purchase your tickets for you. Reach them at ☎ 506/441-7837, rchoice@racsa.co.cr.

■ General Transportation Contacts

Tobías Bolaños Airport at Pavas, San José, ☎ 506/232-2820.

Juan Santamaría International Airport at Alajuela. Flight information, ☎ 506/443-2622.

Daniel Oduber International Airport at Liberia. Administration, ☎ 506/668-1180.

INTERBUS, ☎ 506/283-5573, www.interbusonline.com, interbus@costaricapass.com.

FANTASY BUS, Gray Line, ☎ 506/232-3681, www.graylinecostarica.com, ventas@graylinecostarica.com.

■ Flights

Nature Air, ☎ 506/220-3054, www.natureair.com. Daily departures from the Tobías Bolaños Airport, Pavas, San José to: Barra del Colorado, Drake Bay, Golfito, Liberia, Palmar Sur, Puerto Jiménez, Punta Islita, Quepos, Nosara, Tamarindo, Tambor, Tortuguero and Granada de Nicaragua.

SANSA, ☎ 506/221-9414, www.flysansa.com. Daily departures from the Juan Santamaría International Airport, Alajuela to: Barra del Colorado, Coto 47, Drake Bay, Golfito, Liberia, Nosra, Palmar Sur, Puerto Jiménez, Punta Islita, Quepos, Sámara, Tamarindo, Tambor and Tortuguero.

Buses

It's sad to offer the following advice as we are fans of public transport, but...

- Never leave luggage unattended, even on the bus.
- Be suspicious of people who offer to help. Paranoia anyone?
- Keep photocopies of passports and tickets separate from the originals.
- Always be alert and aware of your surroundings. Which means no booze for lunch.
- If you become lost, find a public place to ask for directions.
- In case of emergency, call 911.

To make a formal complaint for robbery or other criminal acts, call the Office for **Victims of Crime**, ☎ 506/295-3271 or 295-3565. O.I.J Building. Av. 6, Calles 15 and 17, San José.

METRO AREA BUS STOPS BY DESTINATION

BUS	STOPS AT:
Alajuelita	Av 6-8, Calle 8; Av 6, Calle 10
Aserrí	Av 4-6, Calle 7
Barrio Mexico	Av 3, Calle 3
Barrio Luján	Av 2, Calle 5-7
Calle Blancos	Av 5, Calle 1-3
Coronado	Av 7, Calle 0
Curridabat	Av 6, Calle 3-5
Desamparados	Av 4, Calle 5-7
Guadalupe	Av 3, Calle 0
Hatillos	Av 2-6, Calle 6; Av 4, Calle 4-6
Moravia	Av 3, Calle 3-5
Paso Ancho	Av 4-6, Calle 2
Pavas	Av 1, Calle 16-20
Sabana Cementerio	Av 2, Calle 8-10
Sabana Estadio	Av 2, Calle 2-4
Sabanilla	Av 0-2, Calle 9
San Antonio-Escazú	Av 0-1, Calle 16
San Pedro	Av 0, Calle 9-11

METRO AREA BUS STOPS BY DESTINATION	
BUS	**STOPS AT:**
Santa Ana	Av 1-3, Calle 16
Santo Domingo	Av 7-9, Calle 2
Tibás	Av 5-7, Calle 0-2
Tres Ríos	Av Ctrl.-2, Calle l 3
Zapote	Av 2-4, Calle 5

Services from San José to Surrounding Towns

San José to Aerial Tram: (see Guápiles)

San José to Alajuela & Juan Santamariá Airport: Every 10 minutes, 5 am to 10 pm, after 12 pm departures every 30 minutes. Av. 2, Calle 12-14. ☎ 506-449-5141, 30 min or 1+ hr. during rush hours.

San José to Atenas: Daily, 6 am to 10 pm. Every half-hour. Coca-Cola, Calle 16, Av. 1-3. **Coopetransatenas**, ☎ 506-446-5767, 1½ hrs.

San José to Braulio Carillo National Park: (see Guápiles)

San José to Bribrí: Daily at 6, 10, 13:30, 15:30. Calle Ctrl., Av. 13. Terminal del Caribe. ☎ 506-257-8129, Transportation Mepe, 5 hrs.

San José to Cahuita: Daily at 6, 10, 13:30, 15:30. Calle Ctrl., Av. 13. Terminal del Caribe, ☎ 506-257-8129. Transportation Mepe, 4 hrs. Return at 7:30, 9:30, 11:30, 16:30.

San José to Cañas & Upala: Daily at 8:30, 9:50 11:50, 13:40, 15:00, 16:45. Calle 14, Av.1-3. **La Cañera**, ☎ 506-222-3006, 669-0227, 3½ hrs. Return at 4, 5, 9, 14:15.

San José to Cartago: Every10 minutes from 5 am until 12 pm. Av. 4, Calle 9, picks people up at Calle 5, Av. 18 y 20. ☎ 506-233-5350, 50 mins.

San José to Ciudad Quesada: Every half-hour, 5 am to 7:30 pm. Calle 12, Av. 7-9. **Autotransp**. **San José-San Carlos**, ☎ 506-256-8914, 2½ hrs.

San José to Golfito: Daily at 7 am, 3 pm. Av. 5, Calles 14-16. **Tracopa**, ☎ 506-222-2666, 8 hrs. Return at 5 am, 13:30.

San José to Grecia: Every half-hour, 6 am to 10:10 pm. Av. 3, Calles 16-18. **Tuan**, ☎ 506-258-2004, 1½ hrs.

San José to Guápiles, Aerial Tram & Braulio Carillo: Every half-hour, 5:30 am to 10 pm. Calle 13, Av. Central. **Empresarios Guapileños**, ☎ 506-222-0610, 1½ hrs.

San José to Heredia: Every 10 minutes, 5 am to 11 pm. Calle 1, Av. 7-9, 30 mins.

San José to Irazú Volcano: Saturdays, Sundays, Holidays: 8 am. Av. 2, Calle 1-3 in front of Ministerio de Hacienda. ☎ 506-534-4125, 2 hrs. Return at 12:30.

San José to Jacó: Daily, 7:30, 10:30, 13:00, 15:30, 18:30. Coca-Cola, Av. 3, Calle 16. ☎ 506-223-1109, 2½ hrs.

San José to La Fortuna: Daily at 6:15, 8:40, 11:30. Calle 12, Av. 7-9. **Autotransp. San José-San Carlos,** ☎ 506-256-8914, 4½ hrs. Return at 12:30, 14:30.

San José to Liberia: Every half-hour, 6 am to 8 pm (express service at 3 pm, 5 pm). Calle 24, Av. 5-7. **Pulmitan,** ☎ 506-256-9552, 4 hrs.

San José to Limón: Daily every 45 min, 5 am to 7 pm. **Caribeños,** ☎ 506-222-0610. Calle Central, Av. 13, 2½ hrs.

San José to Los Chiles (Nicaragua Border): Daily, 5:30 am, 3:30 pm. Calle 12, Av. 7-9. **Autotransp. San José- San Carlos,** ☎ 506-256-8914, 5 hrs. Return at 5 am and 3 pm.

San José to Manuel Antonio (Direct): Daily at 6, 12:00, 18:00, 19:30, from Coca-Cola. **Transportes Delio Morales,** ☎ 506-223-5567, 3½ hrs. Return at 6, 9:30, 12:00, 17:00. (See also Quepos.)

San José to Monteverde: Daily at 6:30 am, 14:30. Calle 12, Av.7-9. **Transmonteverde,** ☎ 506-222-3854, 5 hrs. Return at 6:30, 14:30.

San José to Nicoya via Bridge: Daily at 6, 8, 10:30, 12:30, 14:00, 16:00, 18:30. Av. 5, Calles 14-16. **Tracopa-Alfaro Bus,** ☎ 506-222-2666, 5 hrs. Return at 3, 4:30, 5:00, 8:30, 9:00, 14:45, 17:00.

San José to Nosara (Direct): Daily at 6 am. Av. 5, Calles 14-16. **Tracopa-Alfaro,** ☎ 506-222-2666, 6 hrs. Return at noon.

San José to Paso Canoas (Panama Border): Daily at 5:00, 7:30, 11:00, 13:00, 16:30, 18:00 and one extra on Fridays 10 am. Av. 5, Calles 14-16. **Tracopa,** ☎ 506-222-2666, 6½ hrs. Return at 3:30, 8:30, 11:30, 15:00.

San José to Playas Fl amingo, Conchal & Potrero: Daily at 8:00 and 15:00 via Liberia; 10:30, via the Bridge (☎ 506-654-4203). Calle 20, Av. 1-3. **Tralapa Bus,** ☎ 506-221-7202, 6 hrs.

San José to Playa Panamá: Daily at 15:00. Calle 20, Av. 1-3. **Tralapa Bus,** ☎ 506-221-7202, 680-0392, 5 hrs.

San José to Poas Volcano: Daily at 8:30. Av. 2, Calle 12-14. **Tuasa Bus,** ☎ 506-222-5325, 2 hrs. Return at 14:30.

San José to Puerto Jiménez: Daily 12 midnight. Calle 14, Av. 9-11. **Transp. Blanco,** ☎ 506-257-4121, 9 hrs. Return at 5 am.

San José to Puerto Viejo de Limón: Daily at 6:00, 10:00, 13:30, 15:30. Terminal del Caribe, Calle 13, Av. Central. **Trans. Mepe**, ☎ 506-257-8129, 5 hrs. Return at 7:00, 9:00, 11:00, 16:00.

San José to Puerto Viejo de Sarapiquí: Daily at 6:30, 7:30, 10:00, 11:30, 13:30, 14:30, 15:30, 16:30, 18:00. Terminal del Caribe, Calle 13, Av. Central. ☎ 506-222-0610, 2 hrs. Return at 5:30, 7:00, 8:00, 11:00, 13:30, 15:00, 17:30.

San José to Puntarenas (Direct): Every half-hour, 6 am to 7 pm. Calle 16, Av. 12. **Empresarios Unidos**, ☎ 506-222-8231, 2 hrs. Returns from 4:15 to 7pm.

San José to Quepos y Manuel Antonio: Daily at 6:00*, 7:00, 10:00, 12:00*, 14:00, 16:00, 18:00*. Coca-Cola, Calle 16, Av. 3. **Transp. Morales**, ☎ 506-223-5567, 3½ hrs. Return at 5:00, 6:00*, 8:00, 9:30*, 12:00*, 14:00, 16:00, 17:00* (* direct).

San José to Sámara: Daily at 12:30. Av. 5, Calles 14-16. **Tracopa-Alfaro Bus**, ☎ 506-222-2666, 5 hrs.

San José to San Isidro del General: With **Tracopa-Alfaro Bus**, ☎ 506-222-2666, 3 hrs. Calles 14-16, Av. 5. Every half-hour, 5 am to 6 pm. Returns, 5:30 to 19:30. With **Musoc Bus**, ☎ 506-222-2422, 3 hrs. Calle Central, Av. 22-24. Daily at 05:30, 06:30, 07:30, 08:30, 09:30, 10:30, 11:30, 12:30, 13:30, 14:30, 15:30, 16:30, 17:30, 18:30. Return at 5:00, 05:30, 06:30, 07:30, 08:30, 09:30, 10:30, 11:30, 12:30, 13:30, 14:30, 15:30, 16:30, 17:30.

San José to Santa Cruz: With **Tracopa-Alfaro Bus**, ☎ 506-222-2666, 5 hrs. Av. 5, Calles 14-16. Daily at 06:30, 8:00, 10:00, 13:30, 15:00, 17:00. Return at 04:30, 05:00, 07:30, 09:30, 12:00, 14:30, 16:30. With **Tralapa Bus**, ☎ 506-221-7202, 5 hrs. Daily at 07:00, 10:00, 10:30, 13:00, 16:00 via Bridge.

San José to Sarchí: Daily at 12:15, 17:30, 18:10. Calle 18, Av. 5-7. **Costado Oeste Abonos Agro**, ☎ 506-258-2004, 1½ hrs. Return at 05:30, 06:00, 06:30.

San José to Siquirres: Every half-hour, 06:30-19:00. Calle 13, Av. 0. **Transportes Caribeños**, ☎ 506-222-0610, 1½ hrs. Returns, 5:30-19:00.

San José to Sixaola (Panama Border) via Cahuita & P.Viejo: Daily at 0:600, 10:00, 13:30, 15:30. Calle 13, Av. 0, 6 hrs. **Transp. Mepe**, ☎ 506-257-8129. Return at 05:00, 07:30, 09:30, 14:30.

San José to Tamarindo: Daily at 11:00, 15:30 via Liberia. Av. 5, Calle 14-16. **Tracopa-Alfaro Bus**, ☎ 506-222-2666, 5½ hrs. Return at 05:45.

San José to Tilarán: Daily at 07:30, 09:30, 12:45, 15:45, 18:30. Calle 14, Av. 9 -11. ☎ 506-222-3854, 4 hrs. Return at 04:45, 07:00, 09:30, 14:00, 17:00.

San José to Turrialba: Every half-hour, 05:15 to 19:00. Calle 13 Av. 6. **Transtusa Bus,** ☎ 506-222-4464, 1¾ hrs. Returns, 05:00 to 17:30.

San José to Zarcero: Daily, 05:00-19:30. Calle 12, Av. 7-9. **Autotransp. San José-San Carlos,** ☎ 506-256-8914, 1½ hrs. Returns, 05:00-18:00.

San José to Zoo Aves: Saturdays & Sundays, every half-hour 08:00-12:00. La Merced Church: Calle 14, Av. 4. **Station Wagon Bus,** ☎ 506-381-5450, 35 mins. Returns, 10:00-15:00.

Inter-City Connections

Drake from Sierpe (by boat): **Transporte Colectivo,** ☎ 506-786-7311, 1½ hrs. $15, minimum four passengers.

La Fortuna from Tilarán: Daily at 07:00, 12:00. Return at 08:00, 17:00. Tilarán downtown, 3½ hrs.

Mal País & Santa Teresa from Cóbano: Daily at 10:30, 14:30, 17:30. Return at 07:00, 11:30, 15:30. Terminal de Buses Cóbano, ☎ 506-200-0204, 45 mins.

Monteverde from Tilarán: Daily at 12:30. Tilarán downtown, 1¾ hrs.

Sarchí from Alajuela: Every half-hour, 05:00-22:00, same return times. Calle 8, Av. 0-1. ☎ 506-494-2139, 1½ hrs.

Zancudo from Golfito (by boat): Daily at 12:00; Return at 07:00.

For more inter-city bus service see www.costaricabybus.com/scheds.html.

International Service

San José to Bocas de Torro, Panama: 09:00 (part trip in boat). In front of Hotel Cocorí near Coca-Cola. **Boca Toreños Bus,** ☎ 506-385-7551, 9 hrs.

San José to David, Panama: Leaves 07:30, arrives 17:00. Av. 5, Calles 14-16. **Tracopa,** ☎ 506-222-2666. Return at 08:30.

San José to Panama City: With **Panaline Bus,** ☎ 506-256-8721, leaves 13:00, arrives 05:00 next day. Calle 16, Av. 3-5. With **Tica Bus,** ☎ 506-221-8954, leaves 22:00, arrives 16:00 next day. Av. 4, Calles 9-11.

San José to Guatemala & Honduras: **Tica Bus,** ☎ 506-221-8954. Daily at 06:00, 07:30, 12:30. Av. 4, Calles 9-11. (Overnight in Managua, hotel not included; stops in San Pedro Sula y Tegucigalpa.)

San José to Managua (Nicaragua): With **Tica Bus,** ☎ 506-221-8954, 11 hrs. Daily at 06:00, 07:30, 12:30. Av. 4, Calles 9-11. With **Transnica Bus,** ☎ 506-223-4242, x 101, 11 hrs. Daily at 05:30, 09:00. Calle 22, Av. 3-5. Return at 07:00, 10:00. With **Central Line,** ☎ 506-248-9129. Daily at 05:00. Calle 12, Av. 7-9 (Estación de San Carlos). Return at 05:00.

San José to San Salvador, El Salvador: **King Quality Bus**, ☎ 506-221-3318, leaves 03:30, arrives 23:00. Calle 12, Av. 3-5 at the Upala bus stop. Breakfast & lunch included, stops in Managua y Tegucigalpa.

■ Ferries

Puntarenas to Playa Naranjo: Daily at 06:00, 10:00, 14:20, 19:00. Dock Puntarenas. **CONATRAMAR**, ☎ 506-661-3834, 1 hr. Return at 07:30, 12:30, 17:00, 21:00.

Puntarenas to Paquera: Daily at 06:30, 10:30, 14:30, 18:30. Dock Puntarenas. **Ferry Peninsular (ADIP)**, ☎ 506-641-0515. Return at 04:30, 08:30, 12:30, 16:30.

Puntarenas to Tambor: Daily at 08:30, 12:30, 16:30, 20:30. Dock Puntarenas. **Ferry Peninsular (ADIP)**, ☎ 506-641-0515, 1¼ hrs. Return at 06:30, 10:30, 14:30, 18:30, 22:00. Note: This route not in service as of this writing; call for update.

Golfito to Puerto Jiménez: Daily at 11:30. Little Dock at Golfito, in Pueblo Civil. **Sr. Chino Castro**, 1½ hrs. Return 06:00.

Index

Index

T

NOTES

NOTES

NOTES

NOTES

NOTES

NOTES

NOTES

NOTES

NOTES

NOTES

NOTES

NOTES

NOTES

NOTES

NOTES